T0360796

Cosmology, Gravitational Waves and Particles

Proceedings of the Conference on

Cosmology, Gravitational Waves and Particles

Nanyang Technological University, Singapore, 6–10 February 2017

Editor

Harald Fritzsch

Ludwig Maximilian University of Munich, Germany

NEW JERSEY • LONDON • SINGAPORE • BEIJING • SHANGHAI • HONG KONG • TAIPEI • CHENNAI • TOKYO

Published by

World Scientific Publishing Co. Pte. Ltd.
5 Toh Tuck Link, Singapore 596224
USA office: 27 Warren Street, Suite 401-402, Hackensack, NJ 07601
UK office: 57 Shelton Street, Covent Garden, London WC2H 9HE

Library of Congress Cataloging-in-Publication Data
Names: Conference on Cosmology, Gravitational Waves and Particles
(2017 : Nanyang Technological University) | Fritzsch, Harald, 1943– editor.
Title: Cosmology, gravitational waves and particles : proceedings of the Conference on Cosmology, Gravitational Waves and Particles, Nanyang Technological University, Singapore, 6-10 February 2017 / editor, Harald Fritzsch (Ludwig Maximilian University of Munich, Germany).
Other titles: Proceedings of the Conference on Cosmology, Gravitational Waves and Particles, Nanyang Technological University, Singapore, 6-10 February 2017
Description: Singapore ; Hackensack, NJ : World Scientific, [2018] | Includes bibliographical references.
Identifiers: LCCN 2017044584| ISBN 9789813231795 (hardcover ; alk. paper) | ISBN 9813231793 (hardcover ; alk. paper)
Subjects: LCSH: Gravitational waves--Congresses. | Cosmology--Congresses. | Particles (Nuclear physics)--Congresses.
Classification: LCC QC179 .C66 2017 | DDC 523.1--dc23
LC record available at https://lccn.loc.gov/2017044584

British Library Cataloguing-in-Publication Data
A catalogue record for this book is available from the British Library.

For any available supplementary material, please visit
http://www.worldscientific.com/worldscibooks/10.1142/10758#t=suppl

Printed in Singapore

Preface

Modern scientific cosmology started in 1917 with Albert Einstein's paper "Cosmological Considerations of the General Theory of Relativity". To obtain a stable universe, Einstein had to introduce a cosmological constant in his field equations. But in 1922 Alexander Friedmann published the idea of an expanding universe. In 1929 Edwin Hubble discovered with the telescope at the Mount Wilson near Pasadena that the distant galaxies move away from the earth with velocities, which are proportional to the distance between the galaxy and the earth. Thus the universe was expanding, and Einstein removed the cosmological term.

The expansion of the universe implies that the expansion started long time ago in a cosmic explosion, the Big Bang. Today we know that the age of our universe is about 14 billion years. Shortly after the Big Bang the universe was very hot and the associated radiation will also exist today. It was discovered in 1964. The temperature of the cosmic background radiation is close to 2.7 K.

In 1933 the Swiss astrophysicist Fritz Zwicky, who worked at the California Institute of Technology, investigated the Coma cluster of galaxies. He discovered that this cluster is only stable, if most of the matter in the cluster is not observed. It must be dark matter, a hypothetical type of matter distinct from baryonic matter (ordinary matter such as protons and neutrons). Although dark matter has not been directly observed, its existence and properties are inferred from its gravitational effects such as the motions of visible matter, gravitational lensing, its influence on the universe's large-scale structure and its effects on the cosmic microwave background. The energy density due to dark matter in our galaxy is about 0.4 $\mathrm{GeV/cm^3}$.

It is still unknown, what is behind the dark matter. It might be provided by a new neutral stable particle, which is not described within the Standard Theory of particle physics. Such particles might be related to a new symmetry, for example supersymmetry. They might also be related to the internal structure of the weak bosons. Those might be composite particles, consisting of two constituents, analogous to the rho mesons, which are bound states of a quark and an antiquark.

The dark matter particles could be produced in pairs in the collisions, studied by the LHC-experiments. The dark matter particles cannot be observed directly, but one can observe the missing energy in those collisions. Thus far nothing has been observed.

Dark matter particles would have an average speed of 300 km/s in our galaxy. A dark matter particle can collide with an atomic nucleus. Such collisions are searched in various experiments at the Gran Sasso laboratory in Italy.

Some time ago one observed by studying supernova explosions, that the expansion of the universe does not become slower in time, but it accelerates in time. To explain this acceleration, general relativity requires, that much of the energy in the universe consists of a component with large negative pressure, dubbed "dark energy". The exact nature and existence of dark energy remains one of the great mysteries of the Big Bang. It might be described by a cosmological constant.

The experiments indicate that our universe consists of 73% dark energy, 23% dark matter, 4.6% regular matter and less than 1% neutrinos. According to the theory, the energy density in matter decreases with the expansion of the universe, but the dark energy density remains constant as the universe expands.

Gravitational waves are ripples in the curvature of space-time that propagate as waves at the speed of light. Predicted in 1916 by Albert Einstein on the basis of his theory of general relativity, gravitational waves transport energy as gravitational radiation, a form of radiant energy similar to electromagnetic radiation.

On 11 February 2016, the LIGO collaboration announced the detection of gravitational waves, from a signal detected on 14 September 2015 of two black holes with masses of 29 and 36 solar masses merging about 1.3 billion light years away. During the final fraction of a second of the merger, it released more than 50 times the power of all the stars in the observable universe combined. The mass of the new black hole was 62 solar masses. The energy equivalent to three solar masses was emitted as gravitational waves.

This book collected contributions by some of the invited speakers at the recent Conference on Cosmology, Gravitational Waves and Particles, 6–10 February 2017, Institute of Advanced Studies, NTU, Singapore. We thank all contributors for their active participation at the conference.

H. Fritzsch
Physics Department
Ludwig-Maximilians-Universität München

Institute of Advanced Studies

Contents

S. Tiwari[8,20], M. Tonelli[40,17], F. Travasso[6,16], M. C. Tringali[46,20], L. Trozzo[45,17], K. W. Tsang[27], N. van Bakel[27], M. van Beuzekom[27], J. F. J. van den Brand[31,27], C. Van Den Broeck[27], L. van der Schaaf[27], J. V. van Heijningen[27], M. Vardaro[38,15], M. Vasúth[29], G. Vedovato[15], D. Verkindt[22], F. Vetrano[47,12], A. Viceré[47,12], J.-Y. Vinet[2], H. Vocca[39,16], R. Walet[27], G. Wang[8,12], M. Was[22], M. Yvert[22], A. Zadrożny[26], T. Zelenova[6], J.-P. Zendri[15]

[1]*APC, AstroParticule et Cosmologie, Université Paris Diderot, CNRS/IN2P3, CEA/Irfu, Observatoire de Paris, Sorbonne Paris Cité, F-75205 Paris Cedex 13, France*

[2]*Artemis, Université Côte d'Azur, Observatoire Côte d'Azur, CNRS, CS 34229, F-06304 Nice Cedex 4, France*

[3]*Astronomical Observatory Warsaw University, 00-478 Warsaw, Poland*

[4]*Janusz Gil Institute of Astronomy, University of Zielona Góra, 65-265 Zielona Góra, Poland*

[5]*Nicolaus Copernicus Astronomical Center, Polish Academy of Sciences, 00-716, Warsaw, Poland*

[6]*European Gravitational Observatory (EGO), I-56021 Cascina, Pisa, Italy*

[7]*ESPCI, CNRS, F-75005 Paris, France*

[8]*Gran Sasso Science Institute (GSSI), I-67100 L'Aquila, Italy*

[9]*Institute of Mathematics, Polish Academy of Sciences, 00656 Warsaw, Poland*

[10]*INAF, Osservatorio Astronomico di Capodimonte, I-80131, Napoli, Italy*

[11]*INAF, Osservatorio Astronomico di Padova, Vicolo dell'Osservatorio 5, I-35122 Padova, Italy*

[12]*INFN, Sezione di Firenze, I-50019 Sesto Fiorentino, Firenze, Italy*

[13]*INFN, Sezione di Genova, I-16146 Genova, Italy*

[14]*INFN, Sezione di Napoli, Complesso Universitario di Monte S.Angelo, I-80126 Napoli, Italy*

[15]*INFN, Sezione di Padova, I-35131 Padova, Italy*

[16]*INFN, Sezione di Perugia, I-06123 Perugia, Italy*

[17]*INFN, Sezione di Pisa, I-56127 Pisa, Italy*

[18]*INFN, Sezione di Roma, I-00185 Roma, Italy*

[19]*INFN, Sezione di Roma Tor Vergata, I-00133 Roma, Italy*

[20]*INFN, Trento Institute for Fundamental Physics and Applications, I-38123 Povo, Trento, Italy*

[21] *LAL, Univ. Paris-Sud, CNRS/IN2P3, Université Paris-Saclay, F-91898 Orsay, France*

[22] *Laboratoire d'Annecy-le-Vieux de Physique des Particules (LAPP), Université Savoie Mont Blanc, CNRS/IN2P3, F-74941 Annecy, France*

[23] *Laboratoire Kastler Brossel, UPMC-Sorbonne Universités, CNRS, ENS-PSL Research University, Collège de France, F-75005 Paris, France*

[24] *Laboratoire des Matériaux Avancés (LMA), CNRS/IN2P3, F-69622 Villeurbanne, France*

[25] *National Astronomical Observatory of Japan, 2-21-1 Osawa, Mitaka, Tokyo 181-8588, Japan*

[26] *NCBJ, 05-400 Świerk-Otwock, Poland*

[27] *Nikhef, Science Park, 1098 XG Amsterdam, The Netherlands*

[28] *Department of Astrophysics/IMAPP, Radboud University Nijmegen, P.O. Box 9010, 6500 GL Nijmegen, The Netherlands*

[29] *Wigner RCP, RMKI, H-1121 Budapest, Konkoly Thege Miklós út 29-33, Hungary*

[30] *Scuola Normale Superiore, Piazza dei Cavalieri 7, I-56126 Pisa, Italy*

[31] *VU University Amsterdam, 1081 HV Amsterdam, The Netherlands*

[32] *University of Białystok, 15-424 Białystok, Poland*

[33] *Università di Camerino, Dipartimento di Fisica, I-62032 Camerino, Italy*

[34] *Università degli Studi di Genova, I-16146 Genova, Italy*

[35] *Université de Lyon, F-69361 Lyon, France*

[36] *Université Claude Bernard Lyon 1, F-69622 Villeurbanne, France*

[37] *Università di Napoli 'Federico II', Complesso Universitario di Monte S.Angelo, I-80126 Napoli, Italy*

[38] *Università di Padova, Dipartimento di Fisica e Astronomia, I-35131 Padova, Italy*

[39] *Università di Perugia, I-06123 Perugia, Italy*

[40] *Università di Pisa, I-56127 Pisa, Italy*

[41] *Institut de Physique de Rennes, CNRS, Université de Rennes 1, F-35042 Rennes, France*

[42] *Università di Roma 'La Sapienza', I-00185 Roma, Italy*

[43] *Università di Roma Tor Vergata, I-00133 Roma, Italy*

[44] *Università di Salerno, Fisciano, I-84084 Salerno, Italy*

[45] *Università di Siena, I-53100 Siena, Italy*

[46] *Università di Trento, Dipartimento di Fisica, I-38123 Povo, Trento, Italy*

[47] *Università degli Studi di Urbino 'Carlo Bo', I-61029 Urbino, Italy*
**j.degallaix@lma.in2p3.fr*

Advanced Virgo is the French-Italian second generation laser gravitational wave detector, successor of the Initial Virgo. This new interferometer keeps only the infrastructure of its predecessor and aims to be 10 times more sensitive, with its first science run planned for 2017. This article gives an overview of the Advanced Virgo design and the technical choices behind it. Finally, the up-to-date progresses and the planned upgrade for the following years are detailed.

Keywords: Gravitational wave; Virgo; laser interferometer.

1. Introduction

The year 2016 will be remembered as the beginning of the gravitational wave astronomy with the announcement by the LIGO and Virgo collaborations of the first direct detection on Earth of the coalescence of two massive black holes. This remarkable achievement was possible thanks to the unprecedented sensitivity of the so-called second generation of gravitational wave detectors. The first gravitational wave signal was recorded by the two Advanced LIGO detectors, first interferometers to be online for this new generation. In 2017, the LIGO detectors will be joined by the Advanced Virgo interferometer, the French-Italian detector which is the topic of this article.

2. The Importance of Advanced Virgo

As mentioned in the introduction, the year 2017 will see the creation of the first network of three gravitational wave detectors: the 2 LIGO detectors and the Virgo one. That is a crucial step for the development of the new astronomy with gravitational waves. More detectors in the network means better redundancy and so longer cumulated observational time, as the typical duty cycle of a detector is around 60% [1]. More important, a tremendous improvement in the sky localisation of the source is expected by adding a third detector to the LIGO network. An accurate sky localisation is essential to enable the multi-messenger astronomy which includes also observations with electromagnetic and neutrino telescopes.

Simulations have shown that even an early version of Advanced Virgo, with a range only 25% of the one from LIGO can reduce the 90% confidence angular area on the sky of the source by a 60% [2].

3. The Advanced Virgo Interferometer

In this section more historical and technical details will be given about the Advanced Virgo interferometer.

Fig. 1. Aerial view of Virgo detector with the two perpendicular 3-km long arm across the Tuscanian countryside. Credit: The Virgo Collaboration.

3.1. *Background*

Advanced Virgo has been built and operated by the Virgo collaboration. The Virgo collaboration is not a new comer in the field since it has been created in 1994 combining the forces from French CNRS and Italian INFN laboratories. At that time, the common goal was to build a laser interferometer gravitational wave detector in the area of Pisa in Italy. This construction of this first generation detector called Virgo was completed in 2003 (see figure 1) and took scientific data intermittently for several years from 2007 to 2011 conjointly with other detectors. This period marks the birth of the global collaboration with exchange of scientific data and technical knowledge between the different continents[3].

In parallel of running the initial Virgo detector, a research effort was started as early as 2005 (with the Advanced Virgo White paper[4]) for a new detector 10 times more sensitive. The Advanced Virgo detector was approved by the French-Italian institutions in 2009 and the major upgrade started on site at the end 2011 after the end of the last science run of the Initial Virgo.

Today the Virgo collaboration, it is 250 people, 20 laboratories spread over 6 European countries, despite its increase in size, the majority of the collaboration still comes from French and Italy institutions.

3.2. *Optical setup and expected sensitivity*

The optical setup of Advanced Virgo is a dual recycled Michelson interferometer with 3 km long Fabry-Perot arm cavities as shown in figure 2. From the point of view of the optical functionality, the interferometer could be decomposed in

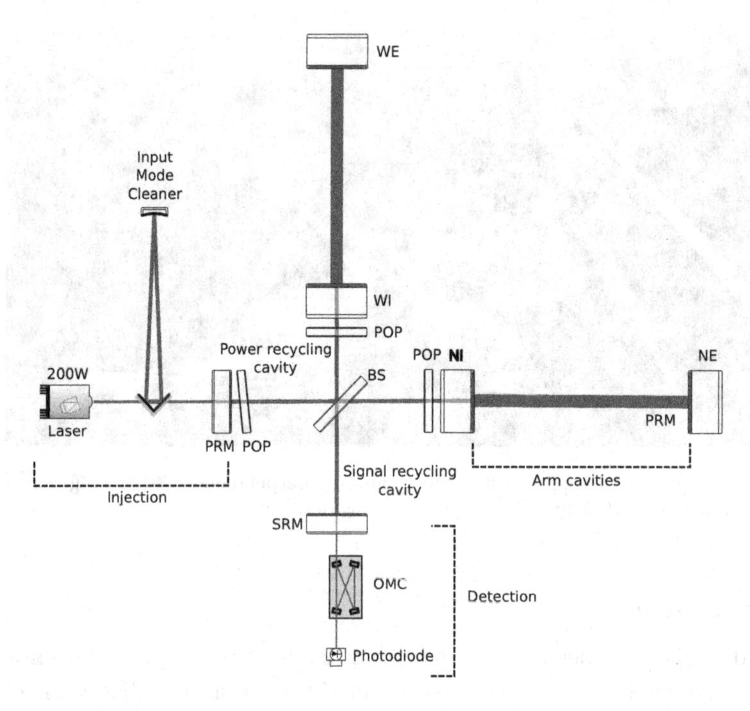

Fig. 2. Schematics of the Advanced Virgo optical setup.

four blocks:

The injection. It is composed by the input light source: a laser at 1064 nm and with a power up to 200 W for the final configuration of the detector and all the components to modulate, to stabilize in amplitude and frequency and also to shape the laser beam. Before the laser and the interferometer, there is also a 3 mirrors 150 m long linear optical cavity to filter the beam and also a 40 times enlarging telescope to achieve the proper beam size on the power recycling mirror.

The detection. This part converts the gravitational signal imprinted in the optical phase of the light to a voltage which can be recorded. Moreover most of the signals used to control the interferometer are derived from the detection side of the interferometer.

The arm cavities. The two perpendicular 3 km long Fabry-Perot cavities are formed by the input and end mirrors. Those mirrors are the most critical optics of the interferometer. Any light lost in the arm cavities is directly synonymous of a degradation of the detector sensitivity.

The recycling cavities. Two additional cavities are formed with the arm cavities and the power recycling mirror (on the injection side) as well as the signal recycling mirror (detection side). The former is called the power recycling cavities and is used to enhance the power circulating in the arm cavities by nearly a factor 40. The latter, the signal recycling cavity, can change the optical response of the interferometer and hence modify the frequency dependent sensitivity curve of the detector.

The goal of Advanced Virgo is to be 10 times more sensitive than the initial Virgo interferometer, enabling a potential volume of detection 1000 times larger than the first generation. Advanced Virgo has kept the same overall infrastructure of the Virgo detector but the core of the interferometer has been largely upgraded.

To ease the commissioning of the detector, it has been decided to have a two steps approach before reaching the final sensitivity of Advanced Virgo[5]:

(1) first, the interferometer will work with a low input power (below 40 W) and without signal recycling mirror. In this configuration, the interferometer is very similar to initial Virgo and so the strategy to handle the interferometer is well known and already experienced. Even with this reduced setup, the astrophysical range for the detection is already more than 80% of the final configuration.
(2) second, in this final step (planned for 2018), the full power of the 200 W laser will be used and the signal recycling mirror will be installed. That will enable the detection full range of Advanced Virgo with recording of black hole merger up to 1 Gpc, with tens of detection per year.

3.3. *Selected key technologies*

To achieve its design sensitivity, Advanced Virgo uses state of the art technologies. It would be too long to detail all the impressive technological aspects of the interferometer, so the author has (arbitrarily) decided to develop only few of them. A more exhaustive view of the various subsystems can be found in Advanced Virgo Technical Design Report[6].

3.3.1. *The large mirrors*

The four mirrors forming the two arm cavities (called NI, WI for the input ones and NE and WE for the end ones as shown in 2) are the most critical optics of the interferometer. Those mirrors have a cylindrical shape with a diameter of 350 mm and a thickness of 200 mm, weighting around 40 kg.

To maximise the interaction between the light and the gravitational wave signal, the various optical losses such as scattering or absorption have to be an kept at an unprecedented low level. In that purpose, the substrate of the mirrors is made of the purest fused silica glass available. The polishing which gives the shape of the surface is done at the atomic level with on the central part, a peak to valley of the height of the mirror surface consistently less than 2 nm. Finally, the very uniform multi-layer coating[7] on the surface allows a reflection of more than 99.999% of the light while absorbing less than 1 part per million[8].

3.3.2. *The mirror payload*

The payload represents the system between the mirrors and the suspension chain. Each arm cavity mirror is suspended using 4 glass fibers of diameter 0.4 mm. The fibers itself are attached to mirror using two glued glassed ears on either side of the mirror barrel. This all glass system is called a monolithic suspension and is a key feature to minimize the mirror displacement induced by thermal noise[9]. Only the arm cavity mirrors are suspended with glass fibers, the other large optics such as the beamsplitter or recycling mirrors are using steel wires.

Compared to Initial Virgo, the payload has been completely redesigned and is now much more complex. For Advanced Virgo, not only the mirror has to be attached though the payload but also additional optics and parts. As an example on the arm cavity input mirror payload, a second large optic is also suspended: the compensation plate, a heating ring (both parts are described in section 3.3.4) and light baffles (see section 3.3.5) are also present.

3.3.3. *The seismic isolation*

The isolation of the mirror from the ground motion is ensured by the super attenuator[10]. The core of the system is a long series of cascaded pendula (8 m high, in vacuum) installed for the Initial Virgo interferometer and was already compliant with the specifications for Advanced Virgo. Thanks to this suspension system, the seismic noise is attenuated by 10 order of magnitude above few Hertz in the six degrees of freedom.

3.3.4. *The thermal compensation*

The combination of very high circulating power (up to 600 kW of light in the arm cavities) and optical absorption (albeit very low, in the order of part per million), leads to the heating of the main optics of the interferometer. The presence of those thermal gradient induces optical aberration jeopardizing the control and sensitivity of the detector. Those negative effects already experienced by the first generation of detector, are much stronger in advanced interferometers. As a result, specific diagnostic tools and actuators are required to mitigate the effect of the absorbed optical power[11].

The distortions induced by the laser beam are monitored in real time by two dedicated light beams probing the input mirror substrates and analyzed through an Hartmann sensor. In parallel, the magnitude and the shape of the interferometer control sidebands (light beams frequency shifted from the main beam) are also been analyzed with phase cameras. Those sidebands are well suited figures of merit about the state of the interferometer since they are extremely sensitive to any defects from the polishing tolerances of the optics, to misalignment or thermal aberrations.

Thanks to those monitoring techniques, error signals can be derived and send to thermal actuators for active aberration control. For example, heating rings have been inserted around the main mirrors to change in-situ their radii of curvature. Controlled heating pattern can also be sent on the Compensation Plates (CP in figure 2) using CO_2 laser beams for near arbitrary correction.

3.3.5. *The diffused light mitigation*

The importance of managing the diffused light has been a hard lesson learned from the first generation of detectors. As some point in time, each detector had their sensitivity limited by diffused light.

All the optics, even the outstanding ones as for Advanced Virgo, always scatter a small amount of the incident light at small and large angles. This light may reach a part of the detector not isolated from the ground, be partially reflected and then recombine to the main laser beam, inducing extra phase noise. Several extensive actions has been taken in Advanced Virgo to mitigate the effect of the diffused light: use of superpolished optics, custom light traps (called baffles) around the optics and all the critical optics for the detection are installed on suspended bench inside vacuum chambers.

4. The Rise of Advanced Virgo

4.1. *A brief timeline*

The installation of Advanced Virgo has started at the end of 2011, right after the last data taking run of the Initial Virgo. After the decommissioning of the previous interferometer, infrastructure works such as the creation of new clean rooms or modification of vacuum pipes have begun. The first subsystem to be ready was the injection which includes the laser and a milestone was achieved with the locking of the Input Mode Cleaner cavity in the middle of 2014. As the large optics of the interferometer were being installed and a stable input beam was available, the first 3 km arm cavity was locked in the middle of 2016. At the beginning of 2017, the final configuration of the first phase of Advanced Virgo was reached with the all the cavities locked at their operating points.

During the installation, several failures of the monolithic suspension occurred with the breaking of the thing glass fibers supporting the arm cavity mirrors. To not delay the commissioning, it has been decided to suspend temporary the mirror with steel wires while the problem was investigated. The decision was not an easy one as using steel wires increases the suspension thermal noise, degrading the interferometer sensitivity at low frequency.

4.2. *Current status*

In this section, we detailed the status of the detector in April 2017. At this date, the interferometer is routinely on the dark fringe with all the cavities reliably locked for several hours. Special care was taken to automate the locking sequence and the interferometer can consistently go to its operating point without any human intervention.

A new phase of the commissioning has started with the first noise budget detailing the noise sources currently limiting the sensitivity. So current effort are dedicated to implement low noise actuators and to engage the auto-alignment loops to close all the degrees of freedom. However, no official sensitivity curve has yet been publicly released.

It was worth noting that the thermal compensation system, a key technology of Advanced Virgo to ensure stability of the recycling cavities and correct the optical aberration is currently being tested and tuned. However, so far, the locking and stability have been achieved without the use of TCS, which is particularly encouraging and simplify the early commissioning.

The short term priority is to be able to join the two LIGO detectors for the second observation run (O2) around the summer of 2017.

In parallel to the progress on the interferometer, a dedicated task force has been created to understand the failure of the monolithic suspension. Extensive investigations has been carried out within and outside the Virgo collaboration to find the origin of the breaking. These investigations were successful and the cause of the failure was found to be linked to the vacuum implementation. The suspending fibers were broken (or deadly fragilized) at the time of venting the tower when exposed to high speed particles coming with the entering flux of air [12].

4.3. *Improvement following O2*

After the second observational run (O2) planned to be completed at the end of August of 2017, a period of around one year, is planned to allow invasive improvements before the next data taking run (O3). This dedicated period of installation and commissioning will be common to the LIGO and Virgo interferometers.

For Advanced Virgo, the priority is to reinstall the monolithic suspension for the arm cavities mirrors to reach the design sensitivity in the low frequency range. This work will also include installation of guards around the fibers and upgrades on the vacuum infrastructure to greatly reduce the amount of projected dust. At this occasion, cleaning of the mirrors and of the vacuum tanks will also be done. This activity is likely to last almost half a year.

A second major work also planned during that period is the installation of a squeezed vacuum light source at the output of the detector [13]. This special light source, provided by the Albert Einstein Institute in Germany, allows to reduce the sensitivity in the high frequency range where the detector is limited by the light shot noise, noise inherent to the quantum nature of light.

Following these two main works, once the state of the interferometer is recovered, commissioning time for several months is required to ensure optimal stability and sensitivity of the detector before considering the next data taking period (O3).

Beyond O3, the final high power laser is expected to be installed as well as the signal recycling mirror with the high reflective coating.

5. The Upgrade Path

For the horizon 2020, large upgrades for Advanced Virgo are also been investigated in the laboratories of the collaboration. The main goal is to contribute in the best way to the international network of detectors both in term of reliability and sensitivity. For that, we must go beyond the design sensitivity of Advanced Virgo while keeping the same infrastructure[14].

Among the promising leads to explore one can quote: new squeezed light source to further reduce the quantum noise at low and high frequencies (called frequency dependent squeezer[15]), cancellation of the Newton noise at very low frequency[16] or even new larger mirrors with advanced coating to reduce the thermal noise[17].

6. Conclusion

Advanced Virgo is a second generation interferometer complementary to the Advanced LIGO detectors. Currently, the detector is in the state of active commissioning and the highest priority is to be able to take data conjointly with the Advanced LIGO interferometers during the summer 2017. After this first observational run, a series of upgrade is already planned to further increase the sensitivity and reliability of the detector promising a bright future for the gravitational wave astronomy.

References

1. B. Abbott, R. Abbott, T. Abbott, M. Abernathy, F. Acernese, K. Ackley, C. Adams, T. Adams, P. Addesso, R. Adhikari *et al.*, Gw150914: The advanced ligo detectors in the era of first discoveries, *Physical review letters* **116**, p. 131103 (2016).
2. L. Singer, Status post-O1: Sources, rates, and localization, LIGO DCC G1601468-v5, (2016).
3. J. Abadie, B. Abbott, R. Abbott, M. Abernathy, T. Accadia, F. Acernese, C. Adams, R. Adhikari, P. Ajith, B. Allen *et al.*, Search for gravitational waves from compact binary coalescence in ligo and virgo data from s5 and vsr1, *Physical Review D* **82**, p. 102001 (2010).
4. R. Flaminio, A. Freise, A. Gennai, P. Hello, P. L. Penna, G. Losurdo, H. Lueck, N. Man, A. Masserot, B. Mours, M. Punturo, A. Spallicci and A. Vicere, Advanced virgo white paper, VIRNOTDIR1390304, (2005).
5. F. Acernese, M. Agathos, K. Agatsuma, D. Aisa, N. Allemandou, A. Allocca, J. Amarni, P. Astone, G. Balestri, G. Ballardin *et al.*, Advanced Virgo: A second-generation interferometric gravitational wave detector, *Classical and Quantum Gravity* **32**, p. 024001 (2014).
6. The Virgo Collaboration, Advanced Virgo Technical Design Report, VIR-0128A-12, (2012).
7. M. Granata, E. Saracco, N. Morgado, A. Cajgfinger, G. Cagnoli, J. Degallaix, V. Dolique, D. Forest, J. Franc, C. Michel *et al.*, Mechanical loss in state-of-the-art amorphous optical coatings, *Physical Review D* **93**, p. 012007 (2016).

8. L. Pinard, C. Michel, B. Sassolas, L. Balzarini, J. Degallaix, V. Dolique, R. Flaminio, D. Forest, M. Granata, B. Lagrange *et al.*, Mirrors used in the LIGO interferometers for first detection of gravitational waves, *Applied Optics* **56**, C11 (2017).
9. P. Amico, L. Bosi, L. Carbone, L. Gammaitoni, F. Marchesoni, M. Punturo, F. Travasso and H. Vocca, Monolithic fused silica suspension for the Virgo gravitational waves detector, *Review of scientific instruments* **73**, 3318 (2002).
10. F. Acernese, F. Antonucci, S. Aoudia, K. Arun, P. Astone, G. Ballardin, F. Barone, M. Barsuglia, T. S. Bauer, M. Beker *et al.*, Measurements of superattenuator seismic isolation by virgo interferometer, *Astroparticle Physics* **33**, 182 (2010).
11. A. Rocchi, E. Coccia, V. Fafone, V. Malvezzi, Y. Minenkov and L. Sperandio, Thermal effects and their compensation in advanced virgo, in *Journal of Physics: Conference Series*, (1)2012.
12. E. Majorana, Advanced Virgo upgrade plans, VIR-0283A-17, (2017).
13. S. Chua, B. Slagmolen, D. Shaddock and D. McClelland, Quantum squeezed light in gravitational-wave detectors, *Classical and Quantum Gravity* **31**, p. 183001 (2014).
14. F. Ricci for the Virgo collaboration, Vision document on the future of Advanced Virgo, VIR-0136B-16, (2016).
15. M. Evans, L. Barsotti, P. Kwee, J. Harms and H. Miao, Realistic filter cavities for advanced gravitational wave detectors, *Physical Review D* **88**, p. 022002 (2013).
16. J. Harms and H. J. Paik, Newtonian-noise cancellation in full-tensor gravitational-wave detectors, *Physical Review D* **92**, p. 022001 (2015).
17. M. Granata, Update on coating research and development at LMA, VIR-0173A-17, (2017).

The Asia-Australia Gravitational Wave Detector Concept

David Blair*, Zong-Hong Zhu†, Li Ju*, Eric Howell*, Chunnong Zhao*,
Hui-Tong Chua*, Simon Anderson*, Leong Cheng Man‡

**The University of Western Australia*
†Beijing Normal University
‡University of Glasgow

The discovery of gravitational waves enabled assessment of the science benefits of new improved gravitational wave detectors. This paper discusses the science benefits of an Asia-Australia Gravitational wave Observatory (AAGO) consisting of a pair of widely spaced gravitational wave detectors on a north-south axis. Initial sensitivity would be ~ 4 times better than the projected sensitivity of Advanced LIGO, but designed for future upgrades to match proposed third generation detectors. AAGO would enable near optimum angular resolution of sources, and signal detections at a rate ~ 1 per hour, sufficient to monitor a substantial fraction of all large mass black hole merger events in the universe. The proposed conceptual design and infrastructure, technical issues and challenges are discussed.

1. Introduction

The detection of gravitational waves from binary black hole coalescence announced in 2016[1] began the era of gravitational wave astronomy. Suddenly we know that our detector technology can actually detect signals. We are now in a position to estimate the benefits of improved detectors.

In 2015, five months before the first detected gravitational wave event GW150914[1], a research program at the Kavli Institute for Theoretical Physics of China entitled The Next Detectors for Gravitational Wave Astronomy[2], considered options for future gravitational wave detectors.

Members of the research program considered various options for creating the next generation of detectors. Most options included increased interferometer arm length. Possibilities included the use of cryogenics, which is currently being pioneered by the Japanese KAGRA project[3]. A second option was to use silicon test masses cooled to 120K and to change the laser wavelength from 1μm to 1.5–2μm, at which silicon is transparent[4,5]. A third option, first proposed informally in China several years ago[6], was to build very long interferometers 10 times the arm length of the current LIGO detectors[7]. The fourth option, which is the topic of this paper, was to scale up Advanced LIGO technology, by doubling the mass of the test masses, the suspension pendulums, and the arm lengths, while maintaining the light intensity on the mirrors, combined with squeezed light technology and Newtonian noise subtraction[8].

Each option presents both advantages and risks. Cryogenics, which is being pioneered by the KAGRA Project reduces test mass thermal noise but requires long cool down times and high thermal conductivity cooling paths to maintain the cryogenic temperature. This can risk the input of mechanical noise and cryogenic contamination of cooled mirrors. The silicon option requires the test masses to be radiatively cooled to 120K where thermoelastic noise falls to zero. This offers the advantage of the very low cryogenic acoustic loss of silicon, which reduces the thermal noise, and minimises of thermal aberrations in the test masses due to the zero thermal expansion coefficient. However it requires the development of low loss silicon optics, including the development of large test masses fabricated from low optical absorption silicon. These technologies have yet to be demonstrated.

The very long interferometer option dilutes the contribution of noise associated with the individual test masses (such as thermal noise) relative to the interferometer arm length, and also allows improved averaging of the thermal noise across a larger laser beam spot size. However it presents difficulties in finding suitable sites, the necessity for substantial civil construction to correct for the curvature of the Earth, and costs associated with the need to use substantially larger vacuum pipe diameters. In addition, the extremely long radius of curvature mirrors ($\sim$20km) become much more sensitive to figure errors which are already specified to the nanometer level in 4km detectors.

Before detection, theoretical predictions of source amplitudes and event rates were very uncertain. In that context the step size for successive generations of detectors was typically chosen to be an order of magnitude. Now that signals have been detected we are in a position, for the detected large mass binary black hole source population, to estimate the benefits of a new detector. We are no longer working in the unknown, and we can make quantitative predictions of the benefits of various design options.

The two-fold scale up option proposed for AAGO, aims to be a judicious balance of arm length dilution of local test mass noise, other technical improvements, costs, risks and development time. It leads to a predicted sensitivity increase of $\sim$4 times the projected sensitivity of Advanced LIGO. This is less than the 10-fold improvement projected for 40km and 10km silicon based options. Before detection this could have seemed inadequate. Since detection, estimates of the reach and the array angular resolution show that there are very strong science benefits. This is reported in a recent document by Howell et al.[9], and is summarized here.

This paper presents a preliminary concept of the Asia-Australia Gravitational Observatory, for which we use the working acronym AAGO. It would be made up of two detectors, one located in China and the other in Australia. Each would use an initial arm length of the order of 8km, but be located on a site where future extension was possible. In its first phase it would be based on technologies proven by the LIGO and VIRGO[10] projects, but would be designed to allow upgrading using whichever of the new approaches discussed above eventually proves to be most effective.

With this staged approach, AAGO would be the first stage of a future “third generation” observatory. The third generation upgrade could be planned to match the time scale for third generation detector projects such as the Einstein-Telescope in Europe and Cosmic Explorer, for which US sites have been considered. The upgrade would probably utilize cooled silicon or longer arms or both. Thus AAGO would contribute not only to the currently operating and planned detectors, but long into the third generation detector era ($\sim$2030 onwards). It allows the southern hemisphere gap in the world-wide array to be filled on a faster time scale.

In section 2 we summarise the science benefits of AAGO using results presented in Howell et al.[9], which include the expected detection rate at both full and reduced sensitivity, and the science outcomes that could be expected.

In section 3 we summarise the preliminary observatory concept and possible locations. We include the facility design, the manufacturing methods, isolator and suspension concepts and the test mass and laser requirements.

In section 4 we discuss some of the technical issues which are faced by the current Advanced LIGO detectors and how they would be addressed in AAGO.

In the conclusion we discuss possible time scales for AAGO and suggest that AAGO could make a pivotal contribution to gravitational wave astronomy in the years before third generation detectors are implemented, thereby paving the way for a full worldwide third generation network.

2. Science Benefits of AAGO

Ground based gravitational wave detectors have always aimed to detect sources in the few Hz to few kHz frequency band. Numerous sources have been predicted including sources associated with neutron stars and stellar mass black holes. These include supernovae, spinning neutron stars with cylindrical asymmetry, and the coalescence of compact binary systems made up of black holes and/or neutron stars. All improved detectors have an increased possibility of detecting the above sources, but quantitative estimates in almost all cases are built on somewhat uncertain theoretical models.

Here we focus specifically on binary black hole sources of the type that have already been detected, because in this case we are able to make reliable predictions based on generally agreed cosmology. The benefits arise because the event rate increases roughly as the sensitivity improvement factor cubed, while the angular resolution improves both from increasing the baselines between detectors and through increasing the signal to noise ratio. The fundamental idea is that with good statistics and good angular resolution we can localize all events to a relatively small volume element in the universe. Thus observations allow us to sample black holes in a large 3D volume of the universe. The data can not only allow the evolution and origins of the BBH population to be determined, but it can provide a completely independent probe of cosmology.

The starting point for estimating event rates and angular resolution is the data on the first three signals, which is summarized in Table I below.

Table I. Summary of the first three binary black hole coalescence events observed during LIGO's first observation run in 2015. The final parameters in the table, sky localization and red shift define the volume localization capability of the two-detector array.

BBH Events	GW150914	GW151226	LVT151012
Masses M⊙	36 - 29	14 -7.5	23 - 13
Final BH Mass M⊙	62	20	35
Radiated Mass M⊙	3	1	1.5
Luminosity distance D_L	420Mpc	440Mpc	1000Mpc
Peak Luminosity W	3.6 x 10^{49}	3.3 x 10^{49}	3.1 x 10^{49}
Inspiral/Final Spin	-0.06, 0.68	0.21, 0.74	0.0, 0.66
Sky Localisation deg^2	230	850	1600
Red Shift	.09 (+.03 -.04)	.09 (+.03 -.04)	0.2 (±.09)

The addition of more detectors which will include Virgo, KAGRA and LIGO-India[11] will improve both the angular resolution and event rates. However the addition of AAGO to the world array has a significant effect on the angular resolution and an even larger effect on the event rates. In particular it allows sufficient sensitivity to be reached that the nearest sources, similar to those in Table I, will be able to be sufficiently localized that a significant fraction will be able to be localized to an individual host galaxy. This enables the Hubble constant to be evaluated using the BBH distance scale which is completely independent from the astronomical distance ladder normally used.

The results below are based on results from Howell et al[9]. Figure 1 shows the event rate for a world wide array with the addition of the AAGO Observatory. We see that the

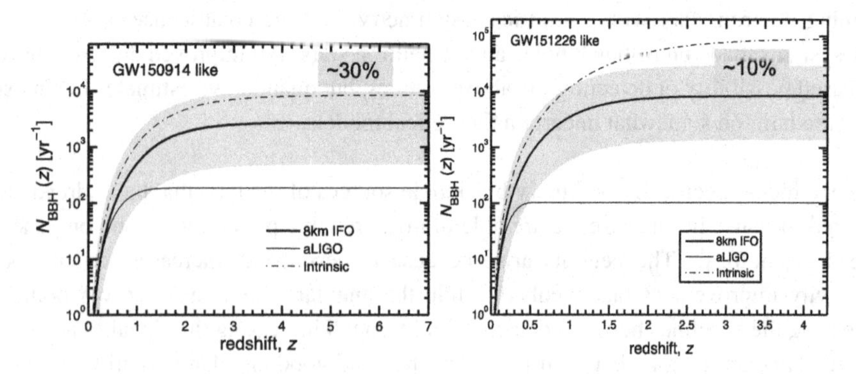

Fig. 1. Preliminary estimate of the total number of events detectable by AAGO as a function of redshift, depending which of the first two events is most typical. The dotted line gives the total number of events in the Hubble volume and the lower curve gives the event rate expected for AdvLIGO. The shaded region is the 95% uncertainty interval. For GW150914-type events AAGO detects $\sim$30% of the universal rate, while for GW151226-type events AAGO detects about 10%.

4-fold improvement over AdvLIGO allows the world array to detect roughly 10–30% of all BBH events in the universe similar to the first two events to be detected. The event rate is between 8 and 24 events per day.

The angular resolution capability is summarized in Tables II and III below. Table II gives the result for 30–30 $M_\odot$ while Table III is for 10–10 $M_\odot$ BBH events. The largest array that includes the two sites for AAGO is the only one for which a substantial fraction (25%) of the nearest high mass BBH sources can be resolved to within 0.1 square degrees. This is roughly the resolution required to identify a substantial fraction of host galaxies at 400–600Mpc.

Tables II and III. Angular resolution results for 30–30 and 10–10 $M_\odot$ BBH events. Here l and H refer to LIGO Livingstone and Hanford, V is Virgo, J is KAGRA, I is LIGO-India, C is China, A is Australia. Arrays with Australia are in all cases superior to other locations.

Table II

Network	Percentage of detections within credible region				
	0.1 deg^2	0.5 deg^2	1 deg^2	5 deg^2	10 deg^2
LHV	0.00	1.11	7.59	44.65	60.50
LHVJ	0.00	16.44	37.59	82.45	91.59
LHVI	0.00	8.26	28.93	80.74	91.30
LHVA	1.05	45.30	71.48	96.07	98.55
LHVC	0.63	32.26	54.02	89.71	95.86
LHVJI	0.00	26.29	51.49	90.39	96.25
LHVJIC	1.58	42.97	66.33	95.09	98.53
LHVJIA	7.31	66.74	88.01	99.71	99.93
LHVJIAC	25.06	81.06	94.63	99.90	100.00

Table III

Network	Percentage of detections within credible region				
	0.1 deg^2	0.5 deg^2	1 deg^2	5 deg^2	10 deg^2
LHV	0.00	0.00	0.00	8.36	21.63
LHVJ	0.00	0.00	1.99	36.75	59.95
LHVI	0.00	0.00	0.00	31.36	56.84
LHVA	0.00	1.20	13.48	73.06	88.59
LHVC	0.00	0.79	9.21	55.24	74.58
LHVJI	0.00	0.00	4.57	51.64	73.49
LHVJIC	0.00	1.90	14.37	67.48	83.24
LHVJIA	0.00	7.74	32.27	88.37	96.24
LHVJIAC	0.00	27.48	52.58	95.20	99.05

For lower mass sources (Table III) source identification is more difficult, but the array with AAGO is the only one capable of achieving 0.5 square degrees for a substantial fraction (27%) of the closest sources. It might be possible to identify the host galaxy using other statistical and astronomical discriminators.

Having 3D localization of $\sim 10^4$ events per year is likely to allow many interesting cosmological studies. Fundamental questions about the origins of the BBH population: are they primordial, co-evolved or dynamically formed will be able to be addressed by studying their spatial distribution. It will be possible to determine how the BBH population relates to large scale structure, baryon acoustic oscillations and dark matter distributions.

3. Design Concept of AAGO

Laser interferometer gravitational wave detectors require a combination of long vacuum pipes, high performance in-vacuum vibration isolators, low loss suspension systems for the very low acoustic loss test masses that also require very high optical performance. The interferometer must be powered by a very high performance high power laser. In addition an extremely complex set of additional optical cavities is required for laser light conditioning and modulation and readout.

The operation of a gravitational wave detector is dependent on an extremely complex set of control systems to lock the optical cavities in the detector arms, control systems for alignment, modulation, frequency and intensity stabilization, and seismic isolation. A substantial amount of the mechanical and optical components must be located and vibration isolated within the high vacuum environment, to prevent acoustic and seismic noise contamination, and due to the combination of high optical power and very high intensity laser light, all materials must be ultrahigh vacuum compatible and free of volatile hydrocarbons which can contaminate optical components.

The proposed parameters for the first stage of AAGO are given in Table IV below.

Table IV.

AAGO Parameters	Component
Configuration	Fabry-Perot dual recycling Michelson interferometer
Readout	8dB Squeezed light
Isolation	Active +Passive
Suspension	1.5m fused silica fibres
Arm Length	8000m
Vacuum Pipe	1.2m diameter stainless steel 316
Corner and end stations	Buried with geothermal low vibration airconditioning
Laser	240W: 1064nm slab with fibre amplifier
Test Masses	87kg fused silica
Newtonian noise	Seismic monitoring and subtraction

In addition to the state of the art optical, mechanical and control systems, the major infrastructure contributes about half of the project cost. This includes the following:

a) Long ultrahigh vacuum stainless steel pipes. AAGO will require 16km of $\sim$1.2m diameter stainless steel pipes. They could be made by spiral welding as used by LIGO, or of corrugated construction as used by the GEO project. The material must be air baked at $\sim$450C to remove hydrogen prior to manufacture.
b) The test masses would be fabricated from high purity fused silica similar to those of LIGO and Virgo. However the diameter would be increased to 50cm, and the total mass would increase to 87kg. The mirror figure error specification would be approximately 0.5 of the Advanced LIGO specification. This would entail the use of figure error correction techniques developed at LMA in France.
c) The isolation system would combine passive very low frequency elements with resonant frequency $\sim$0.05Hz, using techniques developed at Gingin (REF) with isolation chains of $\sim$1hz resonant frequency. Active feedback seismic suppression of microseismic signals using local tilt sensors and seismometer arrays would be used to minimize the input seismic noise, thereby increasing the dynamic range. Early warning of large seismic events would be used to put the system into lockdown a few seconds prior to the disturbance.
d) Fused silica suspensions using laser drawn fibres similar to the LIGO design, but of double the length would be used in a double pendulum reaction mass configuration similar to that of Advanced LIGO.
e) The main laser would be a 1064 nm slab laser with a fibre amplifier. At the power density projected for Advanced LIGO, 240W of pump power is required. Howell et al. considered the case where the cavity power was reduced to power levels similar to those already achieved during AdvLIGO commissioning.

4. Technical Issues

We address here some of the technical issues which have been a problem in existing gravitational wave detectors, and show that new designs can substantially reduce or eliminate many of them.

One of the key technical issues faced by the three long baseline detectors that have operated to date has been locally generated seismic noise from human activity, wind forces and effects of temperature variations. These problems can be overcome by careful design. A team of civil engineers and architects at UWA have spent several years developing a concept for a semi-underground laboratory that is effectively buried by sand berms. This provides excellent thermal insulation combined with high thermal mass that ensures both low cooling requirements and low wind induced building vibration. Figure 2 shows the conceptual design of such a facility.

Fig. 2. Architects proposal for a semi underground laboratory facility.

The low cooling requirement enables airconditioning systems to use low flow velocity, and to have fan and compressor mechanical systems (which generate vibration) located at a large distance. If geothermal energy is available (as it is at the Gingin location in Western Australia) the increased efficiency achievable using ground water heat pumps reduces the compressor power of the air conditioning system[12], which again leads to lower vibration. Air for temperature control can be supplied using underground ducting which minimizes duct vibration which has been frequently observed to be a significant source of noise.

Another significant vibration problem is associated with enhanced microseismic noise. This noise source is generated by ocean waves, and occurs at ~ 0.2Hz. Microseismic noise has long attenuation lengths so is visible at most locations on the earth. However it can be suppressed by using preisolation at frequencies ~ 0.05Hz (considered to be the lowest practical design frequency) and through use of feedback or feedforward techniques based on local measurement with seismometers and tilt meters. Recently LIGO has shown excellent suppression of noise using tilt sensing, and at higher frequency ~ 10Hz, by using geophones to monitor the laboratory floor vibrations. In principle this technique can suppress seismic noise to the noise floor of the measuring instruments. Seismic arrays are being introduced at Virgo and LIGO to develop these techniques further.

Seismic noise and atmospheric density fluctuations can induce gravitational force noise, commonly called Newtonian noise, at frequencies below 10Hz. This noise is induced by time varying gravitational forces which act on individual test masses, and degrade system performance in the 1–5Hz band. This noise can be reduced by monitoring the seismic noise field and the atmospheric density fields and using them to model the gravitational perturbations, and then subtract them from the signal. Gravity gradiometers are under development[13] which could also be used for direct measurement and suppression but to date such devices have not attained sufficient sensitivity. Current detectors are far from reaching limits set by this noise source.

The magnitude of gravitational force is reduced if detectors are constructed say 200m underground as proposed for ET.REF The ability to optimize sensitivity in the 2–5Hz band is likely to be limited by the ability to measure the sources and subtract their signals from the data. Thus precision seismic monitoring will be an important component of AAGO. Studies underway[14,15] are likely to provide better understanding of ground level gravitational force noise suppression in the near future.

Another important technical issue is the problem of parametric instability at high laser power which is induced due to the scattering of laser light from certain acoustic modes of the test masses. This problem can prevent detectors reaching design power. Recently there has been significant progress in designing interferometers for which there is an instability free window in their parameter space[16]. Secondly there are new ideas for low loss dampers that could provide passive suppression of parametric instability[17]. Other more complex solutions have been demonstrated[18,19,20,21,22,23]. For AAGO a significant effort should be undertaken to design for an instability free window, and then to design suitable dampers for back up.

Another challenge is the thermal stability of the test masses. Temperature changes due to absorbed laser light and by ambient temperature fluctuations cause time dependent thermal aberrations, which cause difficulties in all areas from instability control to calibration. Better thermal stability can be achieved, firstly through improved ambient temperature control, which would include sensing and control at the test mass tanks themselves, and secondly, by providing low power laser heating to test masses whenever the main interferometer beam is not active. This can allow the test mass thermal profile to be maintained in a steady state independent of earthquakes or technical failures that can cause loss of laser power on the core optics.

5. Conclusion

The growing prosperity and high levels of education and technology in Asia make AAGO a project that can contribute strongly to national aspirations, scientific literacy, and international cooperation in the Asian region. It can enable the region become a world leader in the fundamental studies of the universe that are made possible by this project. In addition it will expose the region to remarkable technological opportunities based on the ultra-high sensitivity of gravitational wave detectors which are the most sensitive measuring instruments ever created. AAGO would occupy a pivotal role in the world array because of the 7 very long north-south baselines it creates to the other detectors. It will pave the way for the third generation detectors being planned in USA and Europe.

This paper is a preliminary discussion. It will be extended into a more complete discussion document with input from a broader group of gravitational wave scientists.

References

[1] B. P. Abbott et al. (LIGO-VIRGO Collaboration), Observation of Gravitational Waves from a Binary Black Hole Merger, *Phys. Rev. Lett.* **116**, 061102 (2016) 11/02/2016.
[2] http://www.kitpc.ac.cn/?p=ProgDetail&id=PT20150406&i=main.
[3] Y. Aso et al. "Interferometer design of the KAGRA gravitational wave detector", *Phys. Rev. D*, **88**(4), 2013.
[4] LIGO Scientific Collaboration 2015 Instrument Science White Paper https://dcc.ligo.org/LIGO-T1400316/public.
[5] M. Punturo et al., "The Einstein telescope: A third-generation gravitational wave observatory", *Class. Quantum Grav.* **27**, 194, 2010.
[6] Edna Chung Private communication.
[7] S. Dwyer, D. Sigg, S. W. Ballmer, L. Barsotti, N. Mavalvala and M. Evans, "Gravitational wave detector with cosmological reach," *Phys. Rev. D* **91**, 082001, 2015.
[8] D. Blair, L. Ju, C. Zhao, L. Wen, H. Miao, R. Cai, J. Gao, X. Lin, D. Liu, L.-A. Wu, Z. Zhu, G. Hammond, H. J. Paik, V. Fafone, A. Rocchi, C. Blair, Y. Ma, J. Qin and M. Page, "The next detectors for gravitational wave astronomy", *Science China* **58**, 120405 doi:10.1007/s11433-015-5747-7.
[9] E. J. Howell et al., "Host galaxy identification for binary black hole mergers with long baseline gravitational wave detectors", preprint.
[10] F. Acernese et al., "Advanced Virgo: A second-generation interferometric gravitational wave detector", *Class. Quantum Grav.* **32** (2015) 024001.
[11] http://www.gw-indigo.org/tiki-index.php?page=LIGO-India.
[12] P. B. Whittaker, X. Wang, K. Regenauer-Lieb, D. G. Blair and H. T. Chua, "Geothermal air conditioning: Typical applications using deep-warm and shallow-cool reservoirs for cooling in Perth Western Australia", *Int, J. Simu. Multisci. Des. Optim.* **5**, A10, 2014.
[13] D. J. McManus, P. W. F. Forsyth, M. J. Yap, R. L. Ward, D. A. Shaddock, D. E. McClelland and B. J. J. Slagmolen, "Mechanical characterisation of the TorPeDo: A low frequency gravitational force sensor", *Class. Quantum Grav.*, **34**(13):135002, 2017.
[14] J. Harms and K. Venkateswara, "Newtonian-noise cancellation in large-scale interferometric GW detectors using seismic tiltmeters", *Class. Quantum Grav.*, **33**(23):234001, 2016.
[15] MCoughlin, N. Mukund, J. Harms, J. Driggers, R. Adhikari and S. Mitra, 'Towards a first design of a Newtonian-noise cancellation system for advanced LIGO", *Class. Quantum Grav.*, **33**(24):244001, 2016.
[16] J. Zhang, C. Zhao, L. Ju and D. Blair, "Study of parametric instability in gravitational wave detectors with silicon test masses", *Class. Quantum Grav.*, **34**(5):055006, 2017.
[17] L. Ju, H. N. Chen and D. G. Blair, "Multi-stage Dampers for Parametric Instability Suppression", https://dcc.ligo.org/LIGO-G1700852, 2017.
[18] C. Zhao, L. Ju, J. Degallaix, S. Gras and D. G. Blair, "Parametric instabilities and their control in advanced interferometer gravitational-wave detectors", *Phys. Rev. Lett.*, **94**:121102, 2005.
[19] J. Degallaix, C. Zhao, L. Ju and D. Blair, Thermal tuning of optical cavities for parametric instability control", *J. Opt. Soc. Am. B*, **24**(6):1336–1343, Jun 2007.
[20] S. Gras, P. Fritschel, L. Barsotti and M. Evans, "Resonant dampers for parametric instabilities in gravitational wave detectors", *Phys. Rev. D*, **92**, 082001, 2015.
[21] Y. Fan, L. Merrill, C. Zhao, L. Ju, D. Blair, B. Slagmolen, D. Hosken, A. Brooks, P. Veitch and J. Munch, "Testing the suppression of optoacoustic parametric interactions using optical feedback control", *Class. Quantum Grav., Gravity*, **27**(8):084028, 2010.

[22] J. Miller, M. Evans, L. Barsotti, P. Fritschel, M. MacInnis, R. Mittleman, B. Shapiro, J. Soto and C. Torrie. Damping parametric instabilities in future gravitational wave detectors by means of electrostatic actuators. *Physics Letters A*, **375**(3):788–794, 2011.

[23] C. Blair and LSC Instrument Authors, "First demonstration of electrostatic damping of parametric instability at advanced LIGO", *Phys. Rev. Lett.* **118**:151102, Apr 2017.

Gravitational-wave Observations from Ground-based Detectors

Tjonnie G. F. Li

Department of Physics, The Chinese University of Hong Kong, Shatin, N.T., Hong Kong
tgfli@cuhk.edu.hk

Recent detections of gravitational waves by the LIGO detectors herald a new era of observational astronomy. Previously invisible objects and phenomena may now be uncovered through their gravitational interaction. Observation of gravitational waves allows one to explore the extremes of the Universe and study astronomy and fundamental physics like never before. This article gives a brief overview of the detection process, from the production of the data to their physical implications.

Keywords: Gravitational waves; compact binaries; black holes.

1. Introduction

Albert Einstein's theory of general relativity, published in 1916, gave science a radically new way of understanding how space, time and gravity are related. This quickly became science's best tool for understanding phenomena including compact objects such as black holes and neutron stars, the expansion of the Universe and the Big Bang. Last year, scientists from The Laser Interferometer Gravitational-Wave Observatory (LIGO) announced the momentous discovery of tiny ripples in spacetime, better know as *gravitational waves* [1]. So far, the LIGO detectors have confidently detected three gravitational-wave signals, GW150914, GW151226 and GW170104, since the start of its operation in 2015 [1–3].

This article gives a brief overview of the detector, the production of data, the data-analysis process, what has been learned about astrophysics and fundamental physics, and what we may expect in the future.

2. Detecting Gravitational Waves

2.1. *Detector*

Gravitational waves are spacetime disturbances that are typically represented by a strain $h = \Delta L/L$. This gravitational-wave strain can then be measured by detectors, which acts as transducers to convert these space-time perturbations into measurable signals. The LIGO detectors are kilometre-scale Michelson-based interferometers [4]. For Advanced LIGO [5], very pure and homogeneous fused silica mirrors (34 cm diameter, 20 cm thickness, and 40 kg mass) are suspended from monolithic fused silica fibers by multistage pendula [6] and act as freely falling test masses. These multistage pendulum systems are mounted on actively controlled seismic isolation platforms [7,8]. A prestabilised 1064-nm wavelength Nd:YAG laser injects 20W power into the interferometer, which is then converted into 100kW circulated power by coupled optical resonators. The test mass, the suspension system and part of the seismic isolation system are placed in an ultrahigh vacuum system, where the pressure is typically below 1 Pa over the $10,000\,\mathrm{m}^3$ volume. These and other innovations have led to a sensitivity of 10^{23} Hz at 100 Hz in Advanced LIGO's O1 observation run, which is significantly lower strain noise compared Initial LIGO's S6 science run: a factor 50 at 50 Hz and a factor 3 to 4 between 100 Hz and 300 Hz[9].

2.2. *Calibration & detector characterisation*

The output port of the Michelson interferometer is held at an offset from a dark fringe[10]. A differential change in the arm lengths, caused by gravitational waves or environmental noise, will then decrease or increase the amount of light. This light signal is measured and digitized by a photodetector. Besides serving as the strain readout channel, the power fluctuations measured by the photodiodes are also used for feedback control of the test mass actuators to stabilise the operation of the detector[10]. Instead of an unsuppressed change in the differential length, such a control system results in a smaller residual length change. Estimating the gravitational-wave strain therefore requires the calibration of the open loop transfer function, which relates the unsuppressed and the residual length changes. This calibration is mainly done by applying a known varying force by reflecting an auxiliary 1047-nm wavelength laser, modulated in intensity, off the end test mass[11]. The response of the optical transducer is measured by sweeping the modulation frequency through the entire detection band. This way, the calibrated strain readout is computed with less than 10% in magnitude and 10 degree in phase across 20 Hz to 1 kHz[12].

Besides calibration uncertainty, data of the LIGO detectors contain non-Gaussian noise transients introduced by complex interactions between the instruments and their environment. If not dealt with appropriately, these noise transients can be misinterpreted as gravitational-wave events or degrade the significance assigned to an event. Each of the LIGO detectors records over 200,000 auxiliary channels that monitor instrument behavior and environmental conditions. These channels, including those from the physical environment monitor sensors, are used for diagnosing instrument faults and identifying correlations between the strain channel and noise. When a source of noise is identified, it is either resolved on the instrumentation level or is omitted from analyses through vetoes. Careful studies of the sources of transient noise have concluded that the detectors were operating nominally at the time of GW150914 and subsequent detections, and have ruled out environmental influences and non-Gaussian instrument noise as the cause of the observed gravitational-wave signals[13].

2.3. *Data analysis*

Once the data has been acquired, calibrated and potential transient noise sources identified, they are scanned by search algorithms for the presence of gravitational-wave signals. Two classes of search algorithms are typically used to search for gravitational-wave transients: modelled and unmodelled. Modelled searches are based on match filtering where the data is correlated with a bank of known template waveforms [14]. For the searches on data from the O1 observation run, the bank of approximately 250,000 template waveforms are distributed in within the masses $m_1 > 1\,M_\odot, m_2 > 1\,M_\odot$, $M = m_1 + m_2 < 100\,M_\odot$ and spin magnitudes of $0 \leq \chi \leq 0.05$ for neutron stars and $0 \leq \chi \leq 0.9895$ for black holes [15]. Two independent implementations of the matched filtering algorithm are used for compact binaries, namely PyCBC [16,17] and GstLAL [18–20]. Both search implementations calculate the matched-filter SNR and a goodness-of-fit measure similar to a χ^2 test. These two ingredients are then combined to compute a ranking statistic, which is used to rank candidate events (triggers) according to their likelihood of being a gravitational-wave signal. Finally, the significance of triggers (e.g. false alarm rate) is calculated by comparing them to a search background which is either calculated by obtaining a set of coincident triggers through artificially shifting in the timestamps of one detectors triggers or by using the distribution of non-coincident triggers as a proxy. Using these methods, the false alarm rate of GW150914 was estimated to be less than 1 event per 203,000 years, equivalent to a significance greater than $5.1\,\sigma$ [15]. Unmodelled algorithms, which search coherent time-frequency content amongst different detectors, have independently identified GW150914 and subsequent detections [1,3,21].

Information about the properties of the source is provided by computing the posterior probability density function $\Pr(\vec{\theta} \mid d)$ of the unknown parameters $\vec{\theta}$, given the instrument data d. The posterior is proportional to the likelihood $\Pr(d \mid \vec{\theta})$ of the data d given the parameters $\vec{\theta}$ and the prior probability $\Pr(\vec{\theta})$ before the experiment. This posterior and the corresponding evidence (marginalised likelihood) are computed using Markov-chain Monte Carlo [22,23] and Nested Sampling [24,25] methods.

3. Implications

3.1. *Astrophysical implications*

The discovery of pairs of merging black holes have shown the existence of stellar-mass black holes that are considerably heavier than those observed by X-ray observation [1,3], which are in the range of $5 - 29\,M_\odot$ [26]. It is believed that mass loss due to stellar winds are related to the metallicity of the stars through the strength of stellar winds [27]. In particular, sub-solar metallicities are needed to form black holes of $M > 20\,M_\odot$ through regular stellar processes, but that such stellar compositions are relatively rare. Therefore, if the black holes observed by LIGO were formed through stellar processes, then these observations will have profound impact on our models of stellar evolution. Moreover, black-hole binaries are expected to be formed through either dynamical or isolated binary evolution (see e.g. Ref. 3 and references therein). These formation scenarios have their characteristic population properties in terms of the rate of merger, mass and spin of the black holes. So far, the LIGO detections are still consistent with both or a mixture between scenarios[3,28].

3.2. *Test of general relativity*

The merger of binary black holes allows for a unique view of the strong-field dynamics of gravity and to test general relativity in a regime never tested before. In particular, one can consider the dimensionless compactness GM/c^2R, where G is Newton's constant, M is the total mass, R is the typical separation, and c is the speed of light, and the dimensionless speed v/c. For binary pulsars, the compactness and speed are observed up to 10^{-4} and 10^{-3}, respectively, which is still orders of magnitude smaller compared to 0.2 and 0.4 for binary mergers, respectively. LIGO's detections have been used to test general relativity in a variety of ways. For example, searches have been conducted for generic deviations from waveforms that are calibrated against numerical simulations, the inspiral and post-inspiral portions have been tested for their mutual consistency. Moreover, propagation effects such as dispersion have been constrained including one associated to a non-zero graviton mass. These and other tests have not indicated deviations from general relativity, but have put meaningful constraints on alternative theories[3,29].

4. Outlook

In the coming years other second-generation detectors such as Virgo in Italy[30], KAGRA in Japan[31,32] and LIGO-India in India[33], are expected to join the international network of ground-based interferometric detectors. Such a network will have the ability to see more events, provide better estimation of the source parameters and increase the accuracy on the sky localisation[34]. Third generation detectors, such as Einstein Telescope[35] and Cosmic Explorer[36] are expected to further improve the sensitivity by an order of magnitude. These developments can lead to drastic improvements in the existing science that one can do with binary black holes and their populations. Moreover, novel physics such as the detection of binaries with (exotic forms of) matter (e.g. Refs. 37–42), the observation of the black hole quasi-normal modes (e.g. Refs. 43–45), near-horizon physics (e.g. Refs. 46, 47), and precision cosmology (e.g. Refs. 41, 48–50) are expected to change the landscape of astronomy and fundamental physics.

Other forms of gravitational-wave detectors, including space-based interferometers[51] and pulsar timing arrays[52] are expected to open different parts of the gravitational-wave spectrum and uncover different classes of sources. Gravitational-wave detectors will also be supplemented by electromagnetic instruments and particle detectors to unveil different aspects of cataclysmic events such as gamma-ray bursts[53–55] and fast radio bursts[56–61].

Indeed, gravitational astronomy will herald a new era in fundamental physics, cosmology and astrophysics, giving us access to the most energetic phenomena in the Universe, far exceeding all processes but the Big Bang itself.

References

1. B. P. Abbott, R. Abbott, T. D. Abbott, M. R. Abernathy, F. Acernese, K. Ackley, C. Adams, T. Adams, P. Addesso, R. X. Adhikari and et al., Observation of gravitational waves from a binary black hole merger, *Physical Review Letters* **116**, p. 061102 (February 2016).
2. B. P. Abbott, R. Abbott, T. D. Abbott, M. R. Abernathy, F. Acernese, K. Ackley, C. Adams, T. Adams, P. Addesso, R. X. Adhikari and et al., GW151226: Observation of gravitational waves from a 22-solar-mass binary black hole coalescence, *Physical Review Letters* **116**, p. 241103 (June 2016).
3. B. P. Abbott, R. Abbott, T. D. Abbott, F. Acernese, K. Ackley, C. Adams, T. Adams, P. Addesso, R. X. Adhikari, V. B. Adya and et al., GW170104: Observation of a 50-solar-mass binary black hole coalescence at redshift 0.2, *Physical Review Letters* **118**, p. 221101 (June 2017).
4. B. P. Abbott, R. Abbott, R. Adhikari, P. Ajith, B. Allen, G. Allen, R. S. Amin, S. B. Anderson, W. G. Anderson, M. A. Arain and et al., LIGO: The laser interferometer gravitational-wave observatory, *Reports on Progress in Physics* **72**, p. 076901 (July 2009).
5. LIGO Scientific Collaboration, J. Aasi, B. P. Abbott, R. Abbott, T. Abbott, M. R. Abernathy, K. Ackley, C. Adams, T. Adams, P. Addesso and et al., Advanced LIGO, *Classical and Quantum Gravity* **32**, p. 074001 (April 2015).
6. S. M. Aston, M. A. Barton, A. S. Bell, N. Beveridge, B. Bland, A. J. Brummitt, G. Cagnoli, C. A. Cantley, L. Carbone, A. V. Cumming, L. Cunningham, R. M. Cutler, R. J. S. Greenhalgh, G. D. Hammond, K. Haughian, T. M. Hayler, A. Heptonstall, J. Heefner, D. Hoyland, J. Hough, R. Jones, J. S. Kissel, R. Kumar, N. A. Lockerbie, D. Lodhia, I. W. Martin, P. G. Murray, J. O'Dell, M. V. Plissi, S. Reid, J. Romie, N. A. Robertson, S. Rowan, B. Shapiro, C. C. Speake, K. A. Strain, K. V. Tokmakov, C. Torrie, A. A. van Veggel, A. Vecchio and I. Wilmut, Update on quadruple suspension design for advanced LIGO, *Classical and Quantum Gravity* **29**, p. 235004 (December 2012).
7. S. Wen, R. Mittleman, K. Mason, J. Giaime, R. Abbott, J. Kern, B. O'Reilly, R. Bork, M. Hammond, C. Hardham, B. Lantz, W. Hua, D. Coyne, G. Traylor, H. Overmier, T. Evans, J. Hanson, O. Spjeld, M. Macinnis, K. Mailand, D. Ottaway, D. Sellers, K. Carter and P. Sarin, Hydraulic external pre-isolator system for LIGO, *Classical and Quantum Gravity* **31**, p. 235001 (December 2014).
8. F. Matichard, B. Lantz, R. Mittleman, K. Mason, J. Kissel, B. Abbott, S. Biscans, J. McIver, R. Abbott, S. Abbott, E. Allwine, S. Barnum, J. Birch, C. Celerier, D. Clark, D. Coyne, D. DeBra, R. DeRosa, M. Evans, S. Foley, P. Fritschel, J. A. Giaime, C. Gray, G. Grabeel, J. Hanson, C. Hardham, M. Hillard, W. Hua, C. Kucharczyk, M. Landry, A. Le Roux, V. Lhuillier, D. Macleod, M. Macinnis, R. Mitchell, B. O'Reilly, D. Ottaway, H. Paris, A. Pele, M. Puma, H. Radkins, C. Ramet, M. Robinson, L. Ruet, P. Sarin, D. Shoemaker, A. Stein, J. Thomas, M. Vargas, K. Venkateswara, J. Warner and S. Wen, Seismic isolation of Advanced LIGO: Review of strategy, instrumentation and performance, *Classical and Quantum Gravity* **32**, p. 185003 (September 2015).
9. B. P. Abbott, R. Abbott, T. D. Abbott, M. R. Abernathy, F. Acernese, K. Ackley, C. Adams, T. Adams, P. Addesso, R. X. Adhikari and et al., GW150914: The advanced LIGO detectors in the era of first discoveries, *Physical Review Letters* **116**, p. 131103 (April 2016).
10. T. T. Fricke, N. D. Smith-Lefebvre, R. Abbott, R. Adhikari, K. L. Dooley, M. Evans, P. Fritschel, V. V. Frolov, K. Kawabe, J. S. Kissel, B. J. J. Slagmolen and S. J.

Waldman, DC readout experiment in Enhanced LIGO, *Classical and Quantum Gravity* **29**, p. 065005 (March 2012).
11. E. Goetz, P. Kalmus, S. Erickson, R. L. Savage, Jr., G. Gonzalez, K. Kawabe, M. Landry, S. Marka, B. O'Reilly, K. Riles, D. Sigg and P. Willems, Precise calibration of LIGO test mass actuators using photon radiation pressure, *Classical and Quantum Gravity* **26**, p. 245011 (December 2009).
12. B. P. Abbott, R. Abbott, T. D. Abbott, M. R. Abernathy, K. Ackley, C. Adams, P. Addesso, R. X. Adhikari, V. B. Adya, C. Affeldt and et al., Calibration of the Advanced LIGO detectors for the discovery of the binary black-hole merger GW150914, *Physical Review D* **95**, p. 062003 (March 2017).
13. B. P. Abbott, R. Abbott, T. D. Abbott, M. R. Abernathy, F. Acernese, K. Ackley, M. Adamo, C. Adams, T. Adams, P. Addesso and et al., Characterization of transient noise in Advanced LIGO relevant to gravitational wave signal GW150914, *Classical and Quantum Gravity* **33**, p. 134001 (July 2016).
14. B. Allen, W. G. Anderson, P. R. Brady, D. A. Brown and J. D. E. Creighton, FINDCHIRP: An algorithm for detection of gravitational waves from inspiraling compact binaries, *Physical Review D* **85**, p. 122006 (June 2012).
15. B. P. Abbott, R. Abbott, T. D. Abbott, M. R. Abernathy, F. Acernese, K. Ackley, C. Adams, T. Adams, P. Addesso, R. X. Adhikari and et al., GW150914: First results from the search for binary black hole coalescence with Advanced LIGO, *Physical Review D* **93**, p. 122003 (June 2016).
16. T. Dal Canton, A. H. Nitz, A. P. Lundgren, A. B. Nielsen, D. A. Brown, T. Dent, I. W. Harry, B. Krishnan, A. J. Miller, K. Wette, K. Wiesner and J. L. Willis, Implementing a search for aligned-spin neutron star-black hole systems with advanced ground based gravitational wave detectors, *Physical Review D* **90**, p. 082004 (October 2014).
17. S. A. Usman, A. H. Nitz, I. W. Harry, C. M. Biwer, D. A. Brown, M. Cabero, C. D. Capano, T. Dal Canton, T. Dent, S. Fairhurst, M. S. Kehl, D. Keppel, B. Krishnan, A. Lenon, A. Lundgren, A. B. Nielsen, L. P. Pekowsky, H. P. Pfeiffer, P. R. Saulson, M. West and J. L. Willis, The PyCBC search for gravitational waves from compact binary coalescence, *Classical and Quantum Gravity* **33**, p. 215004 (November 2016).
18. K. Cannon, R. Cariou, A. Chapman, M. Crispin-Ortuzar, N. Fotopoulos, M. Frei, C. Hanna, E. Kara, D. Keppel, L. Liao, S. Privitera, A. Searle, L. Singer and A. Weinstein, Toward early-warning detection of gravitational waves from compact binary coalescence, *The Astrophysical Journal* **748**, p. 136 (April 2012).
19. S. Privitera, S. R. P. Mohapatra, P. Ajith, K. Cannon, N. Fotopoulos, M. A. Frei, C. Hanna, A. J. Weinstein and J. T. Whelan, Improving the sensitivity of a search for coalescing binary black holes with nonprecessing spins in gravitational wave data, *Physical Review D* **89**, p. 024003 (January 2014).
20. C. Messick, K. Blackburn, P. Brady, P. Brockill, K. Cannon, R. Cariou, S. Caudill, S. J. Chamberlin, J. D. E. Creighton, R. Everett, C. Hanna, D. Keppel, R. N. Lang, T. G. F. Li, D. Meacher, A. Nielsen, C. Pankow, S. Privitera, H. Qi, S. Sachdev, L. Sadeghian, L. Singer, E. G. Thomas, L. Wade, M. Wade, A. Weinstein and K. Wiesner, Analysis framework for the prompt discovery of compact binary mergers in gravitational-wave data, *Physical Review D* **95**, p. 042001 (February 2017).
21. B. P. Abbott, R. Abbott, T. D. Abbott, M. R. Abernathy, F. Acernese, K. Ackley, C. Adams, T. Adams, P. Addesso, R. X. Adhikari and et al., Observing gravitational-wave transient GW150914 with minimal assumptions, *Physical Review D* **93**, p. 122004 (June 2016).
22. C. Röver, R. Meyer and N. Christensen, Bayesian inference on compact binary inspiral gravitational radiation signals in interferometric data, *Classical and Quantum Gravity* **23**, 4895 (August 2006).

23. M. van der Sluys, V. Raymond, I. Mandel, C. Röver, N. Christensen, V. Kalogera, R. Meyer and A. Vecchio, Parameter estimation of spinning binary inspirals using Markov chain Monte Carlo, *Classical and Quantum Gravity* **25**, p. 184011 (September 2008).
24. J. Veitch and A. Vecchio, Bayesian coherent analysis of in-spiral gravitational wave signals with a detector network, *Physical Review D* **81**, p. 062003 (March 2010).
25. J. Skilling, Nested sampling for general bayesian computation, *Bayesian Anal.* **1**, 833 (12 2006).
26. M. C. Miller and J. M. Miller, The masses and spins of neutron stars and stellar-mass black holes, *Physics Reports* **548**, 1 (January 2015).
27. K. Belczynski, T. Bulik, C. L. Fryer, A. Ruiter, F. Valsecchi, J. S. Vink and J. R. Hurley, On the maximum mass of stellar black holes, *The Astrophysical Journal* **714**, 1217 (May 2010).
28. B. P. Abbott, R. Abbott, T. D. Abbott, M. R. Abernathy, F. Acernese, K. Ackley, C. Adams, T. Adams, P. Addesso, R. X. Adhikari and et al., Astrophysical implications of the binary black-hole merger GW150914, *The Astrophysical Journal Letters* **818**, p. L22 (February 2016).
29. B. P. Abbott, R. Abbott, T. D. Abbott, M. R. Abernathy, F. Acernese, K. Ackley, C. Adams, T. Adams, P. Addesso, R. X. Adhikari and et al., Tests of general relativity with GW150914, *Physical Review Letters* **116**, p. 221101 (June 2016).
30. F. Acernese, M. Agathos, K. Agatsuma, D. Aisa, N. Allemandou, A. Allocca, J. Amarni, P. Astone, G. Balestri, G. Ballardin and et al., Advanced virgo: A second-generation interferometric gravitational wave detector, *Classical and Quantum Gravity* **32**, p. 024001 (January 2015).
31. K. Somiya, Detector configuration of KAGRA-the Japanese cryogenic gravitational-wave detector, *Classical and Quantum Gravity* **29**, p. 124007 (June 2012).
32. Y. Aso, Y. Michimura, K. Somiya, M. Ando, O. Miyakawa, T. Sekiguchi, D. Tatsumi and H. Yamamoto, Interferometer design of the KAGRA gravitational wave detector, *Physical Review D* **88**, p. 043007 (August 2013).
33. C. S. Unnikrishnan, IndIGO and ligo-india scope and plans for gravitational wave research and precision metrology in india, *International Journal of Modern Physics D* **22**, p. 1341010 (January 2013).
34. B. P. Abbott, R. Abbott, T. D. Abbott, M. R. Abernathy, F. Acernese, K. Ackley, C. Adams, T. Adams, P. Addesso, R. X. Adhikari and et al., Prospects for observing and localizing gravitational-wave transients with advanced LIGO and advanced virgo, *Living Reviews in Relativity* **19**, p. 1 (February 2016).
35. Einstein Telescope Science Team, *Einstein Gravitational Wave Telescope Conceptual Design Study*, Tech. Rep. ET-0106C-10 (2011).
36. B. P. Abbott, R. Abbott, T. D. Abbott, M. R. Abernathy, K. Ackley, C. Adams, P. Addesso, R. X. Adhikari, V. B. Adya, C. Affeldt and et al., Exploring the sensitivity of next generation gravitational wave detectors, *Classical and Quantum Gravity* **34**, p. 044001 (February 2017).
37. J. Clark, A. Bauswein, L. Cadonati, H.-T. Janka, C. Pankow and N. Stergioulas, Prospects for high frequency burst searches following binary neutron star coalescence with advanced gravitational wave detectors, *Physical Review D* **90**, p. 062004 (September 2014).
38. M. Agathos, J. Meidam, W. Del Pozzo, T. G. F. Li, M. Tompitak, J. Veitch, S. Vitale and C. Van Den Broeck, Constraining the neutron star equation of state with gravitational wave signals from coalescing binary neutron stars, *Physical Review D* **92**, p. 023012 (July 2015).

39. B. D. Lackey and L. Wade, Reconstructing the neutron-star equation of state with gravitational-wave detectors from a realistic population of inspiralling binary neutron stars, *Physical Review D* **91**, p. 043002 (February 2015).
40. J. A. Clark, A. Bauswein, N. Stergioulas and D. Shoemaker, Observing gravitational waves from the post-merger phase of binary neutron star coalescence, *Classical and Quantum Gravity* **33**, p. 085003 (April 2016).
41. W. Del Pozzo, T. G. F. Li and C. Messenger, Cosmological inference using only gravitational wave observations of binary neutron stars, *Physical Review D* **95**, p. 043502 (February 2017).
42. V. Cardoso, E. Franzin, A. Maselli, P. Pani and G. Raposo, Testing strong-field gravity with tidal love numbers, *Physical Review D* **95**, p. 084014 (April 2017).
43. I. Kamaretsos, M. Hannam and B. S. Sathyaprakash, Is black-hole ringdown a memory of its progenitor?, *Physical Review Letters* **109**, p. 141102 (October 2012).
44. S. Gossan, J. Veitch and B. S. Sathyaprakash, Bayesian model selection for testing the no-hair theorem with black hole ringdowns, *Physical Review D* **85**, p. 124056 (June 2012).
45. J. Meidam, M. Agathos, C. Van Den Broeck, J. Veitch and B. S. Sathyaprakash, Testing the no-hair theorem with black hole ringdowns using TIGER, *Physical Review D* **90**, p. 064009 (September 2014).
46. V. Cardoso, E. Franzin and P. Pani, Is the gravitational-wave ringdown a probe of the event horizon?, *Physical Review Letters* **116**, p. 171101 (April 2016).
47. C. Barceló, R. Carballo-Rubio and L. J. Garay, Gravitational wave echoes from macroscopic quantum gravity effects, *Journal of High Energy Physics* **5**, p. 54 (May 2017).
48. W. Del Pozzo, Inference of cosmological parameters from gravitational waves: Applications to second generation interferometers, *Physical Review D* **86**, p. 043011 (August 2012).
49. S. R. Taylor, J. R. Gair and I. Mandel, Cosmology using advanced gravitational-wave detectors alone, *Physical Review D* **85**, p. 023535 (January 2012).
50. S. R. Taylor and J. R. Gair, Cosmology with the lights off: Standard sirens in the Einstein Telescope era, *Physical Review D* **86**, p. 023502 (July 2012).
51. The LISA Consortium, *Laser Interferometer Space Antenna*, tech. rep. (2017), https://www.elisascience.org/files/publications/LISA_L3_20170120.pdf.
52. G. Hobbs, A. Archibald, Z. Arzoumanian, D. Backer, M. Bailes, N. D. R. Bhat, M. Burgay, S. Burke-Spolaor, D. Champion, I. Cognard, W. Coles, J. Cordes, P. Demorest, G. Desvignes, R. D. Ferdman, L. Finn, P. Freire, M. Gonzalez, J. Hessels, A. Hotan, G. Janssen, F. Jenet, A. Jessner, C. Jordan, V. Kaspi, M. Kramer, V. Kondratiev, J. Lazio, K. Lazaridis, K. J. Lee, Y. Levin, A. Lommen, D. Lorimer, R. Lynch, A. Lyne, R. Manchester, M. McLaughlin, D. Nice, S. Oslowski, M. Pilia, A. Possenti, M. Purver, S. Ransom, J. Reynolds, S. Sanidas, J. Sarkissian, A. Sesana, R. Shannon, X. Siemens, I. Stairs, B. Stappers, D. Stinebring, G. Theureau, R. van Haasteren, W. van Straten, J. P. W. Verbiest, D. R. B. Yardley and X. P. You, The international pulsar timing array project: using pulsars as a gravitational wave detector, *Classical and Quantum Gravity* **27**, p. 084013 (April 2010).
53. R. A. M. J. Wijers, J. S. Bloom, J. S. Bagla and P. Natarajan, Gamma-ray bursts from stellar remnants — Probing the universe at high redshift, *Monthly Notices of the Royal Astronomical Society* **294**, L13 (February 1998).
54. C. Porciani and P. Madau, On the association of gamma-ray bursts with massive stars: Implications for number counts and lensing statistics, *The Astrophysical Journal* **548**, 522 (February 2001).
55. X.-F. Cao, Y.-W. Yu, K. S. Cheng and X.-P. Zheng, The luminosity function of Swift

long gamma-ray bursts, *Monthly Notices of the Royal Astronomical Society* **416**, 2174 (September 2011).
56. D. R. Lorimer, M. Bailes, M. A. McLaughlin, D. J. Narkevic and F. Crawford, A Bright Millisecond Radio Burst of Extragalactic Origin, *Science* **318**, p. 777 (November 2007).
57. D. Thornton, B. Stappers, M. Bailes, B. Barsdell, S. Bates, N. D. R. Bhat, M. Burgay, S. Burke-Spolaor, D. J. Champion, P. Coster, N. D'Amico, A. Jameson, S. Johnston, M. Keith, M. Kramer, L. Levin, S. Milia, C. Ng, A. Possenti and W. van Straten, A population of fast radio bursts at cosmological distances, *Science* **341**, 53 (July 2013).
58. L. G. Spitler, P. Scholz, J. W. T. Hessels, S. Bogdanov, A. Brazier, F. Camilo, S. Chatterjee, J. M. Cordes, F. Crawford, J. Deneva, R. D. Ferdman, P. C. C. Freire, V. M. Kaspi, P. Lazarus, R. Lynch, E. C. Madsen, M. A. McLaughlin, C. Patel, S. M. Ransom, A. Seymour, I. H. Stairs, B. W. Stappers, J. van Leeuwen and W. W. Zhu, A repeating fast radio burst, *Nature* **531**, 202 (March 2016).
59. K. Masui, H.-H. Lin, J. Sievers, C. J. Anderson, T.-C. Chang, X. Chen, A. Ganguly, M. Jarvis, C.-Y. Kuo, Y.-C. Li, Y.-W. Liao, M. McLaughlin, U.-L. Pen, J. B. Peterson, A. Roman, P. T. Timbie, T. Voytek and J. K. Yadav, Dense magnetized plasma associated with a fast radio burst, *Nature* **528**, 523 (December 2015).
60. E. F. Keane, S. Johnston, S. Bhandari, E. Barr, N. D. R. Bhat, M. Burgay, M. Caleb, C. Flynn, A. Jameson, M. Kramer, E. Petroff, A. Possenti, W. van Straten, M. Bailes, S. Burke-Spolaor, R. P. Eatough, B. W. Stappers, T. Totani, M. Honma, H. Furusawa, T. Hattori, T. Morokuma, Y. Niino, H. Sugai, T. Terai, N. Tominaga, S. Yamasaki, N. Yasuda, R. Allen, J. Cooke, J. Jencson, M. M. Kasliwal, D. L. Kaplan, S. J. Tingay, A. Williams, R. Wayth, P. Chandra, D. Perrodin, M. Berezina, M. Mickaliger and C. Bassa, The host galaxy of a fast radio burst, *Nature* **530**, 453 (February 2016).
61. V. Ravi, R. M. Shannon, M. Bailes, K. Bannister, S. Bhandari, N. D. R. Bhat, S. Burke-Spolaor, M. Caleb, C. Flynn, A. Jameson, S. Johnston, E. F. Keane, M. Kerr, C. Tiburzi, A. V. Tuntsov and H. K. Vedantham, The magnetic field and turbulence of the cosmic web measured using a brilliant fast radio burst, *Science* **354**, 1249 (December 2016).

Mass Loss Due to Gravitational Waves with $\Lambda > 0$

Vee-Liem Saw
Department of Mathematics and Statistics, University of Otago,
Dunedin 9016, New Zealand
VeeLiem@maths.otago.ac.nz

The theoretical basis for the energy carried away by gravitational waves that an isolated gravitating system emits was first formulated by Hermann Bondi during the '60s. Recent findings from the observation of distant supernovae revealed that the rate of expansion of our universe is accelerating, which may be well explained by sticking a positive cosmological constant into the Einstein field equations for general relativity. By solving the Newman–Penrose equations (which are equivalent to the Einstein field equations), we generalize this notion of Bondi mass–energy and thereby provide a firm theoretical description of how an isolated gravitating system loses energy as it radiates gravitational waves, in a universe that expands at an accelerated rate. This is in line with the observational front of LIGO's first announcement in February 2016 that gravitational waves from the merger of a binary black hole system have been detected.

Keywords: Gravitational waves; mass-loss formula; Bondi–Sachs mass; cosmological constant; de Sitter; null infinity.

The notion that gravitational waves carry energy away from an isolated system of masses was first established by Bondi and his coworkers, leading to the well-known mass-loss formula.[1,2] This assumed that spacetime is asymptotically flat, i.e. the system of masses is confined within a bounded region and spacetime gets ever closer to being Minkowskian towards large distances away from the source. Bondi enunciated a metric ansatz describing an axially symmetric spacetime, and solved the vacuum Einstein equations (together with the Bianchi identities) in the region far away from the source. The mass-loss formula is then obtained from one of the "supplementary conditions" that arose from the Bianchi identities.

Around that same period in the '60s, an equivalent formulation of general relativity was worked out by Newman and Penrose, making use of quantities called *spin coefficients* (which are essentially the connection coefficients).[3] This led to a collection of 38 (mostly linear) differential equations which are equivalent to the Einstein field equations and the Bianchi identities. By solving these equations at large distances for asymptotically flat spacetimes, Newman and Unti obtained the general asymptotic solutions.[4] One of the relationships from the Bianchi identities, integrated over a 2-sphere of constant u at null infinity $\mathcal{I}$, is in fact the Bondi mass-loss formula

$$\frac{dM_B}{du} = -\frac{1}{A}\oint |\dot{\sigma}^o|^2\, d^2S\,, \tag{1}$$

where $M_B = -\frac{1}{A}\oint(\Psi_2^o + \sigma^o\dot{\bar{\sigma}}^o)d^2S$ is the Bondi mass, and A is the area of that 2-sphere of constant u on $\mathcal{I}$. Dot is derivative with respect to u, which is a retarded null coordinate (and may be interpreted as "time"). The term Ψ_2^o is the leading order term of Ψ_2 when expanded over large distances from the source (where Ψ_2 is one of the dyad components of the Weyl spinor), and σ^o is the leading order term of the complex spin coefficient σ (under the large distance expansion). The presence of $\dot{\sigma}^o$ indicates gravitational waves being emitted by the system, so the mass of the system decreases due to the energy carried away by those gravitational waves.

The purpose of this brief review is to present the key results of a generalization to the above mass-loss formula for a universe with a positive cosmological constant $\Lambda > 0$, as was recently reported in Ref. 5. This work is motivated by the fact that we have observed our universe to be expanding at an *accelerated* rate,[6,7] which spawned intense research over the past few years (see Ref. 8 for a more extensive review). A positive cosmological constant $\Lambda > 0$ provides a simple explanation for the accelerated rate of expansion.

By solving the Newman–Penrose equations with a cosmological constant Λ à la Newman–Unti[4] in Ref. 5, the mass-loss formula with $\Lambda > 0$ (obtained from that same Bianchi identity as in the case for $\Lambda = 0$) is

$$\frac{dM_\Lambda}{du} = -\frac{1}{A}\oint\left(|\dot{\sigma}^o|^2 + \frac{\Lambda}{3}|\eth'\sigma^o|^2 + \frac{2\Lambda^2}{9}|\sigma^o|^4 + \frac{\Lambda^2}{18}\,\mathrm{Re}(\bar{\sigma}^o\Psi_0^o)\right)d^2S\,. \tag{2}$$

Here, the definition of the mass with $\Lambda > 0$ is proposed to be

$$\begin{aligned} M_\Lambda := M_B &+ \frac{1}{A}\int\left(\oint(\Psi_2^o + \sigma^o\dot{\bar{\sigma}}^o)\frac{\partial}{\partial u}(d^2S)\right)du \\ &+ \frac{\Lambda}{3A}\int\left(\oint K|\sigma^o|^2\,d^2S\right)du\,, \end{aligned} \tag{3}$$

which ensures that this mass M_Λ strictly decreases whenever σ^o is nonzero (in the absence of incoming gravitational radiation from elsewhere, or $\Psi_0^o = 0$), i.e. the mass of an isolated gravitating system *strictly decreases due to energy carried away by gravitational waves.*[5] (The term with K in Eq. (3) will be explained below, shortly.)

As opposed to the case where $\Lambda = 0$, here $\mathcal{I}$ is *non-conformally flat* for $\Lambda > 0$ — its Cotton–York tensor is nonzero, when those outgoing gravitational waves carry energy away from the source.[5,9] As a result, the integration is not carried out over a round 2-sphere — but instead over a *topological 2-sphere* of constant u on $\mathcal{I}$.[5,10,a]

[a] See Ref. 10 which describes the foliation of the conformally rescaled $\mathcal{I}$ by topological 2-spheres.

These compact 2-surfaces of constant u on $\mathcal{I}$ being topological instead of round 2-spheres give rise to the second term on the right-hand side of Eq. (3), because the surface element d^2S in general has a u-dependence.[b] The Gauss curvature for these topological 2-spheres on $\mathcal{I}$ is K, which in general depends on σ^o when $\Lambda \neq 0$. If $\sigma^o = 0$, then $K = 1$ — indicating a round 2-sphere of constant curvature.

We see that with the presence of Λ, then σ^o itself would contribute to the mass-loss formula in Eq. (2). (Recall that for asymptotically flat spacetimes in Eq. (1), one needs a variation of σ^o with u to give rise to a decrease in mass.) Apart from that, Ψ_0^o (which is the leading order term of the dyad component of the Weyl spinor Ψ_0) represents the incoming radiation and shows up in the mass-loss formula — due to $\mathcal{I}$ being a space-like hypersurface. In a universe expanding at an accelerated rate, the isolated system has a cosmological horizon. Such incoming radiation from elsewhere that lies beyond the cosmological horizon would not affect the system, but *will arrive at $\mathcal{I}$ and get picked up by the mass-loss formula.*

The asymptotic solutions with a cosmological constant have been extended to include Maxwell fields.[11] The full mass-loss formula is

$$\frac{dM_\Lambda}{du} = -\frac{1}{A}\oint\left(|\dot{\sigma}^o|^2 + k|\phi_2^o|^2 + \frac{\Lambda}{3}|\eth'\sigma^o|^2 + \frac{2\Lambda^2}{9}|\sigma^o|^4 + \frac{\Lambda^2}{18}\,\mathrm{Re}(\bar{\sigma}^o\Psi_0^o) - \frac{k\Lambda^2}{36}|\phi_0^o|^2\right)d^2S \tag{4}$$

and if there are only Maxwell fields but no gravitational radiation, i.e. $\sigma^o = 0$, then

$$\frac{dM_\Lambda}{du} = -\frac{k}{A}\oint\left(|\phi_2^o|^2 - \frac{\Lambda^2}{36}|\phi_0^o|^2\right)d^2S\,, \tag{5}$$

where $M_\Lambda = M_B$ (since the compact 2-surface of constant u on $\mathcal{I}$ is a round sphere when $\sigma^o = 0$, so the surface element d^2S does not have a u-dependence). The terms ϕ_2^o and ϕ_0^o (which are the leading order terms of the dyad components of the Maxwell spinor ϕ_2 and ϕ_0, respectively) represent outgoing and incoming electromagnetic radiations, respectively. The Maxwell fields do not affect the structure of $\mathcal{I}$, unlike the gravitational counterpart which would lead to its non-conformal flatness when the outgoing gravitational waves carry energy away from the isolated system. From Eq. (5), it is clear that outgoing electromagnetic radiation that the source emits carries energy away from it, whilst the incoming electromagnetic radiation from elsewhere would increase the isolated system's total mass–energy.

Full details and extensive elaborations are given in Ref. 5 for the gravitational case, with the extension to include Maxwell fields found in Ref. 11. In relation to this, a discussion on the peeling property with a cosmological constant is given in Ref. 12.

[b]This crops up due to a process of interchanging the order of taking a u-derivative and integrating over the topological 2-sphere.

Acknowledgments

V.-L.S. is working on this research project under the University of Otago Doctoral Scholarship. Travel support to the Nanyang Technological University (NTU), Singapore for this Conference was provided by the Division of Science, University of Otago. Furthermore, I am very grateful for the hospitality shown by the Institute of Advanced Studies, NTU.

References

1. H. Bondi, *Nature* **186**, 535 (1960).
2. H. Bondi, M. G. J. van der Burg and A. W. K. Metzner, *Proc. R. Soc. Lond. A: Math. Phys. Eng. Sci.* **269**, 21 (1962), doi:10.1098/rspa.1962.0161.
3. E. T. Newman and R. Penrose, *J. Math. Phys.* **3**, 566 (1962), doi:http://dx.doi.org/10.1063/1.1724257.
4. E. T. Newman and T. W. J. Unti, *J. Math. Phys.* **3**, 891 (1962), doi:http://dx.doi.org/10.1063/1.1724303.
5. V.-L. Saw, *Phys. Rev. D* **94**, 104004 (2016), doi:10.1103/PhysRevD.94.104004.
6. A. G. Riess *et al.*, *Astron. J.* **116**, 1009 (1998), doi:10.1086/300499.
7. S. Perlmutter *et al.*, *Astrophys. J.* **517**, 565 (1999).
8. V.-L. Saw, arXiv:1706.00160.
9. A. Ashtekar, B. Bonga and A. Kesavan, *Class. Quantum Grav.* **32**, 025004 (2015).
10. L. B. Szabados and P. Tod, *Class. Quantum Grav.* **32**, 205011 (2015).
11. V.-L. Saw, *Phys. Rev. D* **95**, 084038 (2017), doi:10.1103/PhysRevD.95.084038.
12. V.-L. Saw, arXiv:1705.00435.

Exploring Fundamental Physics with Gravitational Waves

Archil Kobakhidze

ARC Centre of Excellence for Particle Physics at the Terascale,
School of Physics, The University of Sydney,
Sydney, NSW 2006, Australia
archil.kobakhidze@sydney.edu.au
http://sydney.edu.au/science/people/archil.kobakhidze.php

The direct discovery of gravitational waves opens of new avenues for exploring fundamental microscopic physics. In this presentation, I will illustrate this by obtaining the stringent bounds on the scale of quantum fuzziness of hypothetical noncommutative space-time. I also discuss how future observations of stochastic gravitational wave background can shed light on the nature of the Higgs mechanism of the electroweak symmetry breaking and the related cosmological phase transition.

Keywords: Gravitational waves; quantum space-times; Higgs mechanism; cosmological phase transition.

1. Introduction

The direct observation of gravitational waves from binary black hole mergers[1] represents a triumphal confirmation of Einstein's general covariant theory of gravitation. It has oppened a new era in multimessenger astronomy. In this presentation I would like to argue that current and future observations of gravitational waves also provides with new opportunities for exploring fundamental physical phenomena at microscopic scales, some of which may not be accessible by other means. I discuss two examples of this. In Section 2, I will demonstrate how the observed gravitational wave signals put the stringent constraints on the scale of quantum fuzziness of a hypothetical non-commutative space-time[2]. In Section 3, I will discuss a potential for future observations of a stochastic gravitational wave background from the electroweak phase transition to shed light on the fundamental mechanism of mass generation in particle physics and the properties of the Higgs boson[3,4].

2. Constraints on Quantum Space-time from GW150914

The idea of quantised space-times has first emerged in relation to the short-scale divergences in quantum field theory and traces back to Heisenberg and Pauli. With the advent of the renormalisation program the idea has been forgotten for a while, until in late 90s, when it has been found that quantum space-times emerge in certain field-theoretic limits of the fundametal string theory[5,6]. The space-time in this scenario exhibits fuzziness characterised by a scale ℓ, in analogy with the quantum fuzziness of the phase space in quantum mechanics, defined through Planck's constant $\hbar$. In what follows, I will show that the stringent bound on ℓ follows from the observed gravitational wave signals.

2.1. *Noncommutative space-time*

The quantum space-time with canonical noncommutativity is defined by a set of operator-valued space-time coordinates that obey the following commutation relations:

$$[\hat{x}^\mu, \hat{x}^\nu] = i\,\ell^2 \theta^{\mu\nu}, \tag{1}$$

where $\theta^{\mu\nu}$ is a real and constant antisymmetric tensor.[a] A quantum field theory in noncommutative space-time (1), is convenient to define in terms of ordinary commutative fields with ordinary products replaced by the following nonlocal product (the symmetric Moyal product):

$$f(x)\star g(x) = f(x)g(x) + \sum_{n=1}^{+\infty} \left(\frac{i\ell}{2}\right)^n \frac{1}{n!}\theta^{\alpha_1\beta 1} \dots \theta^{\alpha_n\beta_n}\, \partial_{\alpha_1} \cdots \partial_{\alpha_n} f(x)\, \partial_{\beta_1} \cdots \partial_{\beta_n} g(x). \tag{2}$$

Hence noncommutative field theories can be treated (with sufficient care) as an usual field theories with noncommutative corrections $\mathcal{O}(\ell^n)$ at each *nth* order term in the Moyal product of fields.

Various aspects of noncommutative field theories have been investigated in the past; see Refs. 7 and 8 and references therein. The limits on the noncommutative scale have been obtained from various particle physics processes, including low-energy precision measurements and processes involving Lorentz symmetry violation. Careful considerations[9] show that the scale of noncommutativity is limited to be $\ell \lesssim TeV^{-1}$. Remarkably, this bound can be improved by the 15 orders of magnitude from observations of gravity wave.

2.2. *Bounds on the scale of space-time fuzziness from gravitational waves*

A binary system is commonly approximated in GR by two point masses whose energy-momentum tensor is given by[b]

$$T^{\mu\nu}_{GR}(\mathbf{x},t) = m_1\gamma_1(t)v_1^\mu(t)v_1^\nu(t)\delta^3(\mathbf{x} - \mathbf{y}_1(t)) + 1 \leftrightarrow 2 \tag{3}$$

with m_i the masses; $\mathbf{y}_i(t)$ the positions; and $v_i^\mu(t) = \left(c, \frac{d\mathbf{y}_i(t)}{dt}\right)$ the velocities of the two bodies $i = 1, 2$. The factor γ_1 is expressed through the metric $g_{\mu\nu}$ and its determinant g as

$$\gamma_1 = \frac{1}{\sqrt{g_1 (g_{\alpha\beta})_1 \frac{v_1^\alpha v_1^\beta}{c^2}}}, \tag{4}$$

[a]In the string-theoretic setting rhs of Eq. (1) represents a background value for an antisymmetric B-field.

[b]For an excellent review of the computational formalism see Ref. 10.

and similarly for γ_2. In this expression, the metric and its determinant are evaluated at the location the body 1, namely, $(g_{\alpha\beta})_1 \equiv g_{\alpha\beta}(\mathbf{y}_1(t))$. Thus γ_1 only depends on time. However, the point-mass approximation implies that g_1 and $(g_{\alpha\beta})_1$ are divergent because of the delta functions in Eq. (3). This problem can be solved through the so-called Hadamard regularization whose application to the PN formalism is described in Ref. 10. It is also worth mentioning that the energy-momentum tensor given by Eqs. (3) and (4) reproduces the correct GR equations of motion only up to 2.5PN order. For orders 3PN and higher, the spatial dependence needs to be accounted for in the metric[10]. Specifically, when evaluating the determinant of the metric in Eq. (4) we must use the field value instead of the location of the point mass, namely, g_1 has to be replaced by $g(t, \mathbf{x})$. Since the lowest-order noncommutative correction turns out to occur at 2PN order in the equations of motion, we can safely ignore this technicality in the present article.

In order to compute noncommutative corrections to the energy-momentum tensor (3), we follow the effective field theory formalism which has been used to compute noncommutative corrections to classical black holes in Ref. 11. In this approach, the Schwarzschild BHs are sourced by a massive real scalar field ϕ. Employing the Moyal product 2, we first compute the noncommutative energy-momentum tensor for a real massive scalar field ϕ in the lowest order of nopncommutative parameter ℓ:

$$\begin{aligned}
T^{\mu\nu}_{NC}(x) &= \frac{1}{2}\left(\partial^\mu\phi \star \partial^\nu\phi + \partial^\nu\phi \star \partial^\mu\phi\right) - \frac{1}{2}\eta^{\mu\nu}\left(\partial_\rho\phi \star \partial^\rho\phi - m^2\phi\star\phi\right) \\
&= \partial^\mu\phi\,\partial^\nu\phi - \frac{1}{2}\eta^{\mu\nu}\left(\partial_\rho\phi\,\partial^\rho\phi - m^2\phi^2\right) - \frac{\ell^4}{8}\theta^{\alpha_1\beta_1}\theta^{\alpha_2\beta_2}\Big(\partial_{\alpha_1}\partial_{\alpha_2}\partial^\mu\phi\partial_{\beta_1}\partial_{\beta_2}\partial^\nu\phi \\
&\quad - \frac{1}{2}\eta^{\mu\nu}\partial_{\alpha_1}\partial_{\alpha_2}\partial_\rho\phi\partial_{\beta_1}\partial_{\beta_2}\partial^\rho\phi + \frac{1}{2}\eta^{\mu\nu}m^2\partial_{\alpha_1}\partial_{\alpha_2}\phi\partial_{\beta_1}\partial_{\beta_2}\phi\Big) + \cdots \,, \qquad (5)
\end{aligned}$$

Note that the first two terms in the second line correspond to the usual energy-momentum tensor of a massive scalar field. Next, following the prescription given in Ref. 11, we compute the following energy momentum tensor for binary system including leading noncommutative correction as[2]:

$$\begin{aligned}
T^{\mu\nu}(\mathbf{x}, t) &= m_1\gamma_1(t)v_1^\mu(t)v_1^\nu(t)\delta^3(\mathbf{x} - \mathbf{y}_1(t)) \\
&\quad + \frac{m_1^3 G^2\Lambda^2}{8c^4} v_1^\mu(t)v_1^\nu(t)\theta^k\theta^l\partial_k\partial_l\,\delta^3(\mathbf{x} - \mathbf{y}_1(t)) + 1 \leftrightarrow 2 \,, \qquad (6)
\end{aligned}$$

where we have simplified the notation by introducing $\Lambda\,\theta^i = (\ell M_P)^2\,\theta^{0i}/$, with θ^i representing the components of a three-dimensional unit vector $\boldsymbol{\theta}$, $\theta^i\theta^i = 1$. In this way $\sqrt{\Lambda}$ parametrise the space-time noncommutativite scale measured in units of the inverse Planck mass $1/M_P$ (in natural units).

Equipped with the above energy-momentum tensor, we closely follow the formalism of Ref. 10 and after lengthy calculations obtain the noncommutative corrections, which in the leading order appears in the 2PN phase of the gravitational wave signal[2]:

$$\varphi_4 = \tfrac{15293365}{508032} + \tfrac{27145}{504}\nu + \tfrac{3085}{72}\nu^2 + \tfrac{25}{2}\Lambda^2(1-2\nu) \ . \tag{7}$$

This expression reduces to the standard result in the limit $\Lambda \to 0$, as it should be. Attributing uncertainties in the observe signal, $|\delta\varphi_4| \lesssim 20^1$, to the above noncommutative correction, we immediately derive the bound[2]:

$$\sqrt{\Lambda} \lesssim 3.5 \ . \tag{8}$$

This implies that the observed gravitational wave signal constraints the scale of space-time noncommutativity be around the Planck scale. This is 15 orders of magnitude stronger bound than those which follow from particle physics considerations[9].

3. Higgs Physics and Gravitational Waves from Electroweak Phase Transition

In the remaining of this presentation I will discuss potentially interesting implication of future observations of stochastic gravitational wave background for the nature of electroweak symmetry breaking and the related cosmological phase transition.

Despite a tremendous progress in determining various properties and couplings of the Higgs boson since its discovery at the Large Hadron Collider (LHC), significant uncertainties remain regarding the nature of the electroweak symmetry breaking. Proper understanding of this important issue requires experimental reconstruction of the Higgs potential, i.e., measurements of Higgs self-interaction couplings, which is a notoriously difficult task at the LHC. A detection of the gravitational wave signals, that can potentially be induced during the first-order electroweak phase transition, may provide complimentary information on the nature of the electroweak symmetry breaking. Below I will describe potential gravitational wave signals from the cosmological phase transition within the effective field theory framework of nonlinearly realised electroweak symmetry[12].

3.1. *Standard model with non-linearly realised electroweak symmetry*

Within the conventional SM, the linearly realised $SU(2)_L \times U(1)_Y$ electroweak gauge symmetry is a hidden gauge symmetry with a stability being the group of QED, $U(1)_Q$. Specifically, the theory in the 'broken phase' is invariant under $U(1)_Q$ gauge transformations. Therefore, to make the $SU(2)_L \times U(1)_Y$ symmetry manifest, it is sufficient to gauge only the coset group $SU(2)_L \times U(1)_Y/U(1)_Q$, which can be parameterised in terms of a non-linear field:

$$\mathcal{X}(x) := e^{\frac{i}{2}\pi^i(x)T^i}\begin{pmatrix}0\\1\end{pmatrix} , \tag{9}$$

where $T^i = \sigma^i - \delta^{i3}\mathbb{I}$ are the three broken generators with σ^i being the Pauli matrices. The $\pi^i(x)$ fields are the three would-be Goldstone bosons spanning the $SU(2)_L \times U(1)_Y/U(1)_Q$ coset space. With non-linear realisation of $SU(2)_L \times U(1)_Y$ electroweak gauge invariance the Higgs field h is no longer obliged to form the electroweak doublet irreducible representation. In the minimal scenario the Higgs boson resides in the $SU(2)_L \times U(1)_Y$ singlet field, $\rho(x)$. The standard Higgs doublet then can be identified with the following composite field[c]:

$$H(x) = \frac{\rho(x)}{\sqrt{2}}\mathcal{X}(x)\ . \tag{10}$$

While maintaining $SU(2)_L \times U(1)_Y$ invariance, the non-linear realisation of the electroweak gauge symmetry allows a number of new interactions beyond those present in the SM, including anomalous Higgs-vector boson couplings, flavour and CP-violating Higgs-fermion couplings and anomalous Higgs interactions. A generic model is rather complicated and also severely constrained by the electroweak precision measurements, flavour physics and Higgs data. In this paper we only consider modification of the SM Higgs potential by adding an anomalous cubic coupling:

$$V(\rho) = -\frac{\mu^2}{2}\rho^2 + \frac{\kappa}{3}\rho^3 + \frac{\lambda}{4}\rho^4. \tag{11}$$

We assume that the scalar potential has a global minimum for a non-zero vacuum expectation value of the Higgs field ρ (cf. the next section):

$$\langle\rho\rangle = v, \quad |v| \approx 246 \text{ GeV}. \tag{12}$$

The absolute value of the vacuum expectation value in (12) is fixed to the standard value since the Higgs interactions with the electroweak gauge bosons are assumed to be the same as in SM, i.e.,

$$\frac{\rho^2}{2}(D_\mu\mathcal{X})^\dagger D^\mu\mathcal{X}, \tag{13}$$

where D_μ is an $SU(2)_L \times U(1)_Y$ covariant derivative. The shifted field

$$h(x) = \rho(x) - v, \tag{14}$$

describes the physical excitation associated with the Higgs particle with the tree-level mass squared:

$$m_h^2 = \left.\frac{\partial^2 V}{\partial\rho\partial\rho}\right|_{\rho=v} \approx (125 \text{ GeV})^2\ . \tag{15}$$

Using equations (12) and (15), we find it convenient to rewrite the mass parameter μ^2 and the quartic coupling λ in terms of the (tree level) Higgs mass, $m_h \approx 125$

[c]We note that if $\rho(x)$ field is to be identified with the modulus of the electroweak doublet field, $\rho^2 = H^\dagger H$, it should be restricted to positive ($\rho > 0$) or negative ($\rho < 0$) values only.

GeV, the Higgs vacuum expectation value v and the cubic coupling κ as:

$$\begin{aligned} \mu^2 &= \tfrac{1}{2}\left(m_h^2 + v\kappa\right), \\ \lambda &= \tfrac{1}{2v^2}\left(m_h^2 - v\kappa\right). \end{aligned} \tag{16}$$

The potential must also be bounded from below, that is $\lambda > 0$ and, hence, $v\kappa < m_h^2$.

The presence of the cubic term in the tree-level Higgs potential (11) significantly alters the Higgs vacuum configuration, even without thermal corrections. With this kind of potential one can differentiate the following three cases:

(i) A non-tachyonic mass parameter, i.e., $\mu^2 < 0$ or, equivalently, $v\kappa < -m_h^2$. One of the local minima in this case is at a trivial configuration $\langle\rho\rangle = 0$. We find that the electroweak symmetry breaking minimum (15) is realised as an absolute minimum of the potential if $-3m_h^2 < v\kappa < -m_h^2$.

(ii) A tachyonic mass parameter, i.e., $\mu^2 > 0$, which implies $v\kappa > -m_h^2$. In this case the trivial configuration is a local maximum and the minimum (15) is realised providing $-m_h^2 < v\kappa < 0$.

(iii) For $\mu^2 = 0$ ($v\kappa = -m_h^2$), $v = -\frac{\kappa}{\lambda}$. In this case there are two trivial solutions for the extremum equation, which represent an inflection point of the potential.

Notice that there exists a symmetry of the above vacuum solutions under $\kappa \to -\kappa$ and $v \to -v$. Although the above analysis was done at tree level, we have verified that the one-loop quantum corrections to the tree-level potential does not change the above picture significantly.

In order to calculate the electroweak phase transition dynamics we consider the one-loop finite temperature potential (for a review, see, e.g. Ref. 13). The potential, as a function of temperature, T, can be split into the following parts:

$$V(\rho, T) = V^{(0)}(\rho) + V_{CW}^{(1)}(\rho) + V^{(1)}(\rho, T > 0) + V_{Daisy}(\rho, T > 0), \tag{17}$$

where $V^{(0)}$ is the classical potential and is given in (11), $V_{CW}^{(1)}$ is the Coleman-Weinberg contribution for $T = 0$ and is given by:

$$V_{CW}^{(1)}(\rho) = \sum_{i=W,Z,t,h} n_i \frac{m_i^4(\rho)}{64\pi^2}\left(\ln\left(\frac{m_i^2(\rho)}{v^2}\right) - \frac{3}{2}\right). \tag{18}$$

$V^{(1)}(\rho, T > 0)$ is the finite temperature contribution and is defined via the thermal function J:

$$\begin{aligned} V^{(1)}(\rho, T) &= \frac{T^4}{2\pi^2} \sum_{i=W,Z,t,h} n_i J\left[\frac{m_i^2(\rho)}{T^2}\right], \\ J[m_i^2\beta^2] &:= \int_0^\infty dx\, x^2 \ln\left[1 - (-1)^{2s_i+1} e^{-\sqrt{x^2+\beta^2 m_i^2}}\right], \end{aligned} \tag{19}$$

here s_i corresponds to the spin and n_i to the number of degrees of freedom of the particle species i. $V_{Daisy}(\rho, T > 0)$ are Daisy-corrected terms[13]. The total potential is shown for a range of κ values and temperatures in Fig 1.

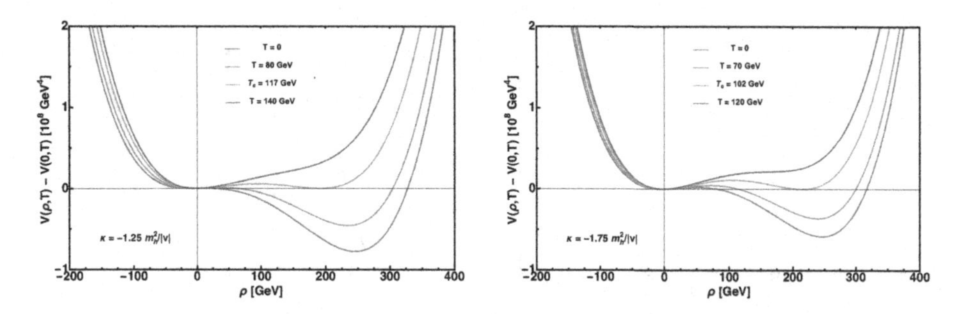

Fig. 1. The finite temperature effective potential for different κ and T.

3.2. *Cosmological phase transition and production of gravity waves*

A salient feature of the effective finite temperature potential (17) is that it contains a single minimum at high temperatures at non-zero values of the Higgs field. Hence, the SM field remains massive in the high-temperature limit. We define this temperature-dependent vacuum state as the false vacuum, $v_T^{(+)}$.

The dynamics of phase transition can be defined using a few key temperatures ($\tilde{T} > T_c > T_n > T_p$) as follows. For $T > \tilde{T}$, $V(\rho, T)$ admits a single minimum at $\rho = v_T^{(+)}$, called false vacuum phase. As the universe cools and reaches $\tilde{T}$, a second minimum, called the true (electroweak) vacuum phase, forms at $\rho = v_T^{(-)}$ with an energy density initially higher than that of the faulse vacuum phase. This energy density then decreases until the two vacua become degenerate at the critical temperature $T = T_c$. For $T < T_c$, the energy density of the true vacuum phase keeps decreasing, causing the false vacuum phase to become metastable: in other words, the Higgs field may tunnel through the potential barrier between $v_T^{(+)}$ and $v_T^{(-)}$. If the decay probability is high enough, bubbles of true vacuum nucleate and expand in the surrounding symmetric phase. The nucleation temperature, T_n is then defined as the temperature at which most of the bubbles are produced. On the other hand, the percolation temperature, T_p corresponds to the instant when a significant volume of the universe (whose value would be specified later on) has been converted from the symmetric to the broken phase. We expect most of bubble-collisions to occur around T_p and not T_n, unlike typical high-temperature (fast) phase transitions.

Numerical analysis show that most of the bubbles of true vacua are produced at $T_n \sim 50$ GeV, which is largelly independent of the coupling κ. However, number of bubbles produced at nucleation temperature expose exponential dependence on κ. On the left panel of Fig. 2 the thermal bounce action is plotted for a range of κ. Since the bubble nucleation rate is $\propto \exp(-S_3/T)$, one can see that larger $|\kappa|$ (larger the barrier between false and true vacua) implies less buubles. Consequently, larger $|\kappa|$ implies smaller percolation temperatures (prolonged phase transition), see the right panel of the same figure.

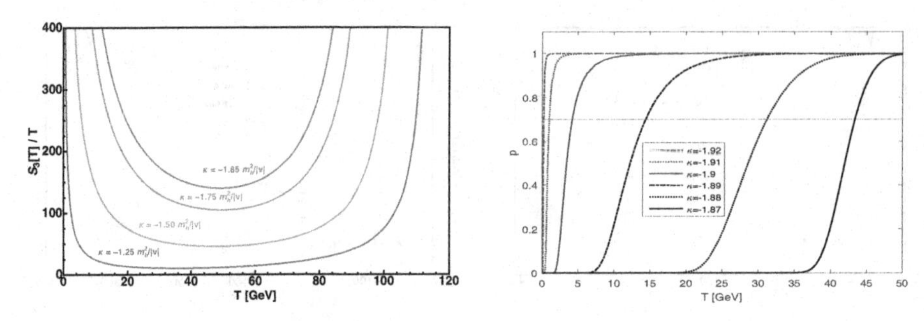

Fig. 2. (Color online) The thermal bounce action (left panel) and probability of false vacuum (right panel) for a range of cubic couplings κ. The percolation temperature is defined as the temperature when $p = 2/3$ (the red horizontal line).

We study dynamics of the bubbles under the reasonable assumption that most of the latent heat released during the phase transition is transferred to the production and growth of bubbles of the true vacua. The final average size of bubbles before percolation critically depends on the duration of phase transition and they typically become of the order of Hubble size at the percolation temperature for the prolonged transition. Violent bubble collisions then disturb the surrounding spacetime by producing gravitational waves. The characteristics of these waves (amplitude and frequency) are defined by the size of colliding bubbles and the energy carried. To compute these characteristics a special care must be taken for the prolonged phase transition. The interested reader can find the details in Ref. 4. Here I only show results of our calculations. On Fig. 3 the gravitational wave spectra are plotted for a fast transition, $T_p \lesssim T_n$ (left panel), and on the prolonged transition, $T_p << T_n$ (right panel). One can see a large spread of possible peak frequences. For a fast transition the gravitational wave frequency peaks in milli-Herz reagion. Future space based experiments, such as eLISA, are sensitive to detect them. The gravitational waves produced during the prolonged phase transition have much smaller peak frequencies. They may be detectable using pulsar timing technique in e.g. Square Kilometer Array observatory (SKA).

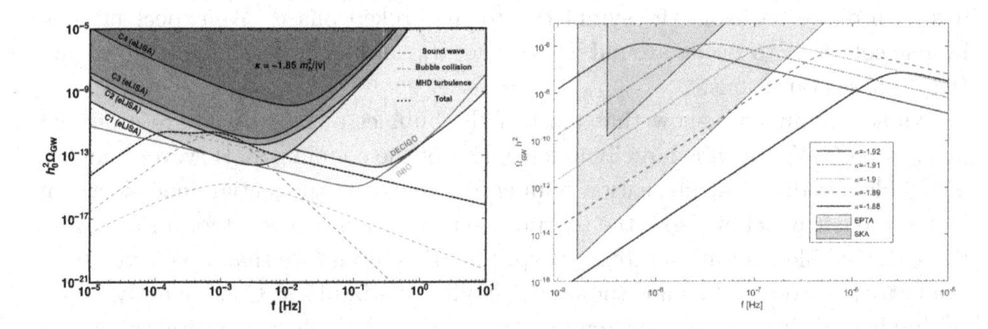

Fig. 3. Gravitational wave spectra for fast (left panel) and slow (right panel) phase transitions.

There is an interesting correlation between the predicted gravitational waves and the trilinear Higgs coupling measurements at LHC. The detection of a stochastic gravitational wave background in one of the above experiments may provide a complimentary information on the nature of electroweak symmetry and the cosmological phase transition. This information will be particularly important since the measurement of the Higgs cubic coupling at high luminosity LHC is feasible only with 30-50% accuracy[14].

Acknowledgments

I would like to thank Harald Fritzsch and other organisers of the Conference on Cosmology Gravitational Waves and Particles for the opportunity to attend the stimulating meeting and present the results reported in this paper. I am mostly indebted to my collaborators Cyril Lagger, Adrian Manning and Jason Yue for their contribution. This work was partially supported by the Australian Research Council.

References

1. B. P. Abbott *et al.* [LIGO Scientific and Virgo Collaborations], Phys. Rev. Lett. **116** (2016) no.6, 061102 doi:10.1103/PhysRevLett.116.061102 [arXiv:1602.03837 [gr-qc]].
2. A. Kobakhidze, C. Lagger and A. Manning, Phys. Rev. D **94** (2016) no.6, 064033 doi:10.1103/PhysRevD.94.064033 [arXiv:1607.03776 [gr-qc]].
3. A. Kobakhidze, A. Manning and J. Yue, arXiv:1607.00883 [hep-ph], Int. J. of Mod. Phys D, 2017, in press.
4. A. Kobakhidze, C. Lagger, A. Manning and J. Yue, arXiv:1703.06552 [hep-ph], Eur. Phys. J. C, 2017, in press.
5. F. Ardalan, H. Arfaei and M. M. Sheikh-Jabbari, JHEP **9902** (1999) 016 doi:10.1088/1126-6708/1999/02/016 [hep-th/9810072].
6. N. Seiberg and E. Witten, JHEP **9909** (1999) 032 doi:10.1088/1126-6708/1999/09/032 [hep-th/9908142].
7. M. R. Douglas and N. A. Nekrasov, Rev. Mod. Phys. **73** (2001) 977 doi:10.1103/RevModPhys.73.977 [hep-th/0106048].
8. R. J. Szabo, Phys. Rept. **378** (2003) 207 doi:10.1016/S0370-1573(03)00059-0 [hep-th/0109162].
9. X. Calmet, Eur. Phys. J. C **41** (2005) 269 doi:10.1140/epjc/s2005-02226-9 [hep-ph/0401097].
10. L. Blanchet, Living Rev. Rel. **17** (2014) 2 doi:10.12942/lrr-2014-2 [arXiv:1310.1528 [gr-qc]].
11. A. Kobakhidze, Phys. Rev. D **79** (2009) 047701 doi:10.1103/PhysRevD.79.047701 [arXiv:0712.0642 [gr-qc]].
12. A. Kobakhidze, arXiv:1208.5180 [hep-ph].
13. M. Quiros, hep-ph/9901312.
14. C. T. Lu, J. Chang, K. Cheung and J. S. Lee, JHEP **1508** (2015) 133 doi:10.1007/JHEP08(2015)133 [arXiv:1505.00957 [hep-ph]].

Galaxy Rotation Curves and the Deceleration Parameter in Weak Gravity

Maurice H. P. M. van Putten

Sejong University, Seoul 143-747, South Korea
mvp@sejong.ac.kr

We present a theory of weak gravity parametrized by a fundamental frequency $\omega_0 = \sqrt{1-q}H$ of the cosmological horizon, where H and q denote the Hubble and, respectively, deceleration parameter. It predicts (i) a C^0 onset to weak gravity across accelerations $\alpha = a_{\rm dS}$ in galaxy rotation curves, where $a_{\rm dS} = cH$ denotes the de Sitter acceleration with velocity of light c, and (ii) fast evolution $Q(z) = dq(z)/dz$ of the deceleration parameter by $\Lambda = \omega_0^2$ satisfying $Q_0 > 2.5$, $Q_0 = Q(0)$, distinct from $Q_0 \lesssim 1$ in ΛCDM. The first is identified in the high resolution data of Lelli *et al.* (2017), the second in the heterogeneous data on $H(z)$ over $0 < z < 2$. A model-independent cubic fit in the second rules out ΛCDM by 4.35σ and obtains $H_0 = 74.0 \pm 2.2$ km s^{-1} Mpc^{-1} consistent with Riess *et al.* [*Astrophys. J.* **826**, 56 (2016)]. Comments on possible experimental tests by the LISA Pathfinder are included.

Keywords: Galaxy rotation curves; deceleration parameter; dark energy; dark matter.

1. Introduction

Modern cosmology shows a universe that is well-described by a three-flat Friedmann–Robertson–Walker (FRW) line-element

$$ds^2 = -dt^2 + a(t)^2(dx^2 + dy^2 + dz^2)\,, \tag{1}$$

that currently experiences accelerated expansion indicated by a deceleration parameter[1,2]

$$q = \frac{1}{2}\Omega_m - \Omega_\Lambda < 0\,, \tag{2}$$

where $q = -H^{-2}\ddot{a}/a$, with Hubble parameter $H = \dot{a}/a$. It points to a finite dark energy density $\Omega_\Lambda = \rho_\Lambda/\rho_c$, where $\rho_c = 3H^2/8\pi G$ is the closure density with Newton's constant G. In this background, galaxies and galaxy clusters formed and evolved by weak gravitational attraction at accelerations α at or below the de Sitter scale

$$a_{\rm dS} = cH\,, \tag{3}$$

where c denotes the velocity of light. As such, (3) sets a scale to *weak gravity* common to cosmological evolution and large scale structure.

Weak gravity surprises us with *more* than is expected by the Newtonian gravitational attraction of the observed baryonic matter content: $\Lambda = 8\pi\rho_\Lambda > 0$ and enhanced acceleration in galaxy dynamics. This joint outcome is anticipated by $[\Lambda] = \mathrm{cm}^{-2}$ and $[\sqrt{\Lambda}] = \mathrm{cm}^{-1}$ in geometrical units, in which Newton's constant and the velocity of light are set equal to 1, where the latter points to a transition radius r_t in rotation curves of galaxies of mass M_b, $4\pi r_t^2\sqrt{\Lambda} \simeq R_g$, $R_g = GM_b/c^2$. Just such scale r_t of a few kpc is observed in a sharp onset to anomalous behavior in high resolution data for $M_b = M_{11}10^{11}M_\odot$ at[3,4]

$$a_{\mathrm{N}} = a_{\mathrm{dS}}\,, \tag{4}$$

where a_{N} refers to the acceleration anticipated based on Newton's law of gravitational attraction associated with M_b inferred from luminous matter. Further out into weak gravity, these rotation curves satisfy the emperical baryonic Tully–Fisher relation or, equivalently, Milgrom's law.[5–7]

We here describe a theory of weak gravity by spacetime holography,[8–10] parameterized by a fundamental frequency of the cosmological horizon[3]

$$\omega_0 = \sqrt{1-q}H\,, \tag{5}$$

defined by the harmonic oscillator of geodesic separation of null-generators of the cosmological horizon. This result reflects compactness of the holographic phase space of spacetime, set by its finite surface area. By holography, (5) induces a dark energy in the $(3+1)$ spacetime (1) by the square

$$\Lambda = \omega_0^2\,. \tag{6}$$

It may be noted that (6) has a vanishing contribution in the radiation dominated era ($q = 1$), whereas it reduces to CDM in a matter dominated era ($q = 1/2$). (In the equation of state $p = w\rho_\Lambda$ for total pressure p, $w = (2q-1)/(1-q)$ vanishes for $q = 1/2$.) In the holographic encoding of particle mass m_0 and positions, inertia is defined by a thermodynamic potential $U = mc^2$ derived from the unitarity of particle propagators.[3,13] In weak gravity, this may incur[3]

$$m < m_0 \quad (\alpha < a_{\mathrm{dS}})\,, \tag{7}$$

with a C^0 onset to inequality by crossing of apparent Rindler and cosmological horizons. In weak gravity, therefore, a particle's inertial mass ("weight") and gravitational mass may appear distinct.[11,12]

According to the above, weak gravity predicts the following. First, the C^0 onset to (7) is at a transition radius[3]

$$r_t = 4.7\ \mathrm{kpc}\, M_{11}^{1/2}(H_0/H)^{1/2}\,. \tag{8}$$

Beyond, asymptotic behavior ($\alpha \ll a_{\mathrm{dS}}$) satisfies Milgrom's law with[3,14]

$$a_0 = \frac{\omega_0}{2\pi}\,. \tag{9}$$

These expressions (8)–(9) may be confronted with data on galaxy rotation curves over an extended redshift range supported by Hubble data $H(z)$. Second, the associated dark energy (6) is relevant to late times, when deceleration $q(z)$ turns negative. At the present epoch, (6) predicts *fast cosmological evolution*, described by $q_0 = q(0)$ and $Q_0 = Q(0)$, $Q(z) = dq(z)/dz$, satisfying

$$q_0 = 2q_{0,\Lambda\mathrm{CDM}} \tag{10}$$

and

$$Q_0 > 2.5\,, \qquad Q_{0,\Lambda\mathrm{CDM}} \lesssim 1\,. \tag{11}$$

In Sec. 2, we discuss the origin of (4) in inertia from entanglement entropy and its confrontation with the recent high resolution data on galaxy rotation curves.[4] In Sec. 3, we discuss (6) and its implications (11) in confrontation with heterogeneous data on $H(z)$ $(0 < z < 2)$.[15,16] Sensitivity of galaxy dynamics in weak gravity to the Hubble parameter $H(z)$ is studied in Sec. 4. We summarize our findings in Sec. 5 and conclude with an estimate of H_0.

2. Weak Gravity in Galaxy Rotation Curves

Galaxy dynamics is of particular interest as a playground for the equivalence principle (EP) (e.g. Ref. 17) at small accelerations on the de Sitter scale a_{dS}.

Einstein's EP asserts that *the gravitational field — wherein photon trajectories appear curved — seen by an observer on the surface of a massive body is indistinguishable from that in an accelerating rocket, at equal weights of its mass.* On this premise, general relativity embeds Rindler trajectories — non-geodesics in Minkowski spacetime — by gravitational attraction as geodesics in curved spacetime around massive bodies, while weight is measured along non-geodesics. With no scale, this embedding is free of surprises, as Rindler accelerations become arbitrarily small. Following such embedding, acceleration vanishes and inertia cancels out in the equations of geodesic motion (ignoring self-gravity). The origin of inertia is hereby not addressed in Einstein's EP or general relativity.

To address inertia for its potential sensitivity to a cosmological background, we take one step back and consider Rindler inertia in non-geodesics of Minkowski spacetime.

Inertia is commonly defined by mass-at-infinity in an asymptotically flat spacetime (Mach's principle). The latter is an overly strong assumption in cosmologies (1), whose Cauchy surfaces are bounded by a cosmological horizon at Hubble radius $R_H = c/H$. Any reference to large distance asymptotics is inevitably perturbed if not defined by the cosmological horizon. In particular, the apparent horizon h of Rindler spacetime at a distance $\xi = c^2/\alpha$ colludes with the latter at sufficiently small accelerations. Thus, h and H colludes at (4) with corresponding transition radius

$$r_t = \sqrt{R_g R_H}\,, \tag{12}$$

giving (8). In what follows, we argue that it sets the onset to reduced inertia (7).

In what follows, h and H refer to apparent horizons associated with radial accelerations. For orbital motions, we appeal to Newton's separation of particle momenta $\mathbf{p}$ and associated forces $d\mathbf{p}/dt$ in radial and azimuthal components, where the former is imparted by the gravitational field of body of mass M_b. (Locally, $d\mathbf{p}/dt$ is measured as curvature of particle trajectories effectively relative to a tangent plane of null-geodesics.) The radial component of $d\mathbf{p}/dt$ hereby carries h as an apparent horizon defined by instantaneous radial acceleration, giving (8). In spiral galaxies, (12) is indeed a typical distance signaling the onset to anomalous galaxy dynamics.[18]

2.1. *Origin of Rindler inertia in entanglement entropy*

Unitarity in encoding particle positions by holographic screens satisfies

$$P_- + P_+ \equiv 1\,, \tag{13}$$

where e.g. $P_\pm$ refer to the probabilities of finding the proverbial cat and, respectively, out of a box, as defined by its quantum mechanical propagator. Satisfying (13) requires *exact arithmetic*, expressing the probability

$$P_+ = 1 - P_-\,, \tag{14}$$

for a cat in the box with no round-off error even when P_- is exponentially small.[13] In a holographic approach, $P_+ = 1 - P_-$ is encoded in information *on* the box as a compact two-surface with

$$P_- \sim e^{-2\Delta\varphi} \tag{15}$$

in terms of a Compton phase,

$$\Delta\varphi = k\xi\,, \tag{16}$$

expressing the cat's distance ξ to the walls by Compton wave number $k = mc/\hbar$ for a mass m. Exact arithmetic on (13) requires an information[13]

$$I = 2\pi\Delta\varphi\,. \tag{17}$$

Here, the factor of 2π in (17) is associated with encoding m on a spherical screen. (For a single flat screen, $I = 2\Delta\varphi$ and for a cubic box $I = 12\Delta\varphi$.)

We next consider a Rindler observer $\mathcal{O}$ of mass m_0, i.e. a non-geodesic trajectory of constant acceleration in $(1+1)$ Minkowski spacetime (t, x). The light cone at the origin delineates an apparent horizon h, $|ct| = x$, to a Rindler observer with worldline $(t, x) = (\sinh(\lambda\tau), \cosh\lambda\tau)$, where $\lambda = \alpha/c$. $\mathcal{O}$'s distance ξ to h is the Lorentz invariant, which can be attributed to an Unruh temperature[19]

$$k_{\rm B}T = \frac{\alpha\hbar}{2\pi c}\,. \tag{18}$$

With h null, the entropy S, putting Boltzmann's constant $k_{\rm B}$ equal to 1,

$$\mathrm{dS} = -dI \tag{19}$$

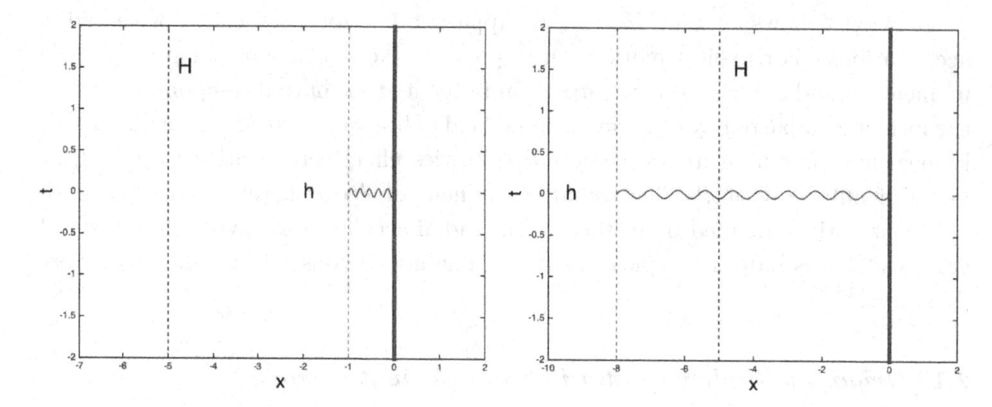

Fig. 1. In three-flat FRW universes with Hubble radius R_H, the Rindler horizon h at distance $\xi = c^2/\alpha$ may fall inside (left) or outside the cosmological the cosmological horizon. Collusion at $\alpha = a_{\rm dS}$ of Rindler and cosmological horizon defines a sharp onset to weak gravity. When H falls within h, it interferes with the phase space of Rindler observers. Rindler inertia m then drops below its Newtonian value set by rest mass energy m_0c^2. (Reprinted from Ref. 3.)

gives rise to a thermodynamic potential

$$dU = -T_U\,{\rm dS} = \left(\frac{\hbar\alpha}{2\pi c}\right)\left(2\pi\frac{mc}{\hbar}d\xi\right). \tag{20}$$

Conforming EP, $\mathcal{O}$ identifies a uniform gravitational field that it may attribute to some massive object well beyond h, featuring a divergent redshift towards h. In this gravitational field and relative to h, $\mathcal{O}$ assumes a potential energy

$$U = \int_0^\xi dU = m_0c^2\,. \tag{21}$$

By a thermodynamical origin to inertia, Rindler observers experience fluctuations therein equivalently to momentum fluctuations by detection of photons from a warm Unruh vacuum.

Emperically, inertia is instantaneous with no associated time scale. Correspondingly, (21) is not an ordinary thermodynamic potential, but one arising from the nonlocal *entanglement entropy* S associated with the apparent horizon h in (19).

In the absence of any length scale in Minkowksi spacetime, (21) establishes a Newtonian identity between mass-energy and inertia. In three-flat cosmology, the resulting $m = m_0$ will hold whenever $\alpha > a_{\rm dS}$, ensuring that h falls within H (Fig. 1).

2.2. C^0 *onset to weak gravity at* $a_{\rm dS}$

In (7), H drops inside h (Fig. 1), and the integral leading to (21) is cutoff at H, leaving U smaller than m_0c^2. By $a_{\rm dS}$, EP is no longer scale free, i.e. inertia m measured by U (gravitational pull in an equivalent gravitational field) may deviate from the Newtonian value m_0. Reduced inertia $m < m_0$ when h falls beyond H

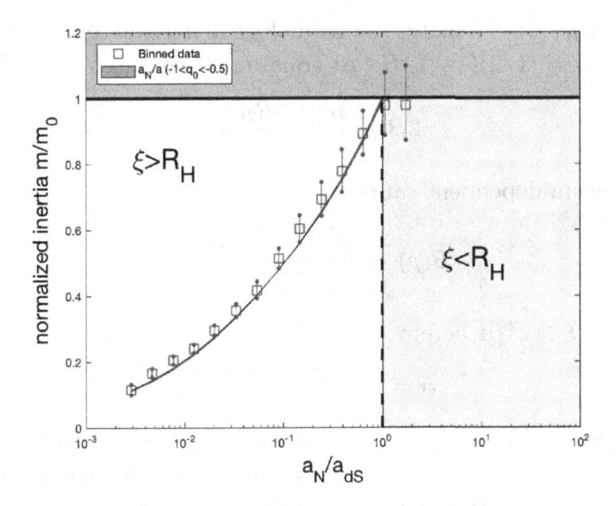

Fig. 2. High resolution data of galaxy rotation curves, here plotted as $m/m_0 = a_N/\alpha$ as a function of a_N/a_{dS}, where α denotes the observed centripetal acceleration and a_N is the expected Newtonian acceleration based on the observed baryonic matter. The results show a sharp onset to weak gravity across $(a_N, m) = (a_{dS}, m_0)$. (Reprinted from Ref. 3.)

in (7) gives rise to an enhanced acceleration at a given Newtonian gravitational forcing $F_N = m_0 M_b / r^2$. (This is prior to a covariant embedding in geodesic motion in curved spacetime, alluded to above.) Arising from the crossing of the apparent Rinder and cosmological horizon surfaces, the onset to $m < m_0$ is C^0 sharp. Just such behavior is apparent in the high resolution galaxy rotation data (Fig. 2).

Figure 2 shows binned rotation curve data from many spiral galaxies, after normalizing the independent variable to a_N (rather than coordinate radius r).[18] Plotting α/a_N as a function of a_N (or a_N/a_{dS}), the results $\alpha > a_N$ are often referred to as the "missing mass problem." In light of our focus on Rindler horizons falling beyond the cosmological horizon, Fig. 2 shows

$$\frac{m}{m_0} = \frac{a_N}{\alpha}. \tag{22}$$

Observationally, the right-hand side of (22) is inferred from the ratio $(V^2/r)/(V_b^2/r) = (V/V_b)^2$ of observed (V) and anticipated (V_c) circular velocities, the latter based on luminous matter. Hence,

$$\frac{E_k}{E_{k,0}} = \frac{mV^2}{m_0 V_b^2} = \frac{m\alpha}{m_0 a_N} = 1\,, \tag{23}$$

showing that the orbital kinetic energies are unchanged.

In weak gravity (7), holographic representations of a particle of mass m involves two low energy dispersion relations of image modes in $(3+1)$ spacetime and of Planck sized surface elements on the cosmological horizon, satisfying[3,14]

$$\hbar\omega = \sqrt{\hbar^2\Lambda + c^2p^2} \quad (r < R_H)\,, \qquad T = \sqrt{T_U^2 + T_H^2} \quad (r = R_H)\,, \tag{24}$$

where $k_B T_H = \hbar a_H/(2\pi c)$ denotes the cosmological horizon temperature at internal surface gravity $a_H = (1/2)(1-q)H$. At equal mode counts, we have[3]

$$2B(p)\frac{m}{m_0} = \frac{R_H}{\xi}, \tag{25}$$

with the momentum-dependent ratio

$$B(p) = \frac{\sqrt{\hbar^2\Lambda + c^2p^2}}{k_B\sqrt{T_U^2 + T_H^2}}. \tag{26}$$

Rindler's relation $\xi = c^2/\alpha$ hereby obtains

$$\alpha = \sqrt{2B(p)a_{dS}\, a_N}, \tag{27}$$

where a_N refers to the Newtonian acceleration $a_N = M_b/r^2$. An effective description of weak gravity (7) now follows, upon taking a momentum average $\langle B(p)\rangle$ of (26) in (27). Averaging over a thermal distribution[3] obtains the green curve in Fig. 2.

In the asymptotic regime, $a_N \ll a_{dS}$, (27) reduces to Milgrom's law[7,18]

$$\alpha = \sqrt{a_0 a_N}, \tag{28}$$

with Milgrom's parameter (9) directly associated with the background cosmology as anticipated based on the dimensional analysis in geometrical units.[3,14] In light of the q-gradients Q_0 in (11), $A_0 = A(0)$, $A(z) = a_0^{-1}\, da_0(z)/dz$, satisfies[3,14]

$$A_0 \simeq -0.5, \qquad A_{0,\Lambda\text{CDM}} > 0. \tag{29}$$

This discrepant outcome of (6) and ΛCDM may be tested observationally in surveys of rotation curves covering a finite redshift range.

3. Accelerated Expansion from Cosmological Holography

The holographic principle proposes a reduced phase space of spacetime, matter and fields to that of a bounding surface in Planck sized degrees of freedom.[8–10] The latter is generally astronomically large in number, and hence excited at commensurably low energies. In weak gravity, these low energies readily reach a low energy scale of the cosmological vacuum, set by the cosmological horizon at Hubble radius $R_H = c/H$ based on the unit of luminosity $L_0 = c^5/G$.[13]

As a null-sphere, the cosmological horizon features closed null-geodesics. The geodesic deviation of a pair of null-geodesics satisfies a harmonic oscillator equation.[3] By explicit calculation of the Riemann tensor in a three-flat FRW universe with deceleration parameter q, the angular frequency satisfies (5). A holographic extension to spacetime obtains a wave equation with dispersion relation $\omega = \sqrt{k^2 + \Lambda}$, where k refers to the wave number of modes orthogonal to the cosmological horizon and Λ in (6). Dark energy is hereby described in terms of the canonical cosmological parameters (H, q), allowing a detailed comparison of cosmic evolution with (6) versus ΛCDM.

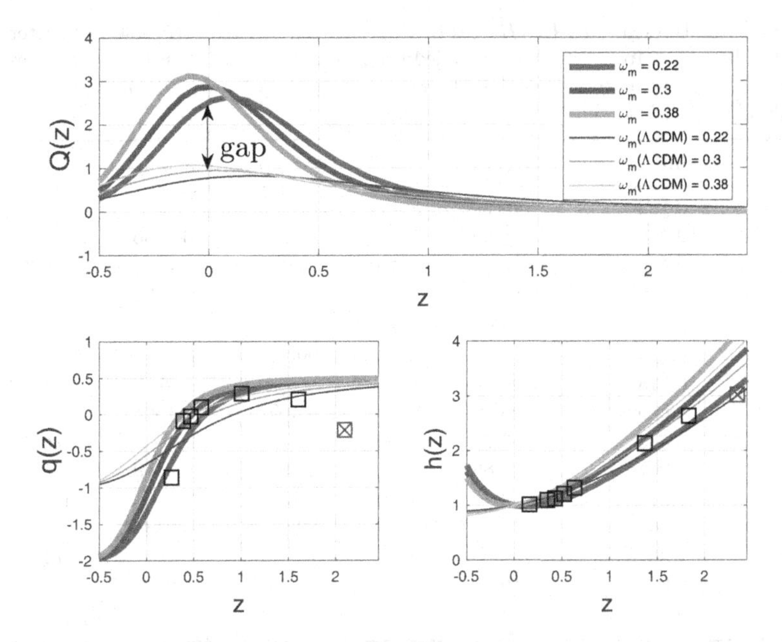

Fig. 3. Evolution of the deceleration parameter $q(z)$ with redshift in a three-flat FRW universe with dark energy $\Lambda = \omega_0^2$ and ΛCDM for various values $q_0 = q(0)$. They show pronounced differences in slope Q_0 at $z = 0$ associated with different curvatures of $h(z) = H(z)/H_0$.

3.1. *Evolution by* $\Lambda = \omega_0^2$ *and* ΛCDM

FRW universes with Λ in (6) leave baryon nucleosynthesis invariant, since $q = 1$ in a radiated dominated epoch. For most of the subsequent evolution of the universe, this dark energy is relatively small compared to matter content. In recent epochs, however, the derivative

$$Q(z) = \frac{dq(z)}{dz} \tag{30}$$

varies rather strongly compared to what is expected in ΛCDM, expressed in (11) and shown in Fig. 3.

Cosmological evolution in the approximation of (1) is described by an ordinary differential equation (ODE) for the scale factor $a(t)$. For (6), this is described by an ODE as a function for $y = \log h$, $h = H/H_0$, as a function of z,

$$y'(z) = 3(1+z)^2 \omega_m e^{-2y} - (1+z)^{-1}, \tag{31}$$

derived from $\ddot{a}(\tau) = a(2h^2 - 3\omega_m a^{-3})$ $(a(0) = 1,\ h(0) = 1)$, $d/dt = -(1+z)Hd/dz$, as a function of time $\tau = tH_0$.[20] For ΛCDM, we have $h(z) = \sqrt{(1-\omega_m) + \omega_m(1+z)^3}$ for ΛCDM. Figure 2 shows illustrative numerical solutions for various values ω_m of dark matter content at $z = 0$. At late times, the dynamical dark energy (6) features a *fast* expansion compared to ΛCDM. In particular, q_0 and Q_0 defined in (10) and (11) are markedly distinct in these two models.

Table 1. Binned data $\{z_k, H(z_k)\}$ $(k = 1, 2, \ldots, 8)$[16] on the Hubble parameter $H(z)$ [km s^{-1} Mpc^{-1}] over an extended range of redshift z, and inferred estimates of $H'(z_{k'})$ and $q(z_{k'})$ at midpoints $z_{k'} = (z_k + z_{k-1})/2$ $(k = 2, 3, \ldots, 8)$, and $Q(z_k)$ $(k = 2, 3, \ldots, 7)$.

k	Redshift z_k, $z_{k'}$	$H(z_k)$	σ_k	$H'(z_{k'})$	$q(z_{k'})$	$Q(z_k)$
1	0.166	75.7	3.35			
	0.2605	78.20	—	26.46	−0.5736	
2	0.355	80.7	1.70	—	—	1.9758
	0.3910	82.65	—	54.17	−0.0884	
3	0.427	84.6	4.80	—	—	0.2659
	0.4725	87.35	—	60.44	−0.0189	
4	0.518	90.1	1.75	—	—	0.1492
	0.5755	93.90	—	66.09	0.1008	
5	0.633	97.7	1.90	—	—	0.0913
	1.0015	127.85	—	81.82	0.2809	
6	1.37	158	8.50	—	—	−0.1470
	1.60	177	—	82.61	0.2135	
7	1.83	196	31.0	—	—	−0.2328
	2.09	210	—	53.85	−0.2077	
8	2.35	224	5.0			

3.2. *Heterogeneous data on* $H(z)$

Measurements of the Hubble parameter $H(z)$ by various methods of observations now extends over an increasingly large redshift range. For recent compilations, see e.g. Sola *et al.*[15] covering $0 < z < 1.936$ and Farooq *et al.*[16] covering $0 < z \leq 2.36$.

Table 1 shows $N = 8$ binned data[16] on $H(z)$ with a mean of normalized standard deviation $\hat{\sigma}_k = \sigma_k / H(z_k)$ satisfying

$$\left\{ \frac{1}{N} \sum_{k=1}^{N} \hat{\sigma}_k^{-2} \right\}^{-\frac{1}{2}} \simeq 3\% . \tag{32}$$

Included are the estimates of $q(z)$ obtained from $H'(z)$ by central differencing.

As heterogeneous data sets, these compilations require tests against physical constraints before they can be used in regression analysis. Different (often unknown) systematics can potentially create trend anomalies that violate essential priors. If so, the data set contains *incompatibilities.*

We recall that the universe entered an essentially matter dominated epoch when z appreciably exceeds the transition redshift z_t defined by the vanishing of the deceleration parameter $q(z)$, when the Hubble flow of galaxies passing through the cosmological horizon changed sign. The constraint $z_t < 1$ seems fairly robust.[16] For $z > 1$, therefore, the positivity conditions

$$H(z) > 0 , \qquad q(z) = -1 + (1 + z) H^{-1}(z) H'(z) > 0 , \tag{33}$$

are *essential physical priors* to any data set.

Table 1 points to a violation of (33) with $q(z) = -0.2077$ at the midpoint $z = 2.09$. Associated with a drop in $H'(z)$, the last data point at $k = 8$ appears to be incompatible with $k = 1, 2, \ldots, 7$ and will be considered incompatible in the following nonlocal analysis.

3.3. *Detecting Q_0 in $H(z)$ data*

Table 2 lists results of nonlinear model regression by our two cosmological models to the Hubble data of Table 1, parametrized by (H_0, ω_m) with the MatLab function

Table 2. Results of nonlinear model regression on the binned compatible data $\{z_k, H(z_k)\}$ $(k = 1, 2, \ldots 7)$ of Table 1.

Model	H_0 [km s^{-1} Mpc^{-1}]	ω_m	q_0	Q_0
$\Lambda = \omega_0^2$	74.1 ± 1.2	0.2821 ± 0.0125	-1.1537 ± 0.0375	2.7990 ± 0.0603
ΛCDM	65.7 ± 1.7	0.3594 ± 0.0384	-0.4609 ± 0.0576	1.0360 ± 0.0486
cubic fit	74.0 ± 2.4	0.2981 ± 0.0566	-1.1057 ± 0.1698	2.2648 ± 0.2910

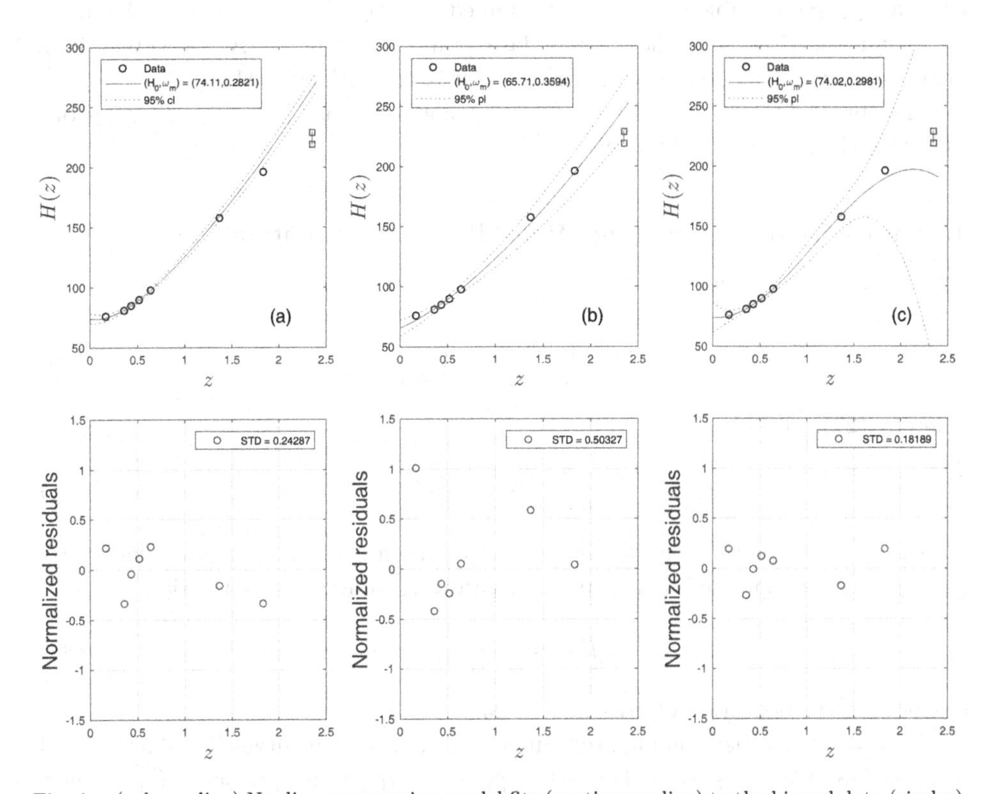

Fig. 4. (color online) Nonlinear regression model fits (continuous line) to the binned data (circles) of Table 1 $(k = 1, 2, \ldots, 7)$ by the MatLab function *fitnlm* to (a) $\Lambda = \omega_0^2$, (b) ΛCDM, (c) a cubic polynomial. Included are the 95% prediction limits (pl, dashed lines; very similar to the 95% confidence limits, not shown). Included is the incompatible data point $k = 8$ (red) in Table 1.

fitnlm using weights $w_k \propto \sigma_k^{-2}$.[16,23] Figure 4 shows the fits and 95% prediction limits (pl, very close but slightly wider than 95% confidence limits).

Figure 4 shows the match of evolution with (6) including the normalized errors, further for ΛCDM and a model-independent cubic fit.

For (6) these errors are essentially the same as obtained by a model-independent cubic fit

$$H(z) = H_0\left(1 + (1+q_0)z + \frac{1}{2}(Q_0 + (1+q_0)q_0)z^2 + b_4 z^3\right) + O(z^4), \qquad (34)$$

with free coefficients (H_0, q_0, Q_0, b_4), and they are about one-half the normalized errors in the fit by ΛCDM.

For the two models, q_0 and Q_0 obtained according to their respective evolution equations, i.e. $\Omega_M = (1/3)(2+q_0)$ for (6), (10) for ΛCDM, and

$$Q_0 = \begin{cases} (2+q_0)(1-2q_0) > 2.5 & (\Lambda = \omega_0^2)\,, \\ (1+q_0)(1-2q_0) \lesssim 1 & (\Lambda\mathrm{CDM})\,, \end{cases} \qquad (35)$$

whereas Q_0 in our cubic fit (34) is determined directly by nonlinear regression.

Included in Fig. 4 is the $k = 8$ data point in Table 1. As expected, it is inconsistent with (6) and only marginally consistent with ΛCDM.

In Table 2, $Q_0 = 2.2648 \pm 0.2910$ of the model-independent cubic fit rules out ΛCDM according to (35) by 4.36σ.

4. Sensitivity to $H(z)$ in High Redshift Galaxy Rotation Curves

By continuity in the onset to weak gravity at $\alpha = a_{\mathrm{dS}}$ (8), we have

$$y_{0,h} = \left(\frac{a_{\mathrm{N}}}{a_{\mathrm{dS}}}\right)_h = \left(\frac{r_t}{R_h}\right)^2 \qquad (36)$$

whereby (27) takes the form

$$\frac{\alpha}{a_{\mathrm{dS}}} = \sqrt{\mu}\,\frac{r_t}{R_h}\,, \qquad (37)$$

with $\mu = 2\langle B(p)\rangle$. In weak gravity $\alpha \lesssim a_{\mathrm{dS}}$, anomalous behavior in galactic dynamics may be expressed by an apparent equivalent dark matter fraction

$$f'_{\mathrm{DM}} = \frac{\alpha - a_{\mathrm{N}}}{\alpha}\,, \qquad (38)$$

conforming the definition of f_{DM} in Ref. 24.

Table 3 lists the data on high redshift sample of rotation curves[24] and associated data on (28)–(38). Figure 5 shows the apparent (f_{DM}) and predicted (f'_{DM}) dark matter fractions. Based on r_t/R_H, this sample of galaxies probes weak gravity (27) in $\alpha < a_{\mathrm{dS}}$ but not the asymptotic regime $\alpha \ll a_{\mathrm{dS}}$, a point recently emphasized by Ref. 25.

Table 3. Analysis on apparent ($f'_{\rm DM}$) versus observed ($f_{\rm DM}$) dark matter fractions in high z rotation curves[24] with baryonic mass $M_b = M_{11}10^{11}\ M_\odot$ and rotation velocities V_c in units of km s^{-1} at the half-light radius R_h.

Galaxy	z	$H(z)/H_0$	$q(z)$	M_{11}	r_t/R_h	μ	$f'_{\rm DM}$	V_c	$f_{\rm DM}$ (95% c.l.)
COS4 01351	0.854	1.5986	0.0853	1.7	0.6740	0.7050	0.2031	276	0.21 ± 0.1
D3a 6397	1.500	2.2883	0.2957	2.3	0.6273	0.6587	0.2342	310	$0.17\,(< 0.38)$
GS4 43501	1.613	2.4246	0.3170	1.0	0.6054	0.6381	0.2497	257	$0.19\,(\pm 0.09)$
zC 406690	2.196	3.1903	0.3996	1.7	0.6060	0.6383	0.2489	301	$0\,(< 0.08)$
zC 400569	2.242	3.2531	0.3919	1.7	0.9992	0.9814	0.0006	364	$0\,(< 0.07)$
D3a 15504	2.383	3.4537	0.4184	2.1	0.5908	0.6239	0.2590	299	$0.12\,(< 0.26)$

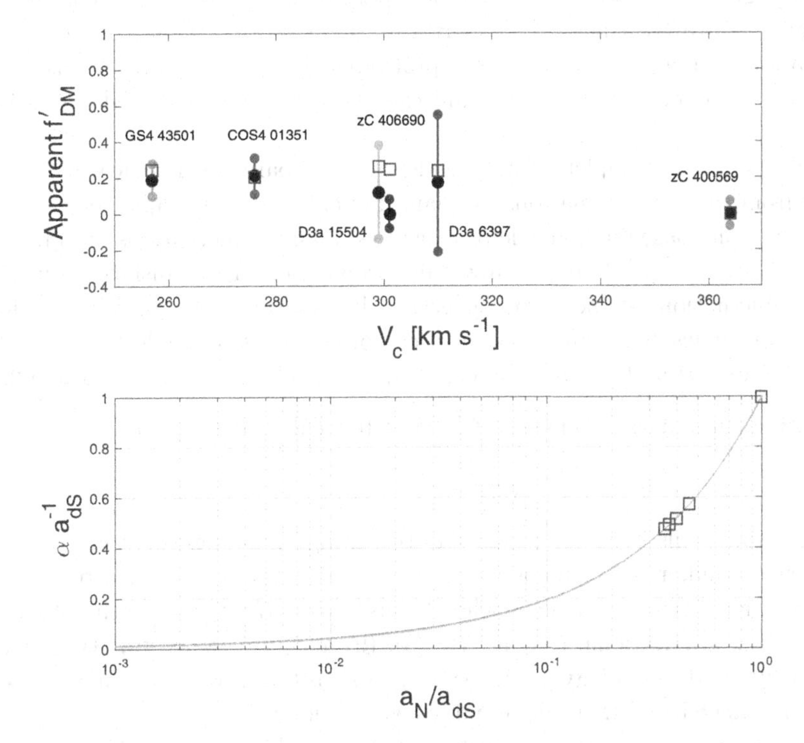

Fig. 5. (color online) (Upper panel.) Apparent dark matter fractions $f'_{\rm DM}$ (blue squares) and observational data on $f_{\rm DM}$. Results agree except for cZ 406690 ($V_c = 301$ km/s). (Lower panel.) Location of the data in Table 3 in weak gravity $a_{\rm N} < a_{\rm dS}$ and associated predicted accelerations $\alpha/a_{\rm dS}$. (Reprinted from Ref. 26.)

Figure 5 shows quantitative agreement of $f_{\rm DM}$ and $f'_{\rm DM}$, except for cZ 4006690 ($V_c = 301$ km s^{-1}). We note that its observed rotation curve is asymmetric, which may indicate systematic errors unique to this galaxy. (The other galaxies all have essentially symmetric rotation curves.) Apart from this particular galaxy, $f'_{\rm DM} \simeq f_{\rm DM}$ in over a broad range of redshifts confirms that (8), and hence weak gravity in galaxy dynamics is co-evolving with $H(z)$ in the background cosmology.

5. Conclusions and Outlook

By volume, structure formation and cosmology represent the most common gravitational interactions at the scale of $a_{\rm dS}$ or less. Relatively strong interactions in our solar system and its general relativistic extension to compact objects are, in fact, the exception. A principle distinction between the two is sensitivity to $a_{\rm dS}$ in weak gravitation, otherwise ignorable in the second. This potential is completely absent in Einstein's EP and classical formulations of general relativity, but becomes a real possibility by causality and compactness of the cosmological horizon.

Inspired by spacetime holography, weak gravity in galactic dynamics and cosmological evolution is parametrized by $a_{\rm dS}$ and the fundamental frequency ω_0 of the cosmological horizon, set by the canonical parameters (H,q) in (1). Observational consequences are (i) a perturbed inertia at small accelerations ($\alpha < a_{\rm dS}$) across a sharp onset at $a_{\rm N} = a_{\rm dS}$ with asymptotic behavior ($a_{\rm N} \ll a_{\rm dS}$) described by a Milgrom's parameter $a_0 = \omega_0/2\pi$; and (ii) $Q_0 > 2.5$ at accelerated expansion by $\Lambda = \omega_0^2$.

Our $\Lambda = \omega_0^2$ is completely independent of the bare cosmological constant Λ_0 arising from vacuum fluctuations in quantum field theory. Complementary to unitary holography based on particle propagators, vacuum fluctuations with no entanglement have a trivial Poincaré invariant propagator. By translation invariance, vacuum fluctuations do not drag spacetime like matter does, whereby Λ_0 has no inertia and carries no gravitational field conform Mach's principle.

Predictions (i) and (ii) are supported by the results of Fig. 1 and Table 2:

- Rotation curve data[4] point to a C^0 onset to weak gravity in Fig. 1 at

$$\left(\frac{a_{\rm N}}{a_{\rm dS}}, \frac{\alpha}{a_{\rm N}}\right) = (1,1)\,. \tag{39}$$

  We attribute this behavior to inertia originating in entanglement entropy, of the apparent Rindler or cosmological horizon, whichever is more nearby. At small accelerations, inertia is reduced when h falls beyond H, enhancing the acceleration for a given gravitational forcing. By (8), (9) and (29), weak gravity is sensitive to background cosmology, manifest in galaxy rotation curves over an extended redshift range listed in Table 3 and shown in Fig. 5.
- A model-independent analysis of the latter by a cubic polynomial identifies $Q_0 \simeq 2.26 \pm 0.29$. ΛCDM with $Q_0 \lesssim 1$ is hereby ruled out at 4.36σ. Normalized errors in the cubic fit and those to $\Lambda = \omega_0^2$ are about one-half of the normalized errors in the fit to ΛCDM.
- Estimates of H_0 in the first two are consistent with recent measurement $H_0 = 73.24 \pm 1.74$ km s^{-1} Mpc^{-1} in local surveys.[21,22] Combining our result for $\Lambda = \omega_0^2$ with the latter obtains relatively fast expansion compared to ΛCDM with

$$H_0 = 73.8 \pm 0.9 \text{ km s}^{-1} \text{ Mpc}\,. \tag{40}$$

The above suggests a strong form of the EP, by insisting on equivalent fluctuations in the inertia of Rindler observers (non-geodesics in Minkowski spacetime) with fluctuations in measuring weight on a scale (non-geodesics in curved spacetime). Momentum fluctuations between two bodies in mutual gravitational attraction herein are fully entangled, ensuring the preservation of total momentum. If so, curvature from the mass must follow very similar rules giving rise to (21). Consequently, any scenario for entropic gravity[27] should include the origin of inertia in entanglement entropy, the details of which remain to be spelled out.

Recently, the LISA Pathfinder has conducted an extensive set of a high precision free fall gravity and inertia experiments at accelerations well below a_{dS}.[28] It uses laser-interferometry and electrostatic forcing to track and perturb test masses about their geodesic motion, in the spacecraft's self-gravitational field. In probing the de Sitter scale of 1 Å s^{-2} or less, we note that (8) conveniently rescales to the size of a small spacecraft, i.e.

$$r_t \simeq 30 \text{ cm } \mathrm{m}_1^{1/2}, \tag{41}$$

about a gravitating mass $m = m_1$ kg, representative for the 2 kg test masses in the LISA Pathfinder mission. It seems worthwhile to look at the data covering $\alpha < a_N$.

Acknowledgments

The author thanks S. S. McGaugh, J. Binney, K. Danzmann, T. Piran, G. Smoot, C. Rubbia, J. Solà, L. Smolin and M. Milgrom for the stimulating discussions. This report was supported in part by the National Research Foundation of Korea under Grant Nos. 2015R1D1A1A01059793 and 2016R1A5A1013277.

References

1. A. Riess *et al.*, *Astrophys. J.* **116**, 1009 (1998).
2. S. Perlmutter *et al.*, *Astrophys. J.* **517**, 565 (1999).
3. M. H. P. M. van Putten, *Astrophys. J.* **837**, 22 (2017).
4. F. Lelli, S. S. McGaugh, J. M. Schombert and M. S. Pawlowski, *Astrophys. J.* **836**, 152 (2017).
5. S. S. McGaugh, *Astron. J.* **143**, 40 (2011).
6. S. S. McGaugh, *Phys. Rev. Lett.* **106**, 121303 (2011).
7. M. Milgrom, *Astrophys. J.* **270**, 365 (1983).
8. J. D. Bekenstein, *Phys. Rev. D* **23**, 287 (1981).
9. G. 't Hooft, arXiv:gr-qc/9310026.
10. L. Susskind, *J. Math. Phys.* **36**, 6377 (1995).
11. M. Milgrom, *Phys. Lett. A* **253**, 273 (1999).
12. L. Smolin, arXiv:1704.00780.
13. M. H. P. M. van Putten, *Int. J. Mod. Phys. D* **4**, 155024 (2015).
14. M. H. P. M. van Putten, *Astrophys. J.* **824**, 43 (2016).
15. J. Solà, A. Gómez-Valent and J. de Cruz Pérez, *Astrophys. J.* **836**, 43 (2017).
16. O. Farooq, F. R. Madiyar, S. Crandall and B. Ratra, *Astrophys. J.* **835**, 26 (2017).
17. A. D. A. M. Spallicci and M. H. P. M. van Putten, *Int. J. Geom. Meth. Mod. Phys.* **13**, 630014 (2016).
18. B. Famae and S. S. McGaugh, *Living Rev.* **15**, 10 (2012).
19. W. G. Unruh, *Phys. Rev. D* **14**, 870 (1976).

20. M. H. P. M. van Putten, *Mon. Not. R. Astron. Soc.* **450**, L48 (2015).
21. A. G. Riess *et al.*, *Astrophys. J.* **826**, 56 (2016).
22. Q.-D. Li *et al.*, *Astrophys. J.* **832**, 103 (2016).
23. S. Podariu, T. Souradeep, J. R. Gott, III, B. Ratra and M. S. Vogeley, *Astrophys. J.* **559**, 9 (2001).
24. R. Genzel *et al.*, *Nature* **543**, 397 (2017).
25. M. Milgrom, arXiv:1703.06110v2.
26. M. H. P. M. van Putten, (2017), in preparation.
27. E. Verlinde, *JHEP* **4**, 29 (2011).
28. M. Armano *et al.*, arXiv:astro-ph/1702.04633v1.

Dark Energy Density in SUGRA Models and Degenerate Vacua

C. D. Froggatt

School of Physics and Astronomy, University of Glasgow, Glasgow, G12 8QQ, UK
Colin.Froggatt@glasgow.ac.uk

H. B. Nielsen

The Niels Bohr Institute, University of Copenhagen, Blegdamsvej 17,
Copenhagen, DK-2100, Denmark
hbech@nbi.dk

R. Nevzorov* and A. W. Thomas

ARC Centre of Excellence for Particle Physics at the Terascale and CSSM,
School of Physical Sciences, The University of Adelaide,
Adelaide, SA 5005, Australia
**roman.nevzorov@adelaide.edu.au*

In $N = 1$ supergravity the tree-level scalar potential of the hidden sector may have a minimum with broken local supersymmetry (SUSY) as well as a supersymmetric Minkowski vacuum. These vacua can be degenerate, allowing for a consistent implementation of the multiple point principle. The first minimum where SUSY is broken can be identified with the physical phase in which we live. In the second supersymmetric phase, in flat Minkowski space, SUSY may be broken dynamically either in the observable or in the hidden sectors inducing a tiny vacuum energy density. We argue that the exact degeneracy of these phases may shed light on the smallness of the cosmological constant. Other possible phenomenological implications are also discussed. In particular, we point out that the presence of such degenerate vacua may lead to small values of the quartic Higgs coupling and its beta function at the Planck scale in the physical phase.

Keywords: Supergravity; cosmological constant; Higgs boson.

1. Introduction

It is expected that at ultra-high energies the Standard Model (SM) is embedded in an underlying theory that provides a framework for the unification of all interactions such as Grand Unified Theories (GUTs), supergravity (SUGRA), String Theory, etc. At low energies this underlying theory could lead to new physics phenomena beyond the SM. Moreover the energy scale associated with the physics beyond the SM is supposed to be somewhat close to the mass of the Higgs boson to avoid a fine-tuning problem related to the need to stabilise the scale where electroweak (EW) symmetry is broken. Despite the successful discovery of the 125 GeV Higgs boson in 2012, no indication of any physics beyond the SM has been detected at the LHC so far. On the other hand, there are compelling reasons to believe that the SM is extremely fine-tuned. Indeed, astrophysical and cosmological observations indicate that there is a tiny energy density spread all over the Universe (the cosmological constant), i.e. $\rho_\Lambda \sim 10^{-123} M_{\rm Pl}^4 \sim 10^{-55} M_Z^4$,[1,2] which is responsible for its acceleration. At the same time much bigger contributions must come from the EW symmetry breaking $(\sim 10^{-67} M_{\rm Pl}^4)$ and QCD condensates $(\sim 10^{-79} M_{\rm Pl}^4)$. Because of the enormous cancellation between the contributions of different condensates to ρ_Λ, which is required to keep ρ_Λ around its measured value, the smallness of the cosmological constant can be considered as a fine-tuning problem.

Here, instead of trying to alleviate fine-tuning of the SM we impose the exact degeneracy of at least two (or even more) vacua. Their presence was predicted by the so-called Multiple Point Principle (MPP).[3] According to the MPP, Nature chooses values of coupling constants such that many phases of the underlying theory should coexist. This corresponds to a special (multiple) point on the phase diagram where these phases meet. At the multiple point these different phases have the same vacuum energy density.

The MPP applied to the SM implies that the Higgs effective potential, which can be written as

$$V_{\rm eff}(H) = m^2(\phi) H^\dagger H + \lambda(\phi)(H^\dagger H)^2\,, \tag{1}$$

where H is a Higgs doublet and ϕ is a norm of the Higgs field, i.e. $\phi^2 = H^\dagger H$, has two degenerate minima. These minima are taken to be at the EW and Planck scales.[4] The corresponding vacua can have the same energy density only if

$$\lambda(M_{\rm Pl}) \simeq 0\,, \quad \beta_\lambda(M_{\rm Pl}) \simeq 0\,, \tag{2}$$

where $\beta_\lambda = \frac{d\lambda(\phi)}{d\log\phi}$ is the beta-function of $\lambda(\phi)$. It was shown that the MPP conditions (2) can be satisfied when $M_t = 173 \pm 5$ GeV and $M_H = 135 \pm 9$ GeV.[4] The application of the MPP to the two Higgs doublet extension of the SM was also considered.[5–7]

The measurement of the Higgs boson mass allows us to determine quite precisely the parameters of the Higgs potential (1). Furthermore using the extrapolation of the SM parameters up to $M_{\rm Pl}$ with full 3-loop precision it was found[8] that

$$\begin{aligned}\lambda(M_{\rm Pl}) = &-0.0143 - 0.0066\left(\frac{M_t}{\rm GeV} - 173.34\right)\\ &+0.0018\left(\frac{\alpha_3(M_Z) - 0.1184}{0.0007}\right) + 0.0029\left(\frac{M_H}{\rm GeV} - 125.15\right).\end{aligned} \tag{3}$$

The computed value of $\beta_\lambda(M_{\rm Pl})$ is also rather small, so that the MPP conditions (2) are basically fulfilled.

The successful MPP predictions for the Higgs and top quarks masses[4] suggest that we may use this idea to explain the tiny value of the cosmological constant as well. In principle, the smallness of the cosmological constant could be related to an almost exact symmetry. Nevertheless, none of the generalisations of the SM provides any satisfactory explanation for the smallness of this dark energy density. An exact global supersymmetry (SUSY) guarantees that the vacuum energy density vanishes in all global minima of the scalar potential. However the nonobservation of superpartners of quarks and leptons implies that SUSY is broken. The breakdown of SUSY induces a huge and positive contribution to the dark energy density which is many orders of magnitude larger than M_Z^4. Here the MPP assumption is adapted to $(N = 1)$ SUGRA models, in order to provide an explanation for the tiny deviation of the measured dark energy density from zero.

2. SUGRA Models Inspired by Degenerate Vacua

The full $(N = 1)$ SUGRA Lagrangian can be specified in terms of an analytic gauge kinetic function $f_a(\phi_M)$ and a real gauge-invariant Kähler function $G(\phi_M, \phi_M^*)$. These functions depend on the chiral superfields ϕ_M. The function $f_a(\phi_M)$ determine, in particular, the gauge couplings $\mathrm{Re}\, f_a(\phi_M) = 1/g_a^2$, where the index a represents different gauge groups. The Kähler function can be presented as

$$G(\phi_M, \phi_M^*) = K(\phi_M, \phi_M^*) + \ln |W(\phi_M)|^2 \,, \tag{4}$$

where $K(\phi_M, \phi_M^*)$ is the Kähler potential while $W(\phi_M)$ is the superpotential of the SUGRA model under consideration. Here we shall use standard supergravity mass units: $\frac{M_{\mathrm{Pl}}}{\sqrt{8\pi}} = 1$.

The SUGRA scalar potential can be written as a sum of F- and D-terms, i.e.

$$\begin{gathered} V(\phi_M, \phi_M^*) = \sum_{M,\bar{N}} e^G \left(G_M G^{M\bar{N}} G_{\bar{N}} - 3\right) + \frac{1}{2}\sum_a (D^a)^2 \,, \\ D^a = g_a \sum_{i,\,j} \left(G_i T^a_{ij} \phi_j\right), \quad G_M \equiv \frac{\partial G}{\partial \phi_M}\,, \quad G_{\bar{M}} \equiv \frac{\partial G}{\partial \phi_M^*}\,, \\ G_{\bar{N}M} \equiv \frac{\partial^2 G}{\partial \phi_N^* \partial \phi_M}\,, \quad G^{M\bar{N}} = G^{-1}_{\bar{N}M}\,. \end{gathered} \tag{5}$$

In Eq. (5) g_a is the gauge coupling associated with the generator T^a. In order to achieve the breakdown of local SUSY in $(N = 1)$ supergravity, a hidden sector is introduced. The superfields of the hidden sector (z_i) interact with the observable ones only by means of gravity. It is expected that at the minimum of the scalar potential (5) hidden sector fields acquire vacuum expectation values (VEVs) so that at least one of their auxiliary fields

$$F^M = e^{G/2} G^{M\bar{P}} G_{\bar{P}} \tag{6}$$

is nonvanishing, leading to the spontaneous breakdown of local SUSY, giving rise to a nonzero gravitino mass $m_{3/2} \simeq \langle e^{G/2} \rangle$. The absolute value of the vacuum energy density at the minimum of the SUGRA scalar potential (5) tends to be of order of $m_{3/2}^2 M_{\rm Pl}^2$. Therefore an enormous degree of fine-tuning is required to keep the cosmological constant in SUGRA models around its observed value.

The successful implementation of the MPP in ($N = 1$) SUGRA models requires us to assume the existence of a supersymmetric Minkowski vacuum.[9,10] According to the MPP this second vacuum and the physical one must be degenerate. Since the vacuum energy density of supersymmetric states in flat Minkowski space vanishes, the cosmological constant problem is solved to first approximation. Such a second vacuum exists if the SUGRA scalar potential (5) has a minimum where

$$W(z_m^{(2)}) = 0\,, \qquad \left.\frac{\partial W(z_i)}{\partial z_m}\right|_{z_m = z_m^{(2)}} = 0\,, \tag{7}$$

where $z_m^{(2)}$ are VEVs of the hidden sector fields in the second vacuum. Equations (7) indicate that an extra fine-tuning is needed to ensure the presence of the supersymmetric Minkowski vacuum in SUGRA models.

The simplest Kähler potential and superpotential that satisfies conditions (7) are given by

$$K(z,\, z^*) = |z|^2\,, \quad W(z) = m_0 (z + \beta)^2\,. \tag{8}$$

The hidden sector of the corresponding SUGRA model involves only one singlet superfield z. If $\beta = \beta_0 = -\sqrt{3} + 2\sqrt{2}$, the corresponding SUGRA scalar potential possesses two degenerate vacua with zero energy density at the classical level. The first minimum associated with $z^{(2)} = -\beta$ is a supersymmetric Minkowski vacuum. In the other minimum, $z^{(1)} = \sqrt{3} - \sqrt{2}$, local SUSY is broken so that it can be identified with the physical vacuum. Varying β around β_0 one can get a positive or a negative contribution from the hidden sector to the total vacuum energy density of the physical phase. Thus parameter β can be always fine-tuned so that the physical and supersymmetric Minkowski vacua are degenerate.

The fine-tuning associated with the realisation of the MPP in ($N = 1$) SUGRA models can be to some extent alleviated within the no-scale inspired SUGRA model with broken dilatation invariance.[10] Let us consider the no-scale inspired SUGRA model that involves two hidden sector superfields (T and z) and a set of chiral supermultiplets φ_σ in the observable sector. These superfields transform differently under the dilatations ($T \to \alpha^2 T$, $z \to \alpha z$, $\varphi_\sigma \to \alpha \varphi_\sigma$) and imaginary translations ($T \to T + i\beta$, $z \to z$, $\varphi_\sigma \to \varphi_\sigma$), which are subgroups of the $SU(1,1)$ group.[10,11] The full superpotential of the model can be written as a sum:[10]

$$\begin{gathered} W(z, \varphi_\alpha) = W_{\rm hid} + W_{\rm obs}\,, \\ W_{\rm hid} = \varkappa \left(z^3 + \mu_0 z^2 + \sum_{n=4}^{\infty} c_n z^n \right), \quad W_{\rm obs} = \sum_{\sigma,\beta,\gamma} \frac{1}{6} Y_{\sigma\beta\gamma} \varphi_\sigma \varphi_\beta \varphi_\gamma\,. \end{gathered} \tag{9}$$

The superpotential (9) includes a bilinear mass term for the superfield z and higher order terms $c_n z^n$ which explicitly break dilatation invariance. A term proportional to z is not included since it can be forbidden by a gauge symmetry of the hidden sector. Here we do not allow the breakdown of dilatation invariance in the superpotential of the observable sector to avoid potentially dangerous terms that may lead to the so-called μ-problem, etc.

The full Kähler potential of the SUGRA model under consideration is given by:[10]

$$K(\phi_M, \phi_M^*) = -3\ln\left[T + \bar{T} - |z|^2 - \sum_\alpha \zeta_\alpha |\varphi_\alpha|^2\right] + \sum_{\alpha,\beta}\left(\frac{\eta_{\alpha\beta}}{2}\varphi_\alpha\varphi_\beta + \text{h.c.}\right) + \sum_\beta \xi_\beta |\varphi_\beta|^2\,, \qquad (10)$$

where ζ_α, $\eta_{\alpha\beta}$, ξ_β are some constants. If $\eta_{\alpha\beta}$ and ξ_β have nonzero values the dilatation invariance is explicitly broken in the Kähler potential of the observable sector. Here we restrict our consideration to the simplest set of terms that break dilatation invariance. Moreover we only allow the breakdown of the dilatation invariance in the Kähler potential of the observable sector, because any variations in the part of the Kähler potential associated with the hidden sector may spoil the vanishing of the vacuum energy density in global minima. When the parameters $\eta_{\alpha\beta}$, ξ_β and $\varkappa$ go to zero, the dilatation invariance is restored, protecting supersymmetry and a zero value of the cosmological constant.

In the SUGRA model under consideration the tree-level scalar potential of the hidden sector is positive definite

$$V_{\text{hid}} = \frac{1}{3(T + \bar{T} - |z|^2)^2}\left|\frac{\partial W_{\text{hid}}(z)}{\partial z}\right|^2, \qquad (11)$$

so that the vacuum energy density vanishes near its global minima. In the simplest case when $c_n = 0$, the SUGRA scalar potential of the hidden sector (11) has two minima, at $z = 0$ and $z = -\frac{2\mu_0}{3}$. At these points V_{hid} attains its absolute minimal value, i.e. zero. In the first vacuum, where $z = -\frac{2\mu_0}{3}$, local SUSY is broken and the gravitino gains mass

$$m_{3/2} = \left\langle \frac{W_{\text{hid}}(z)}{(T + \bar{T} - |z|^2)^{3/2}} \right\rangle = \frac{4\varkappa\mu_0^3}{27\left\langle\left(T + \bar{T} - \frac{4\mu_0^2}{9}\right)^{3/2}\right\rangle}. \qquad (12)$$

All scalar particles get nonzero masses $m_\sigma \sim \frac{m_{3/2}\xi_\sigma}{\zeta_\sigma}$ as well. This minimum can be identified with the physical vacuum. Assuming that ξ_α, ζ_α, μ_0 and $\langle T\rangle$ are all of order unity, a SUSY breaking scale $M_S \sim 1$ TeV can only be obtained when $\varkappa$ is extremely small, i.e. $\varkappa \simeq 10^{-15}$. In the second vacuum, where $z = 0$, the superpotential of the hidden sector vanishes and local SUSY remains intact giving rise to the supersymmetric Minkowski vacuum. If the high order terms $c_n z^n$ are present in Eqs. (9), $V_{\rm hid}$ can have many degenerate vacua, with broken and unbroken SUSY, in which the vacuum energy density vanishes. As a result the MPP conditions are fulfilled without any extra fine-tuning at the tree-level because of the positive-definite form of the scalar potential (11).

It is worth noting that the inclusion of perturbative and nonperturbative corrections to the Lagrangian of the no-scale inspired SUGRA model should spoil the degeneracy of vacua, giving rise to a huge energy density in the minimum of the scalar potential where local SUSY is broken. Furthermore, in the SUGRA model under consideration the mechanism for the stabilisation of the VEV of the hidden sector field T remains unclear. Therefore this model should be considered as a toy example only. It demonstrates that in ($N = 1$) supergravity there might be a mechanism which ensures the vanishing of the vacuum energy density in the physical vacuum. This mechanism can also result in a set of degenerate vacua with broken and unbroken SUSY, leading to the realisation of the MPP.

3. Implications for Cosmology and Collider Phenomenology

3.1. *Model with intermediate SUSY breaking scale*

Now let us assume that the MPP inspired SUGRA model of the type just discussed is realised in Nature. In other words there exist at least two exactly degenerate phases. The first phase is associated with the physical vacuum whereas the second one is identified with the supersymmetric Minkowski vacuum in which the vacuum energy density vanishes in the leading approximation. However nonperturbative effects may give rise to the breakdown of SUSY in the second phase resulting in a small vacuum energy density. This small energy density should be then transferred to our vacuum by the assumed degeneracy.

If SUSY breaking takes place in the second vacuum, it can be caused by the strong interactions in the observable sector. Indeed, the SM gauge couplings g_1, g_2 and g_3, which correspond to $U(1)_Y$, $SU(2)_W$ and $SU(3)_C$ gauge interactions respectively, change with the energy scale. Their evolution obeys the renormalisation group equations (RGEs) that in the one-loop approximation can be written as

$$\frac{d\log\alpha_i(Q)}{d\log Q^2} = \frac{b_i\alpha_i(Q)}{4\pi}, \tag{13}$$

where Q is a renormalisation scale, $i = 1, 2, 3$ and $\alpha_i(Q) = g_i^2(Q)/(4\pi)$. In the pure MSSM $b_3 < 0$ and the gauge coupling $g_3(Q)$ of the $SU(3)_C$ interactions increases in the infrared region. Thus although this coupling can be rather small at high energies it becomes rather large at low energies and one can expect that the role of nonperturbative effects is enhanced.

To simplify our analysis we assume that the SM gauge couplings at high energies are identical in the physical and supersymmetric Minkowski vacua. Consequently for $Q > M_S$, where M_S is a SUSY breaking scale in the physical vacuum, the renormalisation group (RG) flow of these couplings is also the same in both vacua.

When $Q < M_S$ all superparticles in the physical vacuum decouple. Therefore the $SU(3)_C$ beta function in the physical and supersymmetric Minkowski vacua become very different. Because of this, below the scale M_S the values of $\alpha_3(Q)$ in the physical and second vacua ($\alpha_3^{(1)}(Q)$ and $\alpha_3^{(2)}(Q)$) are not the same. For $Q < M_S$ the $SU(3)_C$ beta function in the physical phase $\tilde{b}_3$ coincides with the corresponding SM beta function, i.e. $\tilde{b}_3 = -7$. Using the value of $\alpha_3^{(1)}(M_Z) \approx 0.1184$ and the matching condition $\alpha_3^{(2)}(M_S) = \alpha_3^{(1)}(M_S)$, one can estimate the value of the strong gauge coupling in the second vacuum

$$\frac{1}{\alpha_3^{(2)}(M_S)} = \frac{1}{\alpha_3^{(1)}(M_Z)} - \frac{\tilde{b}_3}{4\pi} \ln \frac{M_S^2}{M_Z^2} \,. \tag{14}$$

In the supersymmetric Minkowski vacuum all particles of the MSSM are massless and the EW symmetry is unbroken. So, in the second phase the $SU(3)_C$ beta function b_3 remains the same as in the MSSM, i.e. $b_3 = -3$. Since the MSSM $SU(3)_C$ beta function exhibits asymptotically free behaviour, $\alpha_3^{(2)}(Q)$ increases in the infrared region. The top quark is massless in the supersymmetric phase and its Yukawa coupling also grows in the infrared with the increasing of $\alpha_3^{(2)}(Q)$. At the scale

$$\Lambda_{\rm SQCD} = M_S \exp\left[\frac{2\pi}{b_3 \alpha_3^{(2)}(M_S)}\right], \tag{15}$$

where the supersymmetric QCD interactions become strong in the supersymmetric Minkowski vacuum, the top quark Yukawa coupling is of the same order of magnitude as the $SU(3)_C$ gauge coupling. So a large value of the top quark Yukawa coupling may give rise to the formation of a quark condensate. This condensate breaks SUSY, resulting in a nonzero positive value for the dark energy density

$$\rho_\Lambda \simeq \Lambda_{\rm SQCD}^4 \,. \tag{16}$$

The dependence of $\Lambda_{\rm SQCD}$ on the SUSY breaking scale M_S is shown in Fig. 1. Since $\tilde{b}_3 < b_3$ the value of the QCD gauge coupling below M_S is larger in the physical phase than in the second one. Therefore the scale $\Lambda_{\rm SQCD}$ is substantially lower than the QCD scale in the SM and decreases with increasing M_S. When the supersymmetry breaking scale in the physical phase is of the order of 1 TeV, we get $\Lambda_{\rm SQCD} = 10^{-26} M_{\rm Pl} \simeq 100$ eV.[10,12,13] This leads to the value of the cosmological constant $\rho_\Lambda \simeq 10^{-104} M_{\rm Pl}^4$, which is enormously suppressed as compared with $v^4 \simeq 10^{-67} M_{\rm Pl}^4$. The measured value of the dark energy density is reproduced when $\Lambda_{\rm SQCD} = 10^{-31} M_{\rm Pl} \simeq 10^{-3}$ eV. The appropriate values of $\Lambda_{\rm SQCD}$ can be obtained only if $M_S \simeq 10^3$–10^4 TeV.[10,12,13] However models with such a large SUSY breaking scale do not lead to a suitable dark-matter candidate and also spoil the unification of the SM gauge couplings.

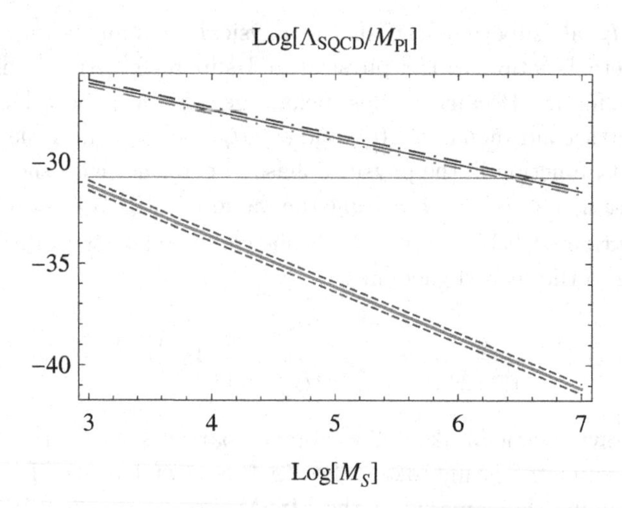

Fig. 1. The value of $\log\left[\Lambda_{\mathrm{SQCD}}/M_{\mathrm{Pl}}\right]$ versus $\log M_S$. The thin and thick solid lines correspond to the pure MSSM and the MSSM with an additional pair of $5+\bar{5}$ supermultiplets, respectively. These lines are obtained for $\alpha_3(M_Z) = 0.1184$. The dashed and dash dotted lines represent the uncertainty in $\alpha_3(M_Z)$. The upper dashed and dash–dotted lines are obtained for $\alpha_3(M_Z) = 0.121$, while the lower ones correspond to $\alpha_3(M_Z) = 0.116$. The horizontal line represents the observed value of $\rho_\Lambda^{1/4}$. The SUSY breaking scale, M_S, is measured in GeV.

3.2. *Split SUSY scenario*

The problems mentioned above can be addressed within the Split SUSY scenario of supersymmetry breaking.[14,15] In other words, let us now assume that in the physical vacuum SUSY is broken so that all scalar bosons gain masses of order of $M_S \gg 10$ TeV, except for a SM-like Higgs boson, whose mass is set to be around 125 GeV. The mass parameters of gauginos and Higgsinos are protected by a combination of an R-symmetry and Peccei–Quinn symmetry so that they can be many orders of magnitude smaller than M_S. To ensure gauge coupling unification all neutralino, chargino and gluino states are chosen to lie near the TeV scale in the Split SUSY scenario.[15] Also a TeV-scale lightest neutralino can be an appropriate dark matter candidate.[15–18]

Thus in the Split SUSY scenario supersymmetry is not used to stabilise the EW scale.[14,15] This stabilisation is expected to be provided by some other mechanism, which may also explain the tiny value of the dark energy density. Therefore in the Split SUSY models M_S is taken to be much above 10 TeV. In the Split SUSY scenario some flaws inherent to the MSSM disappear. The ultra-heavy scalars, whose masses can range from hundreds of TeV up to 10^{13} GeV,[15] ensure the absence of large flavour changing and CP violating effects. The stringent constraints from flavour and electric dipole moment data, that require $M_S > 100 - 1000$ TeV, are satisfied and the dimension-five operators, which mediate proton decay, are also suppressed within the Split SUSY models. Nevertheless, since the sfermions are ultra-heavy the Higgs sector is extremely fine-tuned, with the understanding that the solution to both the hierarchy and cosmological constant problems might not involve natural cancellations, but follow from anthropic-like selection effects.[19] In other words galaxy and star formation, chemistry and biology, are basically impossible without these scales having the values found in our Universe.[19–21] In this case supersymmetry may be just a necessary ingredient in a fundamental theory of Nature like in the case of String Theory.

It has been argued that String Theory can have a huge number of long-lived metastable vacua[22–31] which is measured in googles ($\sim 10^{100}$).[25–31] The space of such vacua is called the "landscape." To analyse the huge multitude of universes, associated with the "landscape" of these vacua a statistical approach is used.[26–31] The total number of vacua in String Theory is sufficiently large to fine-tune both the cosmological constant and the Higgs mass, favouring a high-scale breaking of supersymmetry.[30,31] Thus it is possible for us to live in a universe fine-tuned in the way we find it simply because of a cosmic selection rule, i.e. the anthropic principle.[19]

The idea of the multiple point principle and the landscape paradigm have at least two things in common. Both approaches imply the presence of a large number of vacua with broken and unbroken SUSY. The landscape paradigm and MPP also imply that the parameters of the theory, which results in the SM at low energies, can be extremely fine-tuned so as to guarantee a tiny vacuum energy density and a large hierarchy between $M_{\rm Pl}$ and the EW scale. Moreover the MPP assumption might originate from the landscape of string theory vacua, if all vacua with a vacuum energy density that is too large are forbidden for some reason, so that all the allowed string vacua, with broken and unbroken supersymmetry, are degenerate to very high accuracy. If this is the case, then the breaking of supersymmetry at high scales is perhaps still favoured. Although this scenario looks quite attractive it implies that only a narrow band of values around zero cosmological constant would be allowed and the surviving vacua would obey MPP to the accuracy of the width w of this remaining band. However such accuracy is not sufficient to become relevant for the main point of the present article, according to which MPP "transfers" the vacuum energy density of the second vacuum to the physical vacuum.

In order to estimate the value of the cosmological constant we again assume that the physical and second phases have precisely the same vacuum energy densities and the gauge couplings at high energies are identical in both vacua. This means that the renormalisation group flow of the SM gauge couplings down to the scale M_S is the same in both vacua as before. For $Q < M_S$ all squarks and sleptons in the physical vacuum decouple and the beta functions change. At the TeV scale, the corresponding beta functions in the physical phase change once again due to the decoupling of the gluino, neutralino and chargino. Assuming that $\alpha_3^{(2)}(M_S) = \alpha_3^{(1)}(M_S)$, one finds

$$\frac{1}{\alpha_3^{(2)}(M_S)} = \frac{1}{\alpha_3^{(1)}(M_Z)} - \frac{\tilde{b}_3}{4\pi} \ln \frac{M_g^2}{M_Z^2} - \frac{b_3'}{4\pi} \ln \frac{M_S^2}{M_g^2}\,, \tag{17}$$

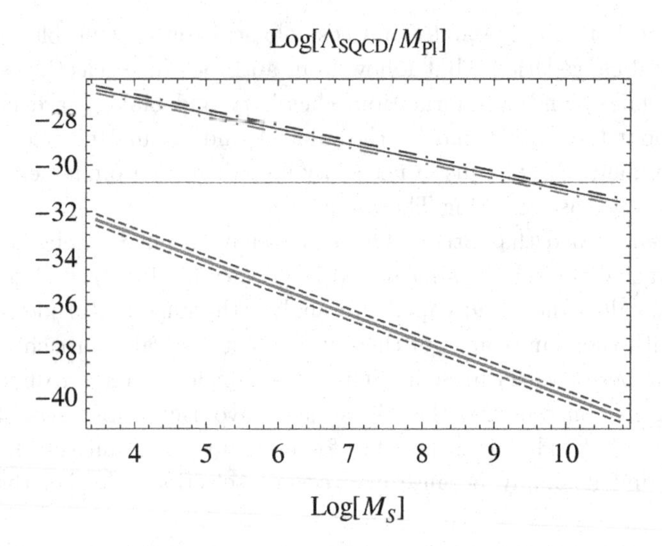

Fig. 2. The value of $\log\left[\Lambda_{SQCD}/M_{Pl}\right]$ versus $\log M_S$ for $M_q = M_g = 3$ TeV. The thin and thick solid lines correspond to the Split SUSY scenarios with the pure MSSM particle content and the MSSM particle content supplemented by an additional pair of $5 + \bar{5}$ supermultiplets respectively. The dashed and dash–dotted lines represent the uncertainty in $\alpha_3(M_Z)$. The thin and thick solid lines are obtained for $\alpha_3(M_Z) = 0.1184$, the upper (lower) dashed and dash–dotted lines correspond to $\alpha_3(M_Z) = 0.116$ ($\alpha_3(M_Z) = 0.121$). The horizontal line represents the observed value of $\rho_\Lambda^{1/4}$. The SUSY breaking scale M_S is measured in GeV.

where M_g is the mass of the gluino and $b'_3 = -5$ is the one-loop beta function of the strong gauge coupling in the Split SUSY scenario. The values of Λ_{SQCD} and ρ_Λ can be estimated using Eqs. (15) and (16), respectively.

In Fig. 2 we explore the dependence of Λ_{SQCD} in the second phase on the SUSY breaking scale M_S in the physical vacuum. In our analysis we set $M_g = 3$ TeV. As before Λ_{SQCD} diminishes with increasing M_S. The observed value of the dark energy density can be reproduced when $M_S \sim 10^9$–10^{10} GeV.[32–34] The value of M_S, which results in the measured cosmological constant, depends on $\alpha_3(M_Z)$ and the gluino mass. However this dependence is rather weak. In particular, with increasing M_g the value of M_S, which leads to an appropriate value of the cosmological constant, decreases. When $\alpha_3(M_Z) = 0.116$–0.121 and $M_g = 500$–2500 GeV, the corresponding value of the SUSY breaking scale varies from $2 \cdot 10^9$ GeV up to $3 \cdot 10^{10}$ GeV.[32–34]

The obtained prediction for M_S can be tested. A striking feature of the Split SUSY model is the extremely long lifetime of the gluino. The gluino decays through a virtual squark to a quark–antiquark pair and a neutralino $\tilde{g} \to q\bar{q} + \chi_1^0$. The large squark masses give rise to a long lifetime for the gluino. This lifetime can be estimated as[35,36]

$$\tau \sim 8\left(\frac{M_S}{10^9\ \text{GeV}}\right)^4 \left(\frac{1\ \text{TeV}}{M_g}\right)^5 s\,. \tag{18}$$

From Eq. (18) it follows that the supersymmetry breaking scale in the Split SUSY models should not exceed 10^{13} GeV.[15] Otherwise the gluino lifetime becomes larger than the age of the Universe. When M_S varies from $2 \cdot 10^9$ GeV ($M_g = 2500$ GeV) to $3 \cdot 10^{10}$ GeV ($M_g = 500$ GeV) the gluino lifetime changes from 1 sec. to $2 \cdot 10^8$ sec. (1000 years). Thus the measurement of the mass and lifetime of gluino should allow one to estimate the value of M_S in the Split SUSY scenario.

3.3. *Models with low SUSY breaking scale*

The observed value of the dark energy density can be also reproduced when the SUSY breaking scale is around 1 TeV. This can be achieved if the MSSM particle content is supplemented by an extra pair of $5 + \bar{5}$ supermultiplets which are fundamental and antifundamental representations of the supersymmetric $SU(5)$ GUT. The additional bosons and fermions would not affect gauge coupling unification in the leading approximation, since they form complete representations of $SU(5)$. In the physical phase states from $5 + \bar{5}$ supermultiplets can gain masses around M_S. The corresponding mass terms in the superpotential can be induced because of the presence of the bilinear terms $[\eta(5 \cdot \bar{5}) + \text{h.c.}]$ in the Kähler potential of the observable sector.[37,38] In the Split SUSY scenario we assume that new bosonic states from $5 + \bar{5}$ supermultiplets gain masses around the supersymmetry breaking scale, whereas their fermion partners acquire masses of the order of the gluino, chargino and neutralino masses. In our numerical studies we set the masses of extra quarks to be equal to the gluino mass, i.e. $M_q \simeq M_g$. In the supersymmetric Minkowski vacuum new bosons and fermions from $5 + \bar{5}$ supermultiplets remain massless. As a consequence they give a substantial contribution to the β functions in this vacuum. Indeed, the one-loop beta function of the strong interaction in the second phase changes from $b_3 = -3$ (the $SU(3)_C$ beta function in the MSSM) to $b_3 = -2$. This leads to a further reduction of Λ_{SQCD}. At the same time, extra fermion states from $5 + \bar{5}$ supermultiplets do not affect much the RG flow of gauge couplings in the physical phase below the scale M_S. For example, in the Split SUSY scenario the one-loop beta function that determines the running of the strong gauge coupling from the SUSY breaking scale down to the TeV scale changes from -5 to $-13/3$. As follows from Figs. 1 and 2 in the case of the SUSY model with extra $5 + \bar{5}$ supermultiplets the measured value of the dark energy density can be reproduced even for $M_S \simeq 1$ TeV.[10,12,13,32–34] Nevertheless, the Split SUSY scenario which was discussed in the previous subsection has the advantage of avoiding the need for any new particles beyond those of the MSSM, provided that $M_S \simeq 10^9$–10^{10} GeV. On the other hand, the MPP scenario with extra $5 + \bar{5}$ supermultiplets of matter and SUSY breaking scale in a few TeV range is easier to verify at the LHC in the near future.

3.4. *The breakdown of SUSY in the hidden sector*

The nonzero value of the vacuum energy density can be also induced if supersymmetry in the second phase is broken in the hidden sector. This can happen if the

SM gauge couplings are sufficiently small in the supersymmetric Minkowski vacuum and by one way or another, only vector supermultiplets associated with unbroken non-Abelian gauge symmetry remain massless in the hidden sector. Then these vector supermultiplets, that survive to low energies, can give rise to the breakdown of SUSY in the second phase. Indeed, at the scale Λ_X, where the gauge interactions that correspond to the unbroken gauge symmetry in the hidden sector become strong in the second phase, a gaugino condensate can be formed. This gaugino condensate does not break global SUSY. Nonetheless if the gauge kinetic function $f_X(z_m)$ has a nontrivial dependence on the hidden sector superfields z_m then the corresponding auxiliary fields F^m can acquire nonzero VEVs

$$F^{z_m} \propto \frac{\partial f_X(z_k)}{\partial z_m}\bar{\lambda}_a\lambda_a + \cdots \,, \tag{19}$$

which are set by $\langle\bar{\lambda}_a\lambda_a\rangle \simeq \Lambda_X^3$. Thus it is only via the effect of a nonrenormalisable term that this condensate causes the breakdown of supersymmetry. Therefore the SUSY breaking scale in the SUSY Minkowski vacuum is many orders of magnitude lower than Λ_X, while the scale Λ_X is expected to be much lower than $M_{\rm Pl}$. As a result a tiny vacuum energy density is induced

$$\rho_\Lambda \sim \frac{\Lambda_X^6}{M_{\rm Pl}^2}\,. \tag{20}$$

The postulated exact degeneracy of vacua implies then that the physical phase has the same energy density as the second phase where the breakdown of local SUSY takes place near Λ_X. From Eq. (20) it follows that in order to reproduce the measured cosmological constant the scale Λ_X has to be somewhat close to $\Lambda_{\rm QCD}$ in the physical vacuum, i.e.

$$\Lambda_X \sim \Lambda_{\rm QCD}/10\,. \tag{21}$$

Although there is no compelling reason to expect that Λ_X and $\Lambda_{\rm QCD}$ should be related, one may naively consider $\Lambda_{\rm QCD}$ and $M_{\rm Pl}$ as the two most natural choices for the scale of dimensional transmutation in the hidden sector.

In the one-loop approximation one can estimate the value of the energy scale Λ_X using the simple analytical formula

$$\Lambda_X = M_{\rm Pl}\exp\left[\frac{2\pi}{b_X\alpha_X(M_{\rm Pl})}\right], \tag{22}$$

where $\alpha_X(M_{\rm Pl}) = g_X^2(M_{\rm Pl})/(4\pi)$, g_X and b_X are the gauge coupling and one-loop beta function of the gauge interactions associated with the unbroken non-Abelian gauge symmetry that survive to low energies in the hidden sector. For the $SU(3)$ and $SU(2)$ gauge groups $b_X = -9$ and -6, respectively. In Fig. 3 we show the dependence of Λ_X on $\alpha_X(M_{\rm Pl})$. As one might expect, the value of the energy scale Λ_X diminishes with decreasing $\alpha_X(M_{\rm Pl})$. The observed value of the dark energy

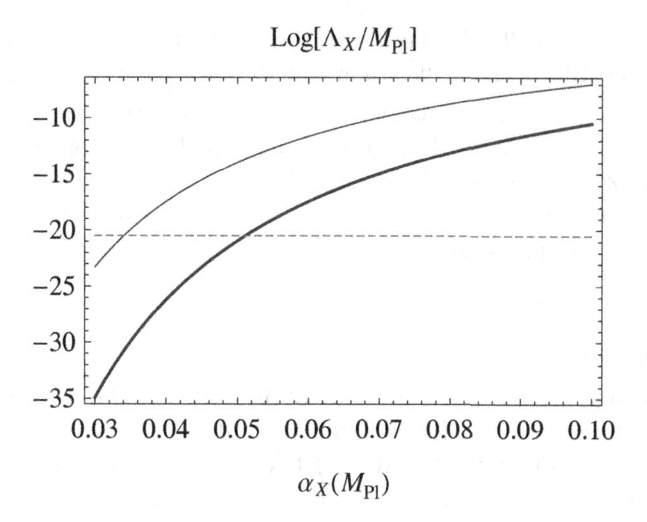

Fig. 3. The value of $\log[\Lambda_X/M_{\rm Pl}]$ as a function of $\alpha_X(M_{\rm Pl})$. The thin and thick solid lines are associated with the $SU(3)$ and $SU(2)$ gauge symmetries, respectively. The horizontal line corresponds to the value of Λ_X that results in the measured cosmological constant.

density is reproduced when $\alpha_X(M_{\rm Pl}) \simeq 0.051$ in the case of the SUSY model based on the $SU(2)$ gauge group and $\alpha_X(M_{\rm Pl}) \simeq 0.034$ in the case of the $SU(3)$ SUSY gluodynamics. It is worth noting that in the case of the $SU(3)$ SUSY model the value of the gauge coupling $g_X(M_{\rm Pl}) \simeq 0.654$, that leads to $\alpha_X(M_{\rm Pl}) \simeq 0.034$, is just slightly larger than the value of the QCD gauge coupling at the Planck scale in the SM, i.e. $g_3(M_{\rm Pl}) = 0.487$.[8]

In this scenario SUSY can be broken at any scale in the physical vacuum. In particular, the breakdown of local supersymmetry can take place near the Planck scale. If this is the case, one can explain the small values of $\lambda(M_{\rm Pl})$ and $\beta_\lambda(M_{\rm Pl})$ by postulating the existence of a third degenerate vacuum. In this third vacuum local SUSY can be broken near the Planck scale while the EW symmetry breaking scale can be just a few orders of magnitude lower than $M_{\rm Pl}$. Since now the Higgs VEV is somewhat close to $M_{\rm Pl}$ one must take into account the interaction of the Higgs and hidden sector fields. Thus the full scalar potential takes the form

$$V = V_{\rm hid}(z_m) + V_0(H) + V_{\rm int}(H, z_m) + \cdots\,, \tag{23}$$

where $V_{\rm hid}(z_m)$ is the part of the full scalar potential associated with the hidden sector, $V_0(H)$ is the part of the scalar potential that depends on the SM Higgs field only and $V_{\rm int}(H, z_m)$ describes the interactions of the SM Higgs doublet with the fields of the hidden sector. Although in general $V_{\rm int}(H, z_m)$ should not be ignored the interactions between H and hidden sector fields can be quite weak if the VEV of the Higgs field is substantially smaller than $M_{\rm Pl}$ (say $\langle H\rangle \lesssim M_{\rm Pl}/10$) and the couplings of the SM Higgs doublet to the hidden sector fields are suppressed. Then the VEVs of the hidden sector fields in the physical and third vacua can be almost identical. As a consequence, the gauge couplings and $\lambda(M_{\rm Pl})$ in the first and third phases should be basically the same and the value of $|m^2|$ in the Higgs effective potential can be still much smaller than $M_{\rm Pl}^2$ and $\langle H^\dagger H\rangle$ in the third vacuum. In this limit $V_{\rm hid}(z_m^{(3)}) \ll M_{\rm Pl}^4$ and the requirement of the existence of the third vacuum with vanishingly small energy density again implies that $\lambda(M_{\rm Pl})$ and $\beta_\lambda(M_{\rm Pl})$ are approximately zero in the third vacuum. Because in this case the couplings in the third and physical phases are basically identical, the presence of such a third vacuum should result in the predictions (2) for $\lambda(M_{\rm Pl})$ and $\beta_\lambda(M_{\rm Pl})$ in the physical vacuum.

Acknowledgments

This work was supported by the University of Adelaide and the Australian Research Council through the ARC Center of Excellence in Particle Physics at the Terascale and through grant LF0 99 2247 (AWT). HBN thanks the Niels Bohr Institute for his emeritus status. CDF thanks Glasgow University and the Niels Bohr Institute for hospitality and support.

References

1. WMAP Collab. (C. L. Bennett *et al.*), *Astrophys. J. Suppl.* **148**, 1 (2003).
2. WMAP Collab. (D. N. Spergel *et al.*), *Astrophys. J. Suppl.* **148**, 175 (2003).
3. D. L. Bennett and H. B. Nielsen, *Int. J. Mod. Phys. A* **9**, 5155 (1994).
4. C. D. Froggatt and H. B. Nielsen, *Phys. Lett. B* **368**, 96 (1996).
5. C. D. Froggatt, L. Laperashvili, R. Nevzorov, H. B. Nielsen and M. Sher, *Phys. Rev. D* **73**, 095005 (2006).
6. C. D. Froggatt, R. Nevzorov, H. B. Nielsen and D. Thompson, *Phys. Lett. B* **657**, 95 (2007).
7. C. D. Froggatt, R. Nevzorov, H. B. Nielsen and D. Thompson, *Int. J. Mod. Phys. A* **24**, 5587 (2009).
8. D. Buttazzo, G. Degrassi, P. P. Giardino, G. F. Giudice, F. Sala, A. Salvio and A. Strumia, *J. High Energy Phys.* **1312**, 089 (2013).
9. C. Froggatt, L. Laperashvili, R. Nevzorov and H. B. Nielsen, *Phys. Atom. Nucl.* **67**, 582 (2004).
10. C. Froggatt, R. Nevzorov and H. B. Nielsen, *Nucl. Phys. B* **743**, 133 (2006).
11. C. Froggatt, L. Laperashvili, R. Nevzorov and H. B. Nielsen, *Bled Workshops Phys.* **5**, 17 (2004).
12. C. D. Froggatt, R. Nevzorov and H. B. Nielsen, *J. Phys. Conf. Ser.* **110**, 072012 (2008).
13. C. D. Froggatt, R. Nevzorov and H. B. Nielsen, *AIP Conf. Proc.* **1200**, 1093 (2010).
14. N. Arkani-Hamed and S. Dimopoulos, *J. High Energy Phys.* **0506**, 073 (2005).
15. G. F. Giudice and A. Romanino, *Nucl. Phys. B* **699**, 65 (2004).
16. N. Arkani-Hamed, S. Dimopoulos, G. F. Giudice and A. Romanino, *Nucl. Phys. B* **709**, 3 (2005).
17. A. Pierce, *Phys. Rev. D* **70**, 075006 (2004).
18. A. Masiero, S. Profumo and P. Ullio, *Nucl. Phys. B* **712**, 86 (2005).
19. S. Weinberg, *Phys. Rev. Lett.* **59**, 2607 (1987).
20. V. Agrawal, S. M. Barr, J. F. Donoghue and D. Seckel, *Phys. Rev. D* **57**, 5480 (1998).
21. C. J. Hogan, *Rev. Mod. Phys.* **72**, 1149 (2000).
22. S. Kachru, R. Kallosh, A. D. Linde and S. P. Trivedi, *Phys. Rev. D* **68**, 046005 (2003).

23. T. Banks, M. Dine and E. Gorbatov, *J. High Energy Phys.* **0408**, 058 (2004).
24. M. Dine, E. Gorbatov and S. D. Thomas, *J. High Energy Phys.* **0808**, 098 (2008).
25. R. Bousso and J. Polchinski, *J. High Energy Phys.* **0006**, 006 (2000).
26. M. R. Douglas, *J. High Energy Phys.* **0305**, 046 (2003).
27. S. Ashok and M. R. Douglas, *J. High Energy Phys.* **0401**, 060 (2004).
28. A. Giryavets, S. Kachru and P. K. Tripathy, *J. High Energy Phys.* **0408**, 002 (2004).
29. J. P. Conlon and F. Quevedo, *J. High Energy Phys.* **0410**, 039 (2004).
30. F. Denef and M. R. Douglas, *J. High Energy Phys.* **0405**, 072 (2004).
31. M. R. Douglas, *Comptes Rendus Physique* **5**, 965 (2004).
32. C. Froggatt, R. Nevzorov and H. B. Nielsen, *PoS* **ICHEP 2010**, 442 (2010).
33. C. Froggatt, R. Nevzorov and H. B. Nielsen, *Int. J. Mod. Phys. A* **27**, 1250063 (2012).
34. C. D. Froggatt, R. Nevzorov and H. B. Nielsen, *AIP Conf. Proc.* **1560**, 300 (2013).
35. S. Dawson, E. Eichten and C. Quigg, *Phys. Rev. D* **31**, 1581 (1985).
36. J. L. Hewett, B. Lillie, M. Masip and T. G. Rizzo, *J. High Energy Phys.* **0409**, 070 (2004).
37. G. F. Giudice and A. Masiero, *Phys. Lett. B* **206**, 480 (1988).
38. J. A. Casas and C. Muñoz, *Phys. Lett. B* **306**, 288 (1993).
39. C. D. Froggatt, R. Nevzorov, H. B. Nielsen and A. W. Thomas, *Phys. Lett. B* **737**, 167 (2014).
40. C. D. Froggatt, R. Nevzorov, H. B. Nielsen and A. W. Thomas, *Nucl. Part. Phys. Proc.* **273-275**, 1465 (2016).
41. R. Nevzorov, H. B. Nielsen, A. W. Thomas and C. Froggatt, *PoS* **EPS-HEP2015**, 380 (2015).

Theory of Dark Matter

Paul H. Frampton

Department of Mathematics and Physics "Ennio de Giorgi",
University of Salento, Lecce, Italy
paul.h.frampton@gmail.com

We discuss the hypothesis that the constituents of dark matter in the galactic halo are primordial intermediate-mass black holes (PIMBHs). The status of axions and weakly interacting massive particles (WIMPs) is discussed, as are the methods for detecting PIMBHs with emphasis on microlensing. The role of the angular momentum $\mathcal{J}$ of the PIMBHs in their escaping previous detection is considered.

Keywords: Dark matter; primordial black holes; wide binaries; CMB distortion; microlensing; Kerr solution.

1. Introduction

Astronomical observations have led to a consensus that the energy make-up of the visible universe is approximately 70% dark energy, 25% dark matter and only 5% normal matter. General discussions of the history and experiments for dark matter are in books authored or edited by Sciama,[1] Sanders,[2] and Bertone.[3] A recent popular book, *The Cosmic Cocktail* by Freese[4] is strong on the panoply of unsuccessful weakly interacting massive particle (WIMP) searches. As we shall see, this lack of success may be due to the fact that WIMPs probably do not exist.

The present ignorance of the dark matter sector is put into perspective by looking at the uncertainty in the values of the constituent mass previously considered. The lightest such candidate is the invisible axion with $M = 1$ μeV. The heaviest such candidate is the intermediate mass black hole (IMBH) with $M = 100{,}000\ M_\odot$ which is a staggering 77 orders of magnitude larger. Our aim is to reduce this uncertainty.

The result of this analysis will be that the number of orders of magnitude uncertainty in the dark matter constituent mass can be reduced to four. We shall conclude, after extensive discussion, that the most viable candidate for the constituent which dominates dark matter is the primordial intermediate mass black hole (PIMBH) with mass in the range

$$20\ M_{\odot} < M_{\mathrm{PIMBH}} < 100{,}000\ M_{\odot}\,. \tag{1}$$

An explanation for the neglect of PIMBHs may be that the literature is confusing. At least one study claimed entirely to rule out Eq. (1). We shall attempt to clarify the situation which actually still permits the whole range in Eq. (1). This talk is, in part, an attempt to redress the imbalance between the few experimental efforts to search for PIMBHs compared to the extensive WIMP searches.

2. Axions

It is worth reviewing briefly the history of the axion particle now believed, if it exists, to lie in the mass range

$$10^{-6}\ \mathrm{eV} < M < 10^{-3}\ \mathrm{eV}\,. \tag{2}$$

The Lagrangian originally proposed for quantum chromodynamics (QCD) was of the simple form, analogous to quantum electrodynamics,

$$\mathcal{L}_{\mathrm{QCD}} = -\frac{1}{4}G^{\alpha}_{\mu\nu}G^{\mu\nu}_{\alpha} - \frac{1}{2}\sum_{i}\bar{q}_{i,a}\gamma^{\mu}D^{ab}_{\mu}q_{i,b} \tag{3}$$

summed over the six quark flavors.

The simplicity of Eq. (3) was only temporary and became more complicated in 1975 by the discovery of instantons which dictated an additional term in the QCD Lagrangian must be added

$$\Delta\mathcal{L}_{\mathrm{QCD}} = \frac{\Theta}{64\pi^2}G^{\alpha}_{\mu\nu}\tilde{G}^{\mu\nu}_{\alpha}\,, \tag{4}$$

where $\tilde{G}_{\mu\nu}$ is the dual of $G_{\mu\nu}$.

When the quark masses are complex, an instanton changes not only Θ but also the phase of the quark mass matrix $\mathcal{M}_{\mathrm{quark}}$ and the full phase to be considered is

$$\bar{\Theta} = \Theta + \arg\det\|\mathcal{M}_{\mathrm{quark}}\|\,. \tag{5}$$

The additional term, Eq. (4), violates P and CP, and contributes to the neutron electric dipole moment whose upper limit provides a constraint

$$\bar{\Theta} < 10^{-9} \tag{6}$$

which fine-tuning is the strong CP problem. The hypothetical axion particle then arises from an ingenious technique to resolve Eq. (6), although as it turns out it may have been too ingenious.

Over 20 years ago, in 1992, there was pointed out a serious objection to the invisible axion. The point is that the invisible potential is so fine-tuned that adding gravitational couplings for weak gravitational fields at the dimension-five level requires tuning of a dimensionless coupling g to be at least as small as $g < 10^{-40}$, more extreme than the tuning of $\bar{\Theta}$ in Eq. (6).

Although a true statement, it is not a way out of this objection to say that we do not know the correct theory of quantum gravity because for weak gravitational fields, as is the case almost everywhere in the visible universe, one can use an effective field theory. To our knowledge, this serious objection to the invisible axion which has been generally ignored since 1992 has not gone away and therefore the invisible axion probably does not exist.

There remains the strong CP problem of Eq. (6). One other solution would be a massless up quark but this is disfavored by lattice calculations. For the moment, Eq. (6) must be regarded as fine-tuning. We recall that the ratio of any neutrino mass to the top quark mass in the Standard Model satisfies

$$\left(\frac{M_\nu}{M_t}\right) < 10^{-12}. \tag{7}$$

3. WIMPs

By WIMP, it is generally meant an unidentified elementary particle with mass in the range, say, between 10 GeV and 1000 GeV and with scattering cross-section with nucleons (N) satisfying, according to the latest unsuccessful WIMP direct searches,

$$\sigma_{\text{WIMP-}N} < 10^{-44}\ \text{cm}^2 \tag{8}$$

which is roughly comparable to the characteristic strength of the known weak interaction.

The WIMP particle must be electrically neutral and be stable or have an extremely long lifetime. In model-building, the stability may be achieved by an *ad hoc* discrete symmetry, for example a Z_2 symmetry under which all the Standard Model particles are even and others are odd. If the discrete symmetry is unbroken, the lightest odd state must be stable and therefore a candidate for a dark matter. In general, this appears contrived because the discrete symmetry is not otherwise motivated.

By far the most popular WIMP example came from electroweak supersymmetry where a discrete R symmetry has the value $R = +1$ for the Standard Model particles and $R = -1$ for all the sparticles. Such an R parity is less *ad hoc* being essential to prevent too-fast proton decay. The lightest $R = -1$ particle is stable and, if not a gravitino which has the problem of too-slow decay in the early universe, it was the neutralino, a linear combination of zino, bino and higgsino. The neutralino provided an attractive candidate.

The big problem with the neutralino is that at the LHC where electroweak supersymmetry not many years ago confidently predicted sparticles (gluinos, etc.) at the weak scale $\sim$250 GeV there is no sign of any additional particle with mass up to at least 2500 GeV and above, so electroweak supersymmetry probably does not exist.

The present run of the LHC is not necessarily doomed if WIMPs and sparticles do not exist. An important question, independent of naturalness but surely related to anomalies, is the understanding of why there are three families of quarks and leptons. For that reason the LHC could discover additional gauge bosons, siblings of the $W^{\pm}$ and Z^{0}, as occur in e.g. the so-called 331-Model.

4. MACHOs

Massive compact halo objects (MACHOs) are commonly defined by the notion of compact objects used in astrophysics as the end products of stellar evolution when most of the nuclear fuel has been expended. They are usually defined to include white dwarfs, neutron stars, black holes, brown dwarfs and unassociated planets, all equally hard to detect because they do not emit any radiation.

This narrow definition implies, however, that MACHOs are composed of normal matter which is too restrictive in the case of black holes. It is here posited[5,a] that black holes of mass up to 100,000 $M_\odot$ (even up to 10^{17} $M_\odot$) can be produced primordially. Nevertheless for the halo the acronym MACHO still nicely applies to dark matter PIMBHs which are massive, compact, and in the halo.

Unlike the axion and WIMP elementary particles which would have a definite mass, the black holes will have a range of masses. The lightest PBH which has survived for the age of the universe has a lower mass limit

$$M_{\rm PBH} > 10^{-18}\ M_\odot \sim 10^{36}\ {\rm TeV} \tag{9}$$

already 36 orders of magnitude heavier than the heaviest would-be WIMP. This lower limit comes from the lifetime formula derivable from Hawking radiation

$$\tau_{\rm BH}(M_{\rm BH}) \sim \frac{G^2 M_{\rm BH}^3}{\hbar c^4} \sim 10^{64} \left(\frac{M_{\rm BH}}{M_\odot} \right)^3 \text{ years}\,. \tag{10}$$

[a]A little history is in order for the identification DM $\equiv$ PIMBH.[5] Our epiphany arrived on July 21, 2015 in the Florentine Duomo during solemn Mass commemorating the quincentennial of Saint Philip Neri. PBHs were first invented in Russia[6] in 1966 then independently in the West[7] in 1974. The idea that PBHs could form the dark matter was first suggested by Chapline[8] in 1975 who, like everybody in the 20th century, assumed the PBHs were much lighter than the Sun. Three decades later, far more massive PBHs, including PIMBHs, were shown to be possible e.g. Refs. 9 and 10 and studies of the entropy of the Universe strongly suggested that far more black holes exist. One of the most convincing arguments in Ref. 5 was the marginalization of the WIMP predicted by the failed theory of weak-scale supersymmetry. We note that Ref. 5 appeared on September 30, 2015, four months before the LIGO announcement[11] on February 11, 2016 of the discovery of gravitational waves from a heavy black hole binary, and five months before a copycat paper involving Adam Riess[12] of March 1, 2016 which now has 92 citations while Ref. 5 has only eight. This is undoubtedly because Riess has a Nobel prize and chose not to cite.[5] It recalls a situation 40 years ago when Sidney Coleman's copycat paper[13] about vacuum decay in quantum field theory appeared on January 24, 1977 totally identical to, but choosing not to cite, our Ref. 14 of September 24, 1976. By this time, Ref. 13 has 1684 citations while Ref. 14 has only 104 which has been called a *kilocite heist.* This case is more important because the theory of dark matter invented in Ref. 5 may fairly soon be tested by, and possibly shown to agree with, experiment. In that case, the *original* seminal papers for the equation DM $\equiv$ PIMBH are therefore, in chronological order.[8,5]

Because of observational constraints the dark matter constituents must generally be another 20 orders of magnitude more massive than the lower limit in Eq. (9).

We assert that most dark matter black holes are in the mass range between ten and 100,000 times the solar mass. The name PIMBHs is appropriate because they lie in mass above stellar-mass black holes and below the supermassive black holes which reside in galactic cores.

Let us discuss three methods (there may be more) which could be used to search for dark matter PIMBHs. While so doing we shall clarify what limits, if any, can be deduced from present observational knowledge.

Before proceeding, it is appropriate first to mention the important Xu–Ostriker upper bound of about a million solar masses from galactic disk stability for any MACHO residing inside the galaxy.

5. Wide Binaries

There exist in the Milky Way pairs of stars which are gravitationally bound binaries with a separation more than 0.1 pc. These wide binaries retain their original orbital parameters unless compelled to change them by gravitational influences, for example, due to nearby IMBHs.

Because of their very low binding energy, wide binaries are particularly sensitive to gravitational perturbations and can be used to place an upper limit on, or to detect, IMBHs. The history of employing this ingenious technique is regretfully checkered. In 2004 a fatally strong constraint was claimed by an Ohio State University group in a paper[15] entitled "End of the MACHO Era" so that, for researchers who have time to read only titles and abstracts, stellar and higher mass constituents of dark matter appeared to be totally excluded.

Five years later in 2009, however, another group[16] this time from Cambridge University reanalyzed the available data on wide binaries and reached a quite different conclusion. They questioned whether *any* rigorous constraint on MACHOs could yet be claimed, especially as one of the important binaries in the earlier sample had been misidentified.

Because of this checkered history, it seems wisest to proceed with caution but to recognize that wide binaries represent a potentially useful source both of constraints on, and the possible discovery of, dark matter IMBHs.

6. Distortion of the CMB

This approach hinges on the phenomenon of accretion of gas onto the PIMBHs. The X-rays emitted by such accretion of gas are downgraded in frequency by cosmic expansion and by Thomson scattering becoming microwaves which distort the CMB, both with regard to its spectrum and to its anisotropy.

One impressive calculation[17] by Ricotti, Ostriker and Mack (ROM) in 2008 of this effect employs a specific model for the accretion, the Bondi–Hoyle model, and carries through the computation all the way up to a point of comparison with data from FIRAS on CMB spectral distortions, where FIRAS was a sensitive device attached to the COBE satellite.

Unfortunately the Bondi–Hoyle model was invented for a static object and assumes spherically symmetric purely s-wave accretion. Studies of the SMBH in the giant galaxy M87 have shown since 2014 that the higher angular momenta strongly dominate, not surprising as the SMBH possesses a gigantic spin angular momentum in natural units.

The results from M87 suggest the upper limits on MACHOs imposed by ROM were too severe by some four or five orders of magnitude and that up to 100% of the dark matter is permitted by arguments about CMB distortion to be in the form of PIMBHs.

7. Microlensing

Microlensing is the most direct experimental method and has the big advantage that it has successfully found examples of MACHOs. The MACHO Collaboration used a method which had been proposed[b] by Paczynski where the amplification of a distant source by an intermediate gravitational lens is observed. The MACHO Collaboration discovered several striking microlensing events whose light curves are exhibited in its 2000 paper. The method certainly worked well for $M < 25\ M_\odot$ and so should work equally well for $M > 25\ M_\odot$ provided one can devise a suitable algorithm and computer program to scan enough sources.

The longevity of a given lensing event is proportional to the square root of the lensing mass and numerically is given by ($\hat{t}$ is longevity)

$$\hat{t} \simeq 0.2 \text{ yr} \left(\frac{M_{\text{lens}}}{M_\odot} \right)^{1/2}, \tag{11}$$

where a transit velocity 200 km/s is assumed for the lensing object.

The MACHO Collaboration[18] investigated lensing events with longevities ranging between about two hours and one year. From Eq. (11) this corresponds to MACHO masses approximately between $10^{-6}\ M_\odot$ and $25\ M_\odot$.

The total number and masses of objects discovered by the MACHO Collaboration could not account for all the dark matter known to exist in the Milky Way. At most 10% could be explained. To our knowledge, the experiment ran out of money and was essentially abandoned in about the year 2000. But perhaps the MACHO Collaboration and its funding agency were too easily discouraged.

What is being suggested is that the other 90% of the dark matter in the Milky Way is in the form of MACHOs which are more massive than those detected by the MACHO Collaboration, and which almost certainly could be detected by a

[b] We have read that such gravitational lensing was later found to have been calculated in unpublished 1912 notes by Einstein who did not publish perhaps because at that time he considered its experimental measurement impracticable.

straightforward extension of their techniques. In particular, the expected microlensing events have a duration ranging up to two centuries.

Microlensing experiments involve systematic scans of millions of distant star sources because it requires accurate alignment of the star and the intermediate lensing MACHO. Because the experiments are already highly computer intensive, it makes us more optimistic that the higher longevity events can be successfully analyzed. Study of an event lasting two centuries should not necessitate that long an amount of observation time. It does require suitably ingenious computer programming to track light curves and distinguish them from other variable sources. This experiment is undoubtedly extremely challenging, but there seems no obvious reason it is impracticable.

8. Intermediate Discussion

Axions probably do not exist for theoretical reasons discovered in 1992. Electroweak supersymmetry probably does not exist for the experimental reason of its non-discovery in Run 1 of the LHC. The idea that dark matter experiences weak interactions (WIMPs) came historically from the appearance of an appealing DM constituent, the neutralino, in the theory of electroweak supersymmetry for which there is no experimental evidence.

The only interaction which we know for certain to be experienced by dark matter is gravity and the simplest assumption is that gravity is the only force coupled to dark matter. Why should the dark matter experience the weak interaction when it does not experience the strong and electromagnetic interactions?

All terrestrial experiments searching for dark matter by either direct detection or production may be doomed to failure.

We began with four candidates for dark matter constituent: (1) axions; (2) WIMPs; (3) baryonic MACHOs; (4) PIMBHs. We eliminated the first two by hopefully persuasive arguments, made within the context of an overview of particle phenomenology including a combination of old and new results. We eliminated the third by the upper limit on baryons imposed by robust Big Bang Nucleosynthesis (BBN) calculations.

We assert that PIMBHs can constitute almost all dark matter while maintaining consistency with the BBN calculations. This is an important point because distinguished astronomers have written an opposite assertion e.g. Begelman and Rees state that black holes cannot form more than 20% of dark matter because the remainder is non-baryonic.

These authors are making an implicit assumption which does not apply to the PIMBHs which we assert comprises almost all dark matters. This assumption is that black holes can be formed only as the result of the gravitational collapse of baryonic stars. We are claiming, on the contrary, that dark matter black holes can be, and the majority must be, formed primordially in the early universe as calculated and demonstrated in FKTY (2010) and independently by CKSY (2010).

Our proposal is that the Milky Way contains between ten million and ten billion massive black holes each with between 20 and 100,000 times the solar mass. Assuming the halo is a sphere of radius a hundred thousand light years the typical separation is between one hundred and one thousand light years which is also the most probable distance of the nearest PIMBH to the Earth. At first sight, it may be surprising that such a huge number of PIMBHs

— the plums in a "*PIMBH plum pudding*" —

(c.f. Thomson, 1904) could remain undetected.

[111 years after Thomson; 31 powers of ten bigger; not replaceable by a nuclear halo.]

However, the mean separation of the plums is at least a hundred light years and the plum size is smaller than the Sun.

9. PIMBHs and PSMBHs

Focusing on the Milky Way halo where we can most easily detect the PBHs, we already know from earlier searches, especially the MACHO Collaboration that masses $M \leq 20\ M_\odot$ can make up not more than 10% of the halo dark matter. At the high mass end, we know from Xu–Ostriker that MACHOs with $M \geq 10^5\ M_\odot$ endanger disk stability. For the Milky Way halo one is led to consider intermediate mass PIMBHs in the mass range

$$20\ M_\odot \leq M_{\mathrm{PIMBH}} \leq 10^5\ M_\odot \tag{12}$$

for the DM constituents. This leads to a *plum pudding* model for the Milky Way halo, named after Thomson's atomic model, where for the DM halo the plums are PIMBHs with masses satisfying Eq. (12) and the pudding is rarefied gas, dust and a few luminous stars.

The formation of PBHs with masses as large as Eq. (12) and much larger is known to be mathematically possible during the radiation era. An existence theorem is provided by hybrid inflationary models. One specific prediction of hybrid inflation is a sharply-peaked PBH mass function. If we need a specific PIMBH mass, we shall use a calligraphic $\mathcal{PIMBH}$ defined by $M_{\mathcal{PIMBH}} \equiv 100\ M_\odot$ exactly. This is merely an example and extension to the whole range of Eq. (12) can also be discussed.

The cosmic time t_{PBH} at which a PBH is formed has been estimated to be

$$t_{\mathrm{PBH}} \simeq \left(\frac{M_{\mathrm{PBH}}}{10^5\ M_\odot}\right)\ \mathrm{s} \tag{13}$$

so that the PIMBHs in Eq. (12) are formed in the time window $0.0002\ \mathrm{s} \leq t_{\mathrm{PIMBH}} \leq 1.0\ \mathrm{s}$ with the special case $t_{\mathcal{PIMBH}} \simeq 0.001$ s. In terms of redshift (Z), this corresponds to

$$5 \times 10^{11} \geq Z_{\mathrm{PIMBH}} \geq 5 \times 10^9 \tag{14}$$

with the special case $Z_{\mathcal{PIMBH}} \simeq 2 \times 10^{11}$.

The formation of BHs which are not primordial, which we shall denote without an initial P or $\mathcal{P}$, necessarily occurs *after* star formation which conservatively occurs certainly only for very different redshifts satisfying

$$Z_{\mathrm{BH}} \leq 100\,. \tag{15}$$

The sharp difference in the redshifts of Eqs. (14) and (15) will become important when we discuss the reasons for previous non-detection, the angular momentum of PIMBHs and BHs, and the central issue of possible CMB distortion by X-rays.

As already mentioned, by using the mathematical models, it is possible to form PBHs not only in the PIMBH mass range of Eq. (12) but also primordial super massive black hole (PSMBHs) in the mass range

$$10^5\, M_\odot \leq M_{\mathrm{PSMBH}} \leq 10^{17}\, M_\odot\,, \tag{16}$$

where the upper limit derives from the formation time t_{PSMBH} given by Eq. (13) staying within the radiation-dominated era. We shall discuss the higher mass range Eq. (16) later in the paper.

Finally for this Introduction, we recall that in a microlensing experiment, e.g. using the LMC or SMC for convenient sources, microlensing by halo PIMBHs, and assuming a typical transit velocity 200 km$\cdot$s^{-1}, the time duration of the microlensing light curve can be estimated to be approximately

$$\tau \simeq \left(\frac{M_{\mathrm{PIMBH}}}{25\, M_\odot}\right)^{\frac{1}{2}} \text{yrs} \tag{17}$$

which we note is close to one year and two years, respectively, for lens masses 25 $M_\odot$ and 100 $M_\odot$. For reference, the highest duration such light curve detected by the MACHO Collaboration which published in the year 2000 corresponded to $M_{\mathrm{PIMBH}} \simeq 20\, M_\odot$.

Nevertheless, if longer duration microlensing light curves can be detected of two years or more, the only known explanation will be the existence of Kerr black holes in the halo with many solar masses.

10. Kerr Metric and Period τ

The PIMBHs are described by a Kerr metric which has the form in Boyer–Lindquist (t, r, θ, ϕ) coordinates, after defining $\alpha = \frac{J}{M}$, $\rho^2 = r^2 + \alpha^2 \cos^2\theta$ and $\Delta = r^2 - 2Mr + \alpha^2$

$$\begin{aligned} ds^2 &= -\left(1 - \frac{2Mr}{\rho^2}\right) dt^2 - \left(\frac{4Mr\alpha \sin^2\theta}{\rho^2}\right) d\phi\, dt + \left(\frac{\rho^2}{\Delta}\right) dr^2 + \rho^2\, d\theta^2 \\ &\quad + \left(r^2 + \alpha^2 + \frac{2Mr\alpha^2 \sin^2\theta}{\rho^2}\right) \sin^2\theta\, d\phi^2\,. \end{aligned} \tag{18}$$

In Eq. (18), there are two free parameters, M and J. Analytic calculations building on Eq. (18) can be difficult, usually leading to numerical techniques.

Table 1.

Astrophysical object	Mass solar masses	Period τ (s)	Angular momentum $\mathcal{J}$ $(\text{kg} \cdot \text{km}^2 \cdot \text{s}^{-1})$
Earth	$M_\oplus = 6 \times 10^{24}$ kg	24 h	1.1×10^{27}
Sun	$M_\odot = 2 \times 10^{30}$ kg	25 days	1.1×10^{36}
PIMBH	20 $M_\odot$	0.013 s	3.0×10^{37}
$\mathcal{PIMBH}$	100 $M_\odot$	0.063 s	7.2×10^{38}
PIMBH	1000 $M_\odot$	0.63 s	7.2×10^{40}
PIMBH	10^4 $M_\odot$	6.3 s	7.2×10^{42}
PIMBH	10^5 $M_\odot$	63 s	7.2×10^{44}
PSMBH (M87)	6×10^9 $M_\odot$	3.8×10^6 s	2.6×10^{54}

In this talk, we shall need only order-of-magnitude estimates for the rotational period τ and, in the next section, for the angular momentum J. These will suffice to make our point about concomitant X-ray emission. The solution is axially symmetric and the radius at the pole $\theta = \frac{\pi}{2}$ is the same as the Schwarzschild radius $R = 2M$. For other values of θ the black hole radius is smaller than the static one and the rest of the static would-be sphere is filled out by an ergosphere whose equatorial radius is also $R = 2M$.

For the primordial black holes of interest, there is no reason to expect that the radiation will collapse in a spherically symmetric fashion to a static Schwarzschild black hole when the PBH formation necessarily occurs in an environment of extreme fluctuations and inhomogeneities. The black holes must be described by the Kerr metric in Eq. (18) with α having a value anything up to the maximal Kerr solution which corresponds to an equatorial speed V equal to the speed of light. The range of V is thus $0 \leq V \leq c$.

We do not know observationally any black hole which is primordial with certainty although many of the observed black holes, including those in the binary coalescences observed by LIGO, could be primordial. For illustration of black hole observations, let us consider the well-studied binary GRS1915+105 of a star and a black hole.

The black hole mass in GRS1915+105 has been established as $M \simeq 13\ M_\odot$ and hence its Schwarzschild radius is $r_s \simeq 39$ km. Its rotation occurs 1150 times per second which translates to an equatorial speed $V \simeq 0.94c$, remarkably close to maximal. We mention this example to show that such high V Kerr black holes are known to exist and although we cannot derive the value of V arising from PBH formation it is to be expected that all values V up to the maximum can occur. For this qualitative purposes, to be conservative, we employ $V = 0.1c$.

To proceed with our estimate we shall therefore take the equatorial velocity of the ergosphere to have magnitude $V = 0.1c$ and use Newtonian mechanics to estimate the rotation period τ as simply

$$\tau = \left(\frac{2\pi R}{V}\right). \tag{19}$$

For the Sun, we have $2\,M_\odot \simeq 3$ km so that for a black hole of mass $M = \eta M_\odot$ and therefore radius $R \simeq 3\eta$ km (Eq. (19)) is, for $V = 0.1c = 3 \times 10^4$ km $\cdot$ s^{-1},

$$\tau = (2 \times 10^{-4}\pi\eta)\ \text{s}\,. \tag{20}$$

Some values of τ, estimated by this method, are shown in the third column of Table 1 and angular momentum $\mathcal{J}$ (discussed later) is in the last column.

11. Angular Momentum $\mathcal{J}$

Let us define the dimensionless angular momentum $\mathcal{J} \equiv J/\text{kg} \cdot \text{km}^2 \cdot \text{s}^{-1}$. We are interested in order of magnitude estimates of $\mathcal{J}$ for the PIMBHs and PSMBHs. The value of $\mathcal{J}$ for astrophysical objects is necessarily a large number, so to set the scene we shall estimate $\mathcal{J}$ for the Earth $\mathcal{J}_\oplus$ and for the Sun $\mathcal{J}_\odot$.

The parameters for the Earth are radius $R_\oplus \simeq 6300$ km, period $\tau_\oplus \simeq 86{,}400$ s, mass $M_\oplus \simeq 6 \times 10^{24}$ kg, hence angular velocity $\omega_\oplus = 2\pi/\tau_\oplus$ and moment of inertia $I_\oplus = \frac{2}{5}M_\oplus R_\oplus^2$ so an estimate is $\mathcal{J}_\oplus \sim I_\oplus\omega_\oplus \simeq 1.1 \times 10^{27}$. For the Sun the similar calculation using $R_\odot \simeq 700{,}000$ km, $\tau_\odot \simeq 25$ days, $M_\odot \simeq 2 \times 10^{30}$ kg gives $\mathcal{J}_\odot \simeq 1.1 \times 10^{36}$.

For the black holes, the value of $\mathcal{J}$ is proportional to η^2 where $\eta = (M/M_\odot)$. A similar estimate to that for the Earth and Sun gives $\mathcal{J} \simeq 7.2 \times 10^{34}\eta^2$, which provides the remaining entries in Table 1.

12. CMB Distortion Revisited

Because of rotational invariance, angular momentum is conserved. The $\mathcal{J}$ of a compact astrophysical object will not change dramatically unless there is an extremely unlikely event like a major collision. For example, the Earth and the Sun in the first two rows of Table 1 were formed 4.6 billion years ago. Their respective angular momenta $\mathcal{J}_\oplus$ and $\mathcal{J}_\odot$ have remained essentially constant all of that time. According to Eq. (13), the PIMBHs listed in the next five rows of Table 1 were all formed at time $t \leq 1$ s and their angular momenta have therefore remained roughly constant for the last 13.8 billion years since then.

In detecting the dark matter, let us focus on the special case $\mathcal{PIMBH}$ with $M = 100\ M_\odot$. The $\mathcal{PIMBH}$ was formed, according to Eq. (13), at time $t = 10^{-3}$ s and rotates with period $t \simeq 63$ ms, thus rotating $\sim$16 times per second and with an absolute angular momentum $\sim 6 \times 10^{11}$ times that of the Earth and $\sim$600 times that of the Sun. There is no known reason that $\mathcal{J}_{\mathcal{PIMBH}}$ would change significantly after its formation.

These remarks about angular momentum are salient to resolving the contradiction between the PIMBH dark matter proposal and the limits on halo MACHOs derived earlier by Ricotti, Ostriker and Mack (ROM)[17] on the basis of X-ray emission and CMB distortion.

The PIMBH proposal was made that the Milky Way dark halo is a plum pudding with, as "plums", PIMBHs in the mass range of Eq. (12) making up 100% of the dark matter. On the other hand, in Fig. 9 of Ref. 17, there is displayed an upper limit of less than 0.01% of the dark matter for this mass range of MACHO. Thus, it would seem that at least one must be incorrect? The conclusion of this talk is that ROM is correct for stellar-collapse black holes but is not applicable to a model which employs primordial black holes.

This ROM upper limit arises from the lack of any observed departure of the CMB spectrum from the predicted black-body curve or of any CMB anisotropy. ROM calculated the accretion of matter on to the MACHOs, the emission of X-rays by the accreted matter and then the downgrading of these X-rays to microwaves by cosmic expansion and more importantly by Compton scattering from electrons.

A crucial assumption made by ROM is that the accretion on to the MACHO can be modeled as if the MACHO has zero angular momentum $J = 0$. The justification for this assumption is based on earlier work by Loeb[19] who studied the collapse of gas clouds at redshifts $200 \leq Z \leq 1400$. Such collapse can form compact objects, eventually black holes, but during the collapse angular momentum is damped out from the electrons by Compton scattering with the CMB.

From Loeb's discussion, the resultant black holes will have $J = 0$ and this appears to underlie why ROM used the Bondi–Hoyle model which presumes spherical symmetry for accretion. This is justified for stellar-collapse black holes by the arguments of Loeb and therefore the upper bounds derived by ROM are applicable.

There is evidence that the Bondi–Hoyle model of accretion is not, by contrast, applicable to spinning PSMBHs, in particular the one at the center of the large galaxy M87. In recent analyses Bondi–Hoyle was used to calculate the number of X-rays expected from the accreted material near M87. In the case of M87 the X-rays are experimentally measured. The conclusion is striking: that the measured X-rays are less by several orders of magnitude than predicted by Bondi–Hoyle theory.

This supports the idea that the SMBHs such as that in M87 are primordial, so we list PSMBH(M87) in the final row of Table 1. The ROM constraints apply to black holes which originate from gravity collapse of baryonic stars. Collecting this fact, together with the ROM limit of $\leq 10^{-4}$ of the dark matter for MACHOs, implies that 99.99% of the dark matter black holes are primordial, formed during the radiation era.

13. Final Discussion

The dark matter and its explanation is a pressing problem which impacts on both high-energy physics and on cosmology. It is indisputable that over 80% of the Milky Way's mass lies in a dark approximately spherical halo surrounding the luminous more planar spiral. The results in this paper strongly support the model involving billions of PIMBHs.

The plum pudding model for the dark halo proposed here arose from a confluence of theoretical threads including study of the entropy of the universe and the knowledge of how to form PBHs with many solar masses as in Eqs. (12) and (16). Nevertheless it was the weakening of the argument for WIMPs which was most decisive.

The strongest objection to the PIMBHs has been based on the X-rays and the CMB distortion as calculated by ROM. In the present talk we have attempted to lay this criticism to rest by noting that ROM assumed $J = 0$ and that the putative PIMBHs have not only many times the Solar mass but also many times the Solar angular momentum. This appears to us to render the ROM constraints inapplicable to the PIMBHs. On the other hand, they do apply to stellar-collapse black holes which implies that almost none ($\leq 0.01\%$) of the dark matter black holes are of that type. To decide whether dark matter really is PIMBHs will require their detection by a dedicated microlensing experiment.

Examples of PSMBHs may already have been observed in galactic cores and quasars. Other PSMBHs can play the role of dark matter in clusters and may well be detectable by other future lensing experiments. There is also the upper mass range contained in Eq. (16). Although masses of PSMBHs up to a few times $10^{10}\ M_\odot$ may have already been observed in quasars, there are what could be called primordial ultra massive black holes (PUMBHs) with masses between 10^{11} and 10^{17} solar masses which might exist within the visible universe.

PUMBHs remain speculative but what can in the near future be examined experimentally is the existence of PIMBHs in the halo. A positive result would solve the 83-year-old problem of the dark matter and explain $\sim$26.7% of the total stress-energy tensor of the visible universe. It would presumably put a stop to searches for WIMPs because the scientific community would accept that WIMPs, like low-energy supersymmetry, do not exist. Searches for axions would perhaps continue but purely within the particle physics domain with no notion that axions, if they exist, can form more than a very tiny fraction of dark matter.

The identification of the dark matter constituents as PIMBHs can revolutionize astronomy and cosmology. To give just one example, the formation of stars which takes place at redshifts $Z \leq 8$ becomes as if only a minor "afterthought" with regard to all the earlier large scale structure formation which would take place in a Universe containing *only* dark matter in the form of PIMBHs. In this sense, the result of this experiment can diminish the cosmological significance of normal matter.

Of the detection methods discussed, extended microlensing observations seem the most promising and an experiment to detect higher longevity microlensing events is being actively pursued. The wide-field telescope must be in the Southern Hemisphere to use the Magellanic Clouds (LMC and SMC) for sources.

The most appropriate active telescope has been identified as the Blanco 4m at Cerro Tololo in Chile. This telescope was named after the late Victor Blanco the Puerto Rican astronomer who was the CTIO Director. A bigger telescope which can confirm the high-duration light curves is the Large Synoptic Survey Telescope (LSST) under construction, also in Chile, expected to take first light in 2022.

References

1. D. W. Sciama, *Modern Cosmology and the Dark Matter* (Cambridge Univ. Press, 2008).
2. R. H. Sanders, *The Dark Matter Problem, A Historical Perspective* (Cambridge Univ. Press, 2014).
3. (Ed.) G. Bertone, *Particle Dark Matter, Observations, Models and Searches* (Cambridge Univ. Press, 2013).
4. K. Freese, *The Cosmic Cocktail, Three Parts Dark Matter* (Princeton Univ. Press, 2014).
5. P. H. Frampton, *Mod. Phys. Lett. A* **31,** 1650093 (2016); G. Chapline and P. H. Frampton,
 J. Cosmol. Astropart. Phys. **11,** 042 (2016).
6. Y. B. Zel'dovich and I. D. Novikov, *Astron. Zh.* **43**, 758 (1966).
7. B. J. Carr and S. W. Hawking, *Mon. Not. R. Astron. Soc.* **168**, 399 (1974).
8. G. F. Chapline, *Nature* **253**, 251 (1975).
9. P. H. Frampton, M. Kawasaki, F. Takahashi and T. T. Yanagida, *J. Cosmol. Astropart. Phys.* **1004**, 023 (2010).
10. S. Clesse and J. Garcia-Bellido, *Phys. Rev. D* **92**, 023524 (2015).
11. LIGO Scientific and Virgo Collabs. (B. P. Abbott *et al.*), *Phys. Rev. Lett.* **116**, 061102 (2016).
12. S. Bird, I. Cholis, J. B. Munoz, Y. Ali-Haimoud, M. Kamionkowski, E. D. Kovetz, A. Raccanelli and A. G. Riess, *Phys. Rev. Lett.* **116**, 201301 (2016).
13. S. Coleman, *Phys. Rev. D* **15**, 2929 (1977).
14. P. H. Frampton, *Phys. Rev. Lett.* **37**, 1378 (1976).
15. J. Yoo, J. Chaname and A. Gould, *Astrophys. J.* **601**, 311 (2004).
16. D. P. Quinn, M. I. Wilkinson, M. J. Irwin, J. Marshall, A. Koch and V. Belokurov, *Mon. Not. R. Astron. Soc.* **396**, 11 (2009).
17. M. Ricotti, J. P. Ostriker and K. J. Mack, *Astrophys. J.* **680**, 829 (2008).
18. C. Alcock *et al.*, *Astrophys. J.* **542**, 281 (2000).
19. A. Loeb, *Astrophys. J.* **403**, 542 (1993).

Supersymmetric versus SO(10) Models of Dark Matter

Keith A. Olive
William I. Fine Theoretical Physics Institute,
School of Physics and Astronomy,
University of Minnesota, Minneapolis, MN 55455, USA
olive@umn.edu

After the results of Run I, can we still "guarantee" the discovery of supersymmetry at the LHC? It is shown that viable dark matter candidates in CMSSM-like models tend to lie in strips (coannihilation, funnel, focus point) in parameter space. The role of grand unification in constructing supersymmetric models is discussed and it is argued that nonsupersymmetric GUTs such as SO(10) may provide alternative solutions to many of the standard problems addressed by supersymmetry.

Keywords: Dark matter; grand unification; supersymmetry.

1. Introduction

For well over 30 years, supersymmetry and the minimal supersymmetric standard model (MSSM) have been prime examples of extensions beyond the standard model (SM) of strong and electroweak interactions. Apart from its algebraic beauty, there are many motivations for supersymmetry. These include the theory's ability to provide gauge coupling unification [1] and address the gauge hierarchy problem. [2] It is well known that the additional fields predicted in the MSSM, if present at low energy, alter the running of the gauge couplings as shown on Fig. 1. At one-loop, the running of the gauge couplings is given by

$$Q\frac{d\alpha_i}{dQ} = \frac{b_i}{2\pi}\alpha_i^2 \,, \tag{1}$$

where the coefficients of the SM beta functions are $b_i = 41/10,\ -19/6,\ -7$ for the U(1), SU(2), and SU(3) gauge couplings respectively. From the left panel of Fig. 1, it is clear that the couplings do not unify at some high energy renormalization scale. However, in a supersymmetric model with new degrees of freedom appearing at the

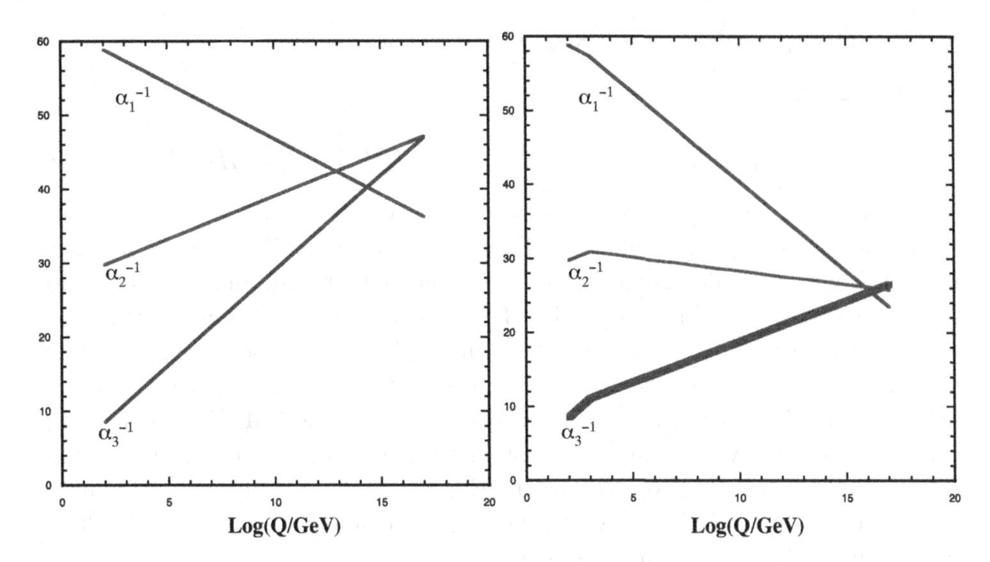

Fig. 1. Running of the gauge couplings in the standard model (left) and in the MSSM (right). $\alpha_i = g_i^2/4\pi$ for each of the gauge couplings associated with $SU(3)_c$, $SU(2)_L$, and $U(1)_Y$. Unification of the couplings becomes possible in the MSSM.

TeV scale (for example), the coefficients become $b_i = 33/5, 1, -3$ and unification is possible as seen on the right panel of Fig. 1.

It is also well known that the quartic coupling, λ, of the Higgs potential

$$V = \frac{\lambda}{4}h^4 + \frac{m^2}{2}h^2 \tag{2}$$

runs negative at a scale $Q \sim 10^{10}$ GeV, with the precise value depending on m_h and the top mass.[3,4] This triggers an instability in the SM as it is run up to high energies. This problem is immediately avoided in the MSSM[3] since the quartic coupling is related to the gauge couplings, $\lambda = \left(g_1^2 + g_2^2\right)/2$ and is positive definite. Furthermore, electroweak symmetry breaking requires that the sign of m^2 in the potential (2) is negative. This radiative electroweak symmetry breaking occurs naturally in the MSSM when running the initially (at the grand unification (GUT) scale) positive soft Higgs masses down to low energies.[5]

In order to ensure that the proton is long-lived, it is common to impose R-parity defined in terms of baryon number, lepton number and spin as $(-1)^{3B+L+2s}$. Models with unbroken R-parity include only those interactions which are direct supersymmetric analogues of SM processes. As an additional consequence, supersymmetric models with R-parity predict the existence of a stable particle which can be a dark matter candidate.[6]

The MSSM can be defined by its superpotential

$$W = \left[y_e H_1 L e^c + y_d H_1 Q d^c + y_u H_2 Q u^c\right] + \mu H_1 H_2 \,, \tag{3}$$

and the soft supersymmetry-breaking part of the Lagrangian

$$\mathcal{L}_{\rm soft} = -\frac{1}{2}M_\alpha \lambda^\alpha \lambda^\alpha - m^2_{ij}\phi^{*i}\phi^j \\ - A_e y_e H_1 L e^c - A_d y_d H_1 Q d^c - A_u y_u H_2 Q u^c - B\mu H_1 H_2 + \text{h.c.} \qquad (4)$$

Many models of supersymmetry breaking (such as gravity mediation[7]) predict universalities among the supersymmetry breaking parameters. For example, at some high energy input scale (usually taken to be the GUT scale), all gaugino masses, M_α, take a common value, $m_{1/2}$, all scalar masses, m^2_{ij}, take a common mass m_0^2, and all trilinear mass terms, A_i, take a common value, A_0. These three parameters, plus the ratio of the two Higgs expectation values, $\tan\beta$, defines what is commonly referred to as the constrained MSSM (CMSSM).[8–11] In the CMSSM, one uses the conditions derived by the minimization of the Higgs potential after radiative electroweak symmetry breaking to solve for the Higgs mixing mass, $|\mu|$ and the bilinear mass term B (or equivalently μ and the Higgs pseudoscalar mass, m_A) for fixed $\tan\beta$, leaving the sign of μ undetermined.

2. Pre-Run I

Before the first results of Run I at the LHC were released, expectations were high for the possible discovery of supersymmetry as supersymmetric models such as the CMSSM provided definite improvements to low energy precision phenomenology and were predicted to lie within the reach of the LHC. The left panel of Fig. 2 shows the results of mastercode[12] — a frequentist Markov Chain Monte Carlo analysis of low energy experimental observables in the context of supersymmetry. At each point sampled in the CMSSM parameter space, mastercode compares the predicted and measured values of a large set of observables, establishing a χ^2 likelihood function across the parameter space. The figure shows the color coded values of $\Delta\chi^2$ relative to the best fit point shown by the white dot at low $m_{1/2}$ and low m_0. Marginalization over A_0 and $\tan\beta$ was performed to produce this $(m_0, m_{1/2})$ plane. The best-fit CMSSM point lies at $m_0 = 60$ GeV, $m_{1/2} = 310$ GeV, $A_0 = 130$ GeV, $\tan\beta = 11$. For a full list of observables used to determine χ^2 see Ref. 12.

There was also considerable optimism for discovering supersymmetric dark matter in direct detection experiments. The left panel of Fig. 3 displays the pre-LHC preferred range of the spin-independent DM scattering cross-section $\sigma_p^{\rm SI}$ (calculated here assuming a relatively large value for the π–N scattering term $\Sigma_N = 64$ MeV) as a function of m_χ.[12] The expected range of $\sigma_p^{\rm SI}$ lied just below the then present experimental upper limits (solid lines).[14,15] As one can see from the successive lower upper limits from later experiments[16–18] shown by the bands, the pre-LHC predicted values for the elastic scattering cross-section was well within reach of current experiments.

For fixed $\tan\beta$, A_0, and $sgn(\mu)$, it is instructive to view the regions of the CMSSM parameter space in the $(m_{1/2}, m_0)$ plane as in Fig. 4. Parameter regions

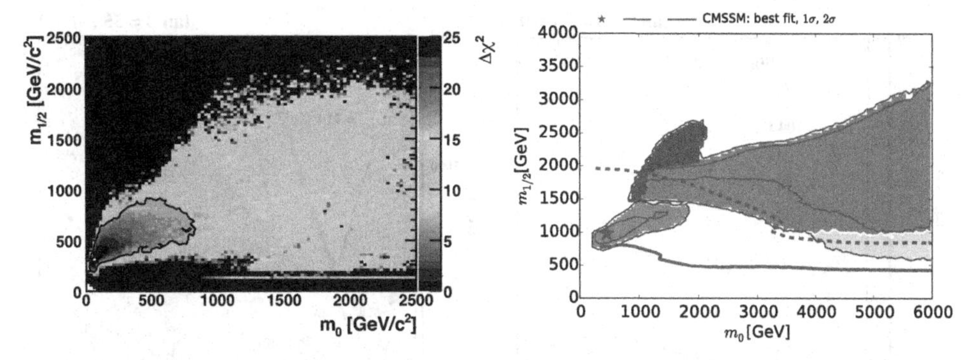

Fig. 2. (Color online) The $\Delta\chi^2$ functions in the $(m_0, m_{1/2})$ planes for the CMSSM from a mastercode frequentist analysis. The pre-LHC result is shown on the left panel.[12] Red and blue contours correspond to 68% and 95% CL contours and the best fit point is depicted by a white dot. The post-LHC result is shown on the right panel[13] using 8 TeV data at 20 fb^{-1}. Here the best fit point is shown by the filled star. The color of the shaded region indicates the dominant annihilation mechanism for obtained the correct relic density: stau coannihilation-pink; A/H funnel-blue; focus point-cyan; and a hybrid region of stau coannihilation and funnel-purple. The solid and dashed purple curves show the Run I reach and the expected Run II reach at 14 TeV at 3000 fb^{-1} respectively. The latter corresponds approximately to the 95% CL exclusion sensitivity with 300/fb at 14 TeV.

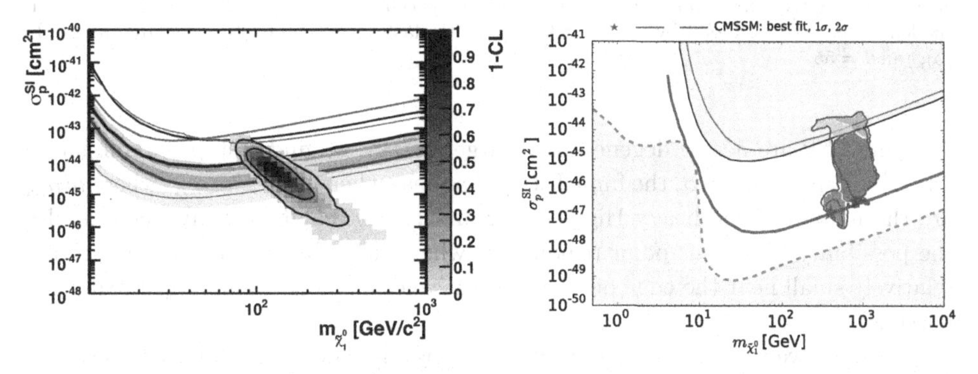

Fig. 3. (Color online) (Left) The pre-LHC prediction for the spin-independent DM scattering cross-section, σ_p^{SI}, versus m_χ in the CMSSM.[12] The solid lines are the pre-LHC experimental upper limits from CDMS[14] and XENON10,[15] while the bands are the more recent limits from XENON100[16,17] and LUX.[18] (Right) The post-Run I likelihood contours for σ_p^{SI}.[13] Shading within the likelihood contours is the same as in Fig. 2, though here we also see a region where chargino coannihilations are dominant (green). The green and black lines show the sensitivities of the XENON100[17] and LUX[18] experiments, respectively, and the solid purple lines show the projected 95% exclusion sensitivity of the LUX-Zeplin (LZ) experiment.[19] The dashed orange line shows the astrophysical neutrino "floor,"[20,21] below which astrophysical neutrino backgrounds dominate (yellow region).

compatible with the cosmic microwave background (CMB) determinations of the relic density,[22,23] are largely found in "strips" of the parameter space, due to necessary relations in the sparticle mass spectrum. For parameters where the stau and LSP are nearly degenerate, we obtain the stau coannihilation strip,[24–26] or when the

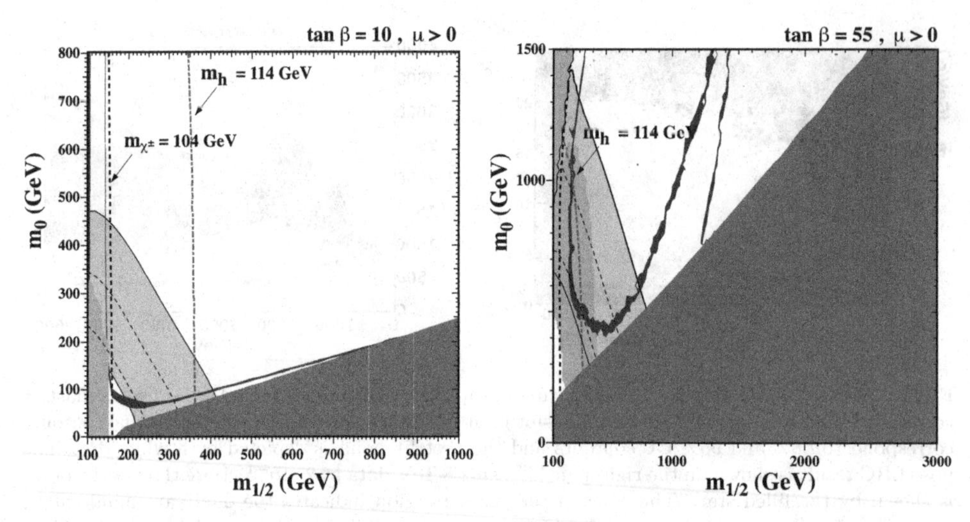

Fig. 4. (Color online) (Left) The $(m_{1/2}, m_0)$ planes for (a) $\tan\beta = 10$ and $\mu > 0$, assuming $A_0 = 0$, $m_t = 173.1$ GeV and $m_b(m_b)^{\overline{MS}}_{SM} = 4.25$ GeV. The near-vertical (red) dot-dashed lines are the contours $m_h = 114$ GeV, and the near-vertical (black) dashed line is the contour $m_{\chi^\pm} = 104$ GeV. The medium (dark green) shaded region is excluded by $b \to s\gamma$,[33] and the dark (blue) shaded area is the cosmologically preferred region. In the dark (brick red) shaded region, the LSP is the charged $\tilde{\tau}_1$. The region allowed by the E821 measurement of $g_\mu - 2$ at the 2σ level,[34] is shaded (pink) and bounded by solid black lines, with dashed lines indicating the 1σ ranges. In (b), $\tan\beta = 55$.

stop and LSP are nearly degenerate at large A_0/m_0, we find a stop coannihilation strip.[27–29] At large $\tan\beta$, the funnel strips[8] appear when $2m_\chi \simeq m_{H,A}$, where $m_{H,A}$ are the masses of the heavy Higgs scalar and pseudoscalar. Finally, there is also the possibility of a focus point region,[30,31] where the value of the μ term becomes relatively small near the edge of where radiative electroweak symmetry breaking is possible.

In Fig. 4, we see directly regions with acceptable relic density and which satisfy other phenomenological constraints. The dark (blue) shaded regions correspond to that portion of the CMSSM plane such that the computed relic density yields is in agreement with that of CMB determinations. On the left panel, with $\tan\beta = 10$, $A_0 = 0$, and $\mu > 0$, there is a bulk region of acceptable relic density at relatively low values of $m_{1/2}$ and m_0. This region tapers off as $m_{1/2}$ is increased becoming the stau-coannihilation strip. At higher values of m_0, annihilation cross-sections are too small to maintain an acceptable relic density and $\Omega_\chi h^2$ is too large. At large $m_{1/2}$, coannihilation processes[32] between the LSP and the next lightest sparticle (in this case the $\tilde{\tau}$) enhance the annihilation cross-section and reduce the relic density. This occurs when the LSP and NLSP are nearly degenerate in mass. The dark (red) shaded region has $m_{\tilde{\tau}} < m_\chi$ and is excluded. On the right panel, $\tan\beta = 55$ and we see clearly a funnel-like region of acceptable relic density. Between the two sides of the funnel, there is a line with $2m_\chi = m_{A,H}$.

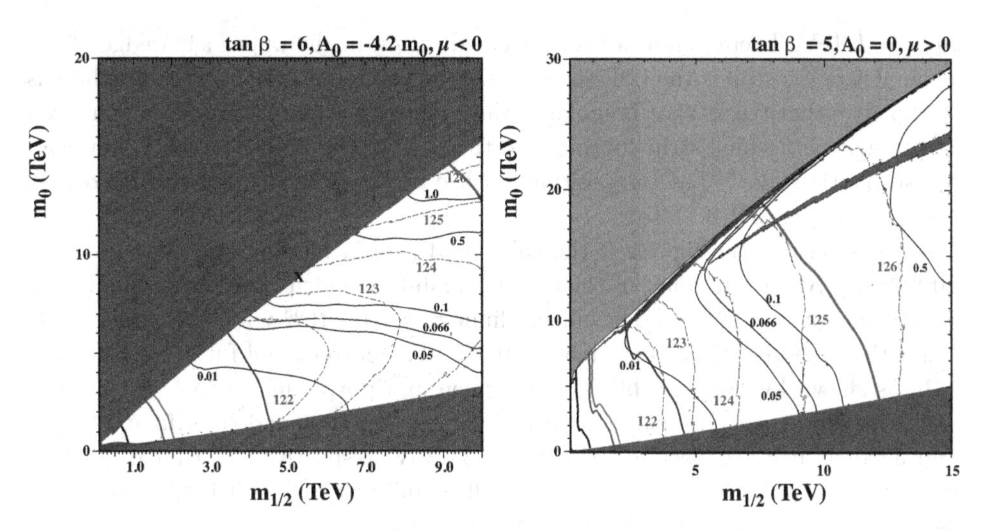

Fig. 5. (Color online) Sample CMSSM $(m_{1/2}, m_0)$ planes showing the stop coannihilation strip with $\tan\beta = 6$ and $A_0 = -4.2m_0$, $\mu < 0$ (left) and the focus-point strip for $\tan\beta = 5$ and $A_0 = 0$ with $\mu > 0$ (right). In the light mauve shaded regions, it is not possible to satisfy the electroweak symmetry breaking conditions. In the brown shaded regions, the LSP is charged and/or colored. The dark blue shaded strips show the areas where $0.06 < \Omega_\chi h^2 < 0.2$ on the right panel and the further enlarged range of $0.01 < \Omega_\chi h^2 < 2.0$ on the left panel. These ranges have been enhanced to improve visibility. The red dot-dashed contours indicate the Higgs mass, labeled in GeV, and the solid black contours indicate the proton lifetime in units of 10^{35} yrs. The bold solid black, blue, green, purple, and red lines in each panel are current and future limits from the LHC at 8 TeV, 300 and 3000 fb^{-1} at 14 TeV, 3000 fb^{-1} with the HE-LHC at 33 TeV, and 3000 fb^{-1} with the FCC-hh at 100 TeV, respectively, taken from the analysis of Ref. 29.

3. Post-Run I

As everyone knows, supersymmetry was not discovered during Run I at the LHC,[35] and the prospects for discovering supersymmetry at the LHC dimmed significantly. On the right panel of Fig. 2, the post-Run I likelihood contours in the $(m_0, m_{1/2})$ plane[13] are shown using 8 TeV results at 20 fb^{-1}.[35] The best fit point based on the 8 TeV data is shown by the filled star at (420,970) GeV with $A_0 = 3000$ GeV and $\tan\beta = 14$, though the likelihood function is quite flat and the exact position of the best point is not very well defined.

The right panel of Fig. 3 shows that there is still hope for direct detection experiments though the new best fit point implies a cross-section of $\sim 10^{-47}$ cm^2, nearly two orders of magnitude below the current upper bound. The likelihood function, however, is rather flat between 10^{-47} cm$^2 \lesssim \sigma_p^{\rm SI} \lesssim 10^{-45}$ cm^2. Note that in this case, a lower value of $\Sigma_{\pi N} = 50 \pm 7$ MeV was used.

In Fig. 5, we show two CMSSM $(m_{1/2}, m_0)$ planes[37] displaying the stop-coannihilation strip with $\mu < 0$ (left) and focus-point strip with $\mu > 0$ (right). Higgs mass contours are shown as red dot-dashed curves labeled by m_h in GeV. As in Fig. 4, there is a stau coannihilation strip at low m_0 just above the brown shaded region where the stau is the LSP. However, it does not extend much past

$m_{1/2} = 1$ TeV. There is now a brown shaded region in the upper left wedge of the plane where the stop is the LSP (or tachyonic). Though it is barely visible, there is a stop strip that tracks that boundary. Since relic density range has been enhanced in the figure, the blue strip continues to the edge of the plot. In reality, however, the stop strip ends [28,29] at the position marked by the **X** in the figure which occurs at $(m_{1/2}, m_0) \simeq (5.2, 8.8)$ TeV.

Also shown in the figures is the calculated proton lifetime labeled in units of 10^{35} years. At the endpoint of the stop coannihilation strip, the proton lifetime is approximately 2×10^{34} yrs. The current limit $\tau_p > 6.6 \times 10^{33}$ yrs [36,38] would exclude the entire area below the curve labeled 0.066. The current and future reach of the LHC is shown by the solid black, blue, green and purple lines which are particle exclusion reaches for $\not{E}_T$ searches with 20/fb at 8 TeV, 300 and 3000/fb at 14 TeV, and 3000/fb at a prospective HE-LHC at 33 TeV, respectively. [29] Unlike the stau strip, it is unlikely that the entire strip will be fully probed as it is seen to extend beyond the reach of a future 33 TeV LHC upgrade.

The right panel of Fig. 5 shows an example of a focus point strip. [30,31] The light mauve shaded region in the parts of the right panel with large $m_0/m_{1/2}$ is excluded because there are no solutions to the EWSB conditions: along this boundary $\mu^2 = 0$. Just below the regions where EWSB fails, there is a narrow dark blue strip where the relic density falls within the range determined by CMB and other experiments. [23] The planes also feature stau-coannihilation strips [25] close to the boundaries of the brown shaded region at low m_0. They extend to $m_{1/2} \simeq 1$ TeV, but are very difficult to see on the scale of this plot, even with our enhancement of the relic density range. There are also "thunderbolt"-shaped brown shaded bands at intermediate $m_0/m_{1/2}$ where the chargino is the LSP.

Though the CMSSM contains viable regions of parameter space, it is useful to consider alternatives. Here, I mention some of these alternatives which may either be more or less constrained than the CMSSM. Among the more restricted set of models is mSUGRA, [7] where the condition that $B_0 = A_0 - m_0$ is applied and hence $\tan\beta$ is no longer free. [7,10,11] This is a 3-parameter model. Pure gravity mediated models [39] are the most restricted set of models and in principle contain only one free parameter, the gravitino mass $m_{3/2}$. [41] In this case, gaugino masses (and A-terms) are generated through anomalies. [40] $m_0 = m_{3/2}$, $B_0 \simeq -m_0$ and the spectrum is reminiscent of split supersymmetry [42] and typically contains a wino LSP. In practice, though, the model is over-restrictive and it is necessary to include a Giudice–Masiero term [43,44] in order to achieve radiative electroweak symmetry breaking. [41] This allows one to take $\tan\beta$ as a free parameter once again. The model is similar in some respects with minimal anomaly mediation (mAMSB) [45] in which gaugino masses and A-terms are determined from $m_{3/2}$, but m_0 is kept as a free parameter which together with $\tan\beta$ defines a 3-parameter theory. An example of the latter is shown in Fig. 6 for $\tan\beta = 5$. [46] The pink triangular region at large m_0 and relatively small $m_{3/2}$ is excluded because there are no consistent solutions to the electroweak vacuum conditions in that region. The prominent near-horizontal

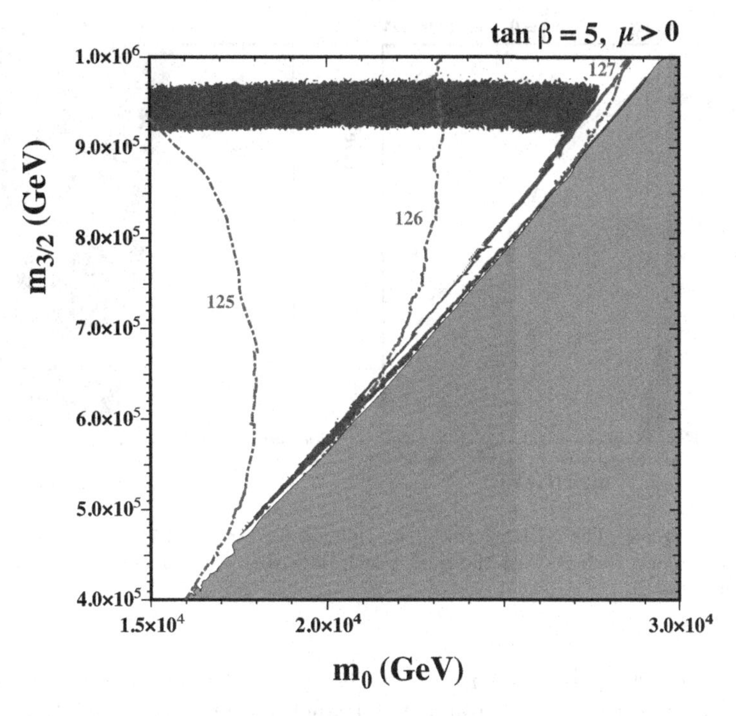

Fig. 6. (Color online) The $(m_0, m_{3/2})$ plane for $\tan\beta = 5$. There are no consistent solutions of the electroweak vacuum conditions in the pink shaded triangular regions at lower right. The neutralino LSP density falls within the range of the CDM density indicated by Planck and other experiments in the dark blue shaded bands. Contours of the Higgs mass are shown as red dashed lines.

band at $m_{3/2} \approx 950$ TeV, shows the region where the relic density agrees with CMB determination and there the LSP is nearly pure wino. We stress that any value of $m_{3/2}$ below this band would also be allowed if the wino provides only a fraction of the total cold dark matter density. There is also a very narrow V-shaped diagonal strip running close to the electroweak vacuum boundary, where the LSP has a large Higgsino component.

There are many possible models which are extensions of the CMSSM. These include models where the Higgs masses are non-universal: the NUHM1 ($m_1 = m_2 \neq m_0$),[10,11,47,48] and NUHM2 ($m_1 \neq m_2 \neq m_0$);[10,11,48,49] subGUT models where the input universality scale differs from the GUT scale with $M_{\rm in} < M_{\rm GUT}$;[10,11,50] super-GUT models with $M_{\rm in} > M_{\rm GUT}$.[37,51] The above-mentioned models all have 1–2 additional parameters relative to the CMSSM. In the case of the NUHM, one may replace m_1, and/or m_2 with μ and/or the Higgs pseudoscalar mass, m_A. Two examples of NUHM1 $(m_{1/2}, m_0)$ planes are shown in Fig. 7 for $\tan\beta = 4.5$, $A_0 = 0$ and $\mu = 1000$ GeV (left) and $\mu = 1050$ GeV (right). On the left panel, we see a small stau LSP region at low m_0 and $m_{1/2}$. Since μ is fixed, the composition of the LSP changes as $m_{1/2}$ is increased. At small $m_{1/2}$ the LSP is mainly bino and the

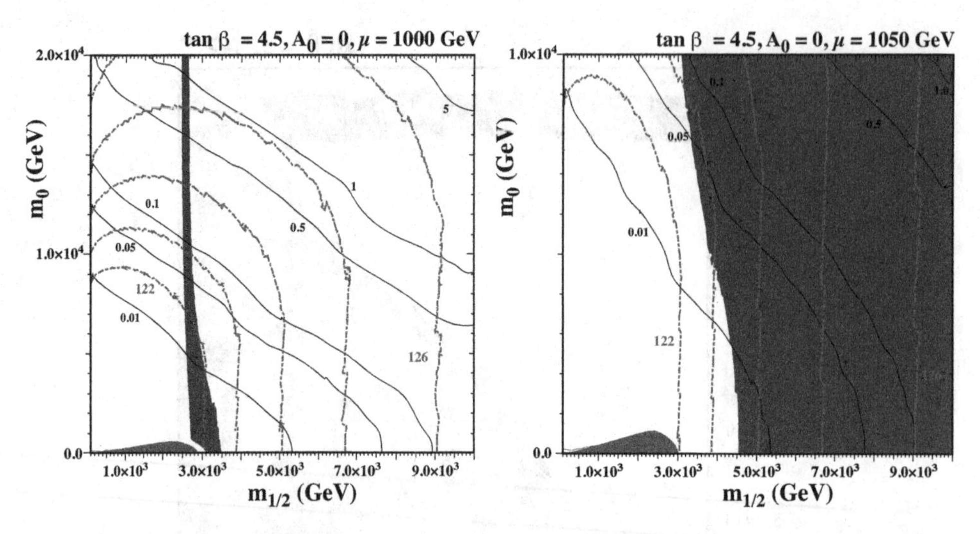

Fig. 7. (Color online) The NUHM1 $(m_{1/2}, m_0)$ planes for $\tan\beta = 4.5$. We take $\mu = 1000$ GeV on the left panel and 1050 GeV on the right panel, both with $A_0 = 0$. The shading and contour types are as in Fig. 5.

relic density is too high. As $m_{1/2}$ is increased, the Higgsino component increases and the relic density passes through the Planck range across a relatively narrow, near-vertical transition strip. At larger $m_{1/2}$ the LSP is a Higgsino with a mass of about 1050 GeV which is slightly low for a Higgsino LSP and, as a result, the relic density is somewhat too small when $m_{1/2} \gtrsim 3$ TeV. In this panel, we see that we obtain an acceptable Higgs mass ($m_h > 124$ GeV) when $m_0 \gtrsim 13$ TeV. The proton lifetime is sufficiently large ($\tau_p \gtrsim 0.25 \times 10^{35}$ yrs) for this value of m_0.

On the right panel of Fig. 7, we have increased μ slightly to 1050 GeV. The most striking feature is that the dark matter region fills the right part of the plane: indeed, it extends infinitely far to the right towards large gaugino masses. In this case, when the gaugino mass is large, the LSP is a nearly pure Higgsino, as is the NLSP. This near-degeneracy facilitates coannihilation that brings the relic density within the acceptable range, with $\Omega_\chi h^2$ being determined predominantly by μ.[52] The Higgsino mass in this case is very close to 1100 GeV, which remains constant at large $m_{1/2}$. Thus there is a very large (infinite) area where the relic density matches the Planck result. At low $m_{1/2}$, the relic density is too large and drops monotonically as the gaugino mass is increased and asymptotes to the Planck density at very large $m_{1/2}$. For $m_0 \lesssim 10$ TeV, when $A_0 = 0$, the Higgs mass contours are nearly vertical and the value of $\tan\beta = 4.5$ was chosen to maximize the area with good relic density and Higgs masses. The area between $m_{1/2} = 5$ TeV and 9 TeV has m_h between 124 GeV and 126 GeV, and increasing (decreasing) $\tan\beta$ by 0.5 would raise (lower) m_h by roughly 1 GeV. Much of this region has $\tau_p \gtrsim 0.05 \times 10^{35}$ yrs: requiring $\tau_p > 5 \times 10^{33}$ yrs implies either $m_{1/2} \gtrsim 7.8$ TeV for small m_0 or $m_0 \gtrsim 8$ TeV for $m_{1/2} \simeq 4$ TeV.

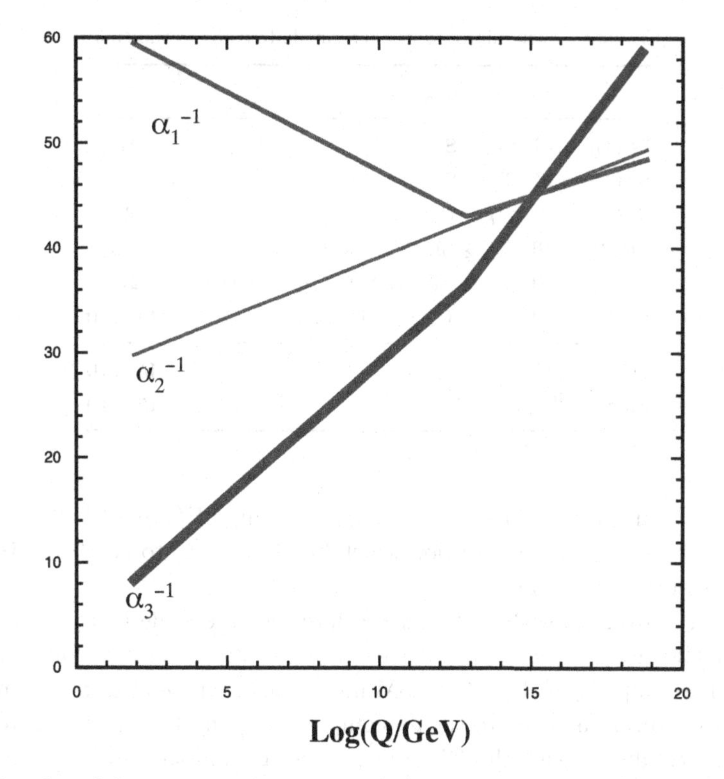

Fig. 8. Running of the gauge couplings in a SO(10) model broken to $SU(4) \times SU(2)_L \times U(1)_R$ at the unification scale, $M_{\rm GUT} = 1.3 \times 10^{15}$ GeV, which is subsequently broken to the standard model at an intermediate scale $M_{\rm int} = 7.8 \times 10^{12}$ GeV.

4. SO(10) GUT Dark Matter

Nonsupersymmetric SO(10)[53] models can also naturally account for the presence of a dark matter candidate. If symmetry breaking to the SM occurs through an intermediate scale gauge group, the gauge couplings may be deflected at the intermediate scale and hence allow for gauge coupling unification[54–59] as shown in Fig. 8.[56] If the intermediate scale gauge group is broken via a **126** dimensional representation of SO(10), a Z_2 discrete symmetry (similar to R-parity) is preserved[60] thus allowing for the existence of a stable dark matter candidate.[56–59,61,62] Furthermore, in models with gauge coupling unification and a stable dark matter candidate, it is also possible to stabilize the electroweak vacuum while at the same time radiatively break the electroweak symmetry.[59]

SO(10) models necessarily involve rather large representations. The gauge multiplet is a **45** which decomposes as $\mathbf{45} = (\mathbf{24},\mathbf{0}) + (\mathbf{10},\mathbf{4}) + (\overline{\mathbf{10}},-\mathbf{4}) + (\mathbf{1},\mathbf{0})$ under $SU(5)\times U(1)$ or $\mathbf{45} = (\mathbf{15},\mathbf{1},\mathbf{1})+(\mathbf{6},\mathbf{2},\mathbf{2})+(\mathbf{1},\mathbf{3},\mathbf{1})+(\mathbf{1},\mathbf{1},\mathbf{3})$ under $SU(4)\times SU(2)\times SU(2)$. Each generation of matter fields lie in a **16** which similarly decomposes as $\mathbf{16} = (\mathbf{10},-\mathbf{1}) + (\bar{\mathbf{5}},\mathbf{3}) + (\mathbf{1},-\mathbf{5})$ or $\mathbf{16} = (\mathbf{4},\mathbf{1},\mathbf{2}) + (\bar{\mathbf{4}},\mathbf{2},\mathbf{1})$. The Higgs representations will depend on the particular symmetry breaking pattern chosen. For

Table 1. Candidates for the intermediate gauge group G_{int}.

G_{int}	R_1
$\mathrm{SU}(4)_C \otimes \mathrm{SU}(2)_L \otimes \mathrm{SU}(2)_R$	**210**
$\mathrm{SU}(4)_C \otimes \mathrm{SU}(2)_L \otimes \mathrm{SU}(2)_R \otimes D$	**54**
$\mathrm{SU}(4)_C \otimes \mathrm{SU}(2)_L \otimes \mathrm{U}(1)_R$	**45**
$SU(3)_C \otimes \mathrm{SU}(2)_L \otimes \mathrm{SU}(2)_R \otimes \mathrm{U}(1)_{B-L}$	**45**
$SU(3)_C \otimes \mathrm{SU}(2)_L \otimes \mathrm{SU}(2)_R \otimes \mathrm{U}(1)_{B-L} \otimes D$	**210**
$SU(3)_C \otimes \mathrm{SU}(2)_L \otimes \mathrm{U}(1)_R \otimes \mathrm{U}(1)_{B-L}$	**45**, **210**
$\mathrm{SU}(5) \otimes \mathrm{U}(1)$	**45**, **210**
Flipped $\mathrm{SU}(5) \otimes \mathrm{U}(1)$	**45**, **210**

example, to obtain the intermediate scale gauge group $\mathrm{SU}(4)_C \otimes \mathrm{SU}(2)_L \otimes \mathrm{SU}(2)_R$, we should take a Higgs representation which breaks SO(10) to be $R_1 = \mathbf{210}$. Other possibilities are given in Table 1.

As noted above, we wish to break the intermediate scale gauge group with a Higgs in a **126** representation in order to preserve the Z_2 discrete for dark matter stability. The coupling of the **126** to SM matter fields embedded in a **16** naturally gives rise to a Majorana mass to the ν_R component of the **16** of order $\langle \mathbf{126} \rangle \sim M_{\text{int}}$ which when combined with the Dirac mass arising from the vev of the SM Higgs (now residing in a 10-plet of SO(10)) gives rise to the seesaw mechanism for light neutrino masses.[63]

The dark matter in SO(10) models may be either fermionic or bosonic. A fermionic DM candidate should be parity even and belong to a **10**, **45**, **54**, **120**, **126**, **210** or **210′** representation, while scalar DM is parity odd and belongs to a **16** or **144** representation. There are of course possibilities for candidates in higher dimensional representation, but the discussion here is limited to representations no larger than the **210**. Table 2 shows the various dark matter candidates labelled by their SM quantum numbers.[58]

Depending on the dark matter and Higgs representation chosen, the three renormalization group equations for the gauge couplings can be used to determine, the GUT scale, the intermediate scale, and the value of the GUT gauge coupling. One such example is a singlet fermion (F_1^0) originating in the $(\mathbf{15}, \mathbf{1}, \mathbf{1})$ representation (in terms of $\mathrm{SU}(4) \otimes \mathrm{SU}(2) \otimes \mathrm{SU}(2)$) included in the **45** of SO(10). The evolution of the gauge couplings in this model was shown in Fig. 8. In this model,[57] $R_1 = \mathbf{54}$, and we have $\log(M_{\text{int}}) = 13.66$, $\log(M_{\text{GUT}}) = 15.87$, and $g_{\text{GUT}} = 0.567$. For further details concerning this model, see Ref. 57.

Given the large number of possible intermediate gauge groups and the large number of possible dark matter representations (scalar or fermionic), one may think that there are a vast number of dark matter models in the SO(10) framework. However, once we demand gauge coupling unification, require that the GUT scale

Table 2. List of $SU(2)_L \otimes U(1)_Y$ multiplets in SO(10) representations that contain an electric neutral color singlet.

Model	$B-L$	$\mathrm{SU}(2)_L$	Y	SO(10) representations
$\mathrm{F}_{\mathbf{1}}^{0}$		**1**	0	**45**, **54**, **210**
$\mathrm{F}_{\mathbf{2}}^{1/2}$		**2**	1/2	**10**, **120**, **126**, **210′**
$\mathrm{F}_{\mathbf{3}}^{0}$	0	**3**	0	**45**, **54**, **210**
$\mathrm{F}_{\mathbf{3}}^{1}$		**3**	1	**54**
$\mathrm{F}_{\mathbf{4}}^{1/2}$		**4**	1/2	**210′**
$\mathrm{F}_{\mathbf{4}}^{3/2}$		**4**	3/2	**210′**
$\mathrm{S}_{\mathbf{1}}^{0}$		**1**	0	**16**, **144**
$\mathrm{S}_{\mathbf{2}}^{1/2}$	1	**2**	1/2	**16**, **144**
$\mathrm{S}_{\mathbf{3}}^{0}$		**3**	0	**144**
$\mathrm{S}_{\mathbf{3}}^{1}$		**3**	1	**144**
$\hat{\mathrm{F}}_{\mathbf{1}}^{0}$		**1**	0	**126**
$\hat{\mathrm{F}}_{\mathbf{2}}^{1/2}$	2	**2**	1/2	**210**
$\hat{\mathrm{F}}_{\mathbf{3}}^{1}$		**3**	1	**126**

is high enough to ensure a sufficiently long proton lifetime, and further require that $M_{\rm int} < M_{\rm GUT}$, there are in fact only a handful of viable models. We also require that we can split the SO(10) multiplets in such a way to leave only the DM candidate (and perhaps weak partners) with weak scale masses. The resulting acceptable models for scalar dark matter candidates is shown in Table 3. For more information on these models see Ref. 58.

The scalar singlet models (SA) have a phenomenology similar to that of so-called Higgs portal models.[66] To roughly estimate the favored mass region for a scalar singlet, consider the quartic interaction between the singlet DM ϕ and the SM Higgs field: $\lambda_{H\phi}\phi^2|H|^2/2$. Through this coupling, the singlet DM particles annihilate into a pair of the SM Higgs bosons and weak gauge bosons. The annihilation cross-section is

$$\sigma_{\rm ann} v_{\rm rel} \simeq \frac{\lambda_{H\phi}^2}{16\pi m_{\rm DM}^2}, \tag{5}$$

assuming that the DM mass $m_{\rm DM}$ is much larger than the SM Higgs mass m_h and neglecting terms proportional to v^2. The DM relic abundance is, on the other hand, related to the annihilation cross-section by

$$\Omega_{\rm DM} h^2 \simeq \frac{3\times 10^{-27}\ {\rm cm}^3\ {\rm s}^{-1}}{\langle \sigma_{\rm ann} v_{\rm rel}\rangle}. \tag{6}$$

Table 3. One-loop result for $M_{\rm GUT}$, $M_{\rm int}$, $\alpha_{\rm GUT}$, and proton lifetimes for scalar dark matter models. The DM mass is set to be $m_{DM} = 1$ TeV. The mass scales are given in GeV and the proton lifetimes are in units of years. These models evade the proton decay bound, $\tau(p \to e^+\pi^0) > 1.4 \times 10^{34}$ yrs.[64,65]

Model	$R_{\rm DM}$	$\log_{10} M_{\rm GUT}$	$\log_{10} M_{\rm int}$	$\alpha_{\rm GUT}$	$\log_{10} \tau_p(p \to e^+\pi^0)$
		$G_{\rm int} = {\rm SU}(4)_C \otimes {\rm SU}(2)_L \otimes {\rm SU}(2)_R$			
SA422 ($\mathtt{S}_{\mathbf{1}}^{0}$)	$\mathbf{4}, \mathbf{1}, \mathbf{2}$	16.33	11.08	0.0218	36.8 ± 1.2
SB422 ($\mathtt{S}_{\mathbf{2}}^{1/2}$)	$\mathbf{4}, \mathbf{2}, \mathbf{1}$	15.62	12.38	0.0228	34.0 ± 1.2
		$G_{\rm int} = {\rm SU}(3)_C \otimes {\rm SU}(2)_L \otimes {\rm SU}(2)_R \otimes {\rm U}(1)_{B-L}$			
SA3221 ($\mathtt{S}_{\mathbf{1}}^{0}$)	$\mathbf{1}, \mathbf{1}, \mathbf{2}, 1$	16.66	8.54	0.0217	38.1 ± 1.2
SB3221 ($\mathtt{S}_{\mathbf{2}}^{1/2}$)	$\mathbf{1}, \mathbf{2}, \mathbf{1}, -1$	16.17	9.80	0.0223	36.2 ± 1.2
SC3221 ($\mathtt{S}_{\mathbf{2}}^{1/2}$)	$\mathbf{1}, \mathbf{2}, \mathbf{3}, -1$	15.62	9.14	0.0230	34.0 ± 1.2
		$G_{\rm int} = {\rm SU}(3)_C \otimes {\rm SU}(2)_L \otimes {\rm SU}(2)_R \otimes {\rm U}(1)_{B-L} \otimes D$			
SA3221D ($\mathtt{S}_{\mathbf{1}}^{0}$)	$\mathbf{1}, \mathbf{1}, \mathbf{2}, 1$	15.58	10.08	0.0231	33.8 ± 1.2
SB3221D ($\mathtt{S}_{\mathbf{2}}^{1/2}$)	$\mathbf{1}, \mathbf{2}, \mathbf{1}, -1$	15.40	10.44	0.0233	33.1 ± 1.2

To account for the observed DM density $\Omega_{\rm DM}h^2 = 0.12$,[23] the DM mass should be $m_{\rm DM} \lesssim 3$ TeV for $\lambda_{H\phi} \lesssim 1$. This gives us a rough upper bound for the DM mass. The other scalar DM candidates are ${\rm SU}(2)_L \otimes {\rm U}(1)_Y$ multiplets, which can interact with SM particles through gauge interactions in addition to the quartic coupling mentioned above. In particular, $\mathtt{S}_{\mathbf{2}}^{1/2}$ is similar to the dark matter model known as the Inert Higgs Doublet Model and has been widely studied in the literature.[67,68] As one can see from Table 3, there are relatively few phenomenologically viable models.

An example of a fermionic singlet model was discussed above, and its relic density is mainly determined by the reheat temperature after inflation (see Refs. 56 and 57 for more details). Nonsinglet fermions behaving as well studied wimps are also possible. From Table 2, we see that the only candidate without hypercharge is the weak (wino-like) triplet $\mathtt{F}_{\mathbf{3}}^{0}$. One example of a triplet candidate is given in Table 4. As one can see, this state is a singlet under both ${\rm SU}(4)_C$ and ${\rm SU}(2)_R$ and originates in a **45** of SO(10). While the intermediate scale is relatively low, the GUT scale is quite high and hence the proton lifetime is unobservably long.

It is also possible that a fermionic dark matter candidate carries hypercharge and in this case, it may be either a weak doublet (Higgsino-like) or triplet. Some examples are shown in Table 5. In this case, we must introduce another representation (at the intermediate scale) to mix with $R_{\rm DM}$ in order to induce some splitting in the DM multiplet to evade current DM detection experimental results. These are denoted $R'_{\rm DM}$ in the table.

Table 4. The one-loop results for M_{GUT}, M_{int}, α_{GUT}, and proton lifetimes for a real triplet fermionic DM models. Here the DM mass is set to be 1 TeV. The mass scales and proton decay lifetime are in units of GeV and years, respectively.

Model	R_{DM}	$\log_{10} M_{\text{int}}$	$\log_{10} M_{\text{GUT}}$	α_{GUT}	$\log_{10} \tau_p(p \to e^+\pi^0)$
		$G_{\text{int}} = \text{SU}(4)_C \otimes \text{SU}(2)_L \otimes \text{SU}(2)_R$			
$\mathbf{F}_\mathbf{3}^0$	$(\mathbf{1}, \mathbf{3}, \mathbf{1})$	6.54	17.17	0.0252	39.8 ± 1.2

Table 5. Possible hypercharged fermionic DM models that are not yet excluded by current proton decay experiments. The quantum numbers are labeled in the same order as G_{int}. The subscripts D and W refer to Dirac and Weyl respectively. The numerical results are calculated for DM mass of 1 TeV. The mass scales and proton decay lifetime are in unit of GeV and years, respectively.

Model	R_{DM}	R'_{DM}	$\log_{10} M_{\text{int}}$	$\log_{10} M_{\text{GUT}}$	α_{GUT}	$\log_{10} \tau_p$
		$G_{\text{int}} = \text{SU}(4)_C \otimes \text{SU}(2)_L \otimes \text{U}(1)_R$				
FA421 ($\mathbf{F}_\mathbf{2}^{1/2}$)	$(\mathbf{1}, \mathbf{2}, 1/2)_D$	$(\mathbf{15}, \mathbf{1}, 0)_W$	3.48	17.54	0.0320	40.9 ± 1.2
		$G_{\text{int}} = \text{SU}(4)_C \otimes \text{SU}(2)_L \otimes \text{SU}(2)_R$				
FA422 ($\mathbf{F}_\mathbf{2}^{1/2}$)	$(\mathbf{1}, \mathbf{2}, \mathbf{2})_W$	$(\mathbf{1}, \mathbf{3}, \mathbf{1})_W$	9.00	15.68	0.0258	34.0 ± 1.2
FB422 ($\mathbf{F}_\mathbf{2}^{1/2}$)	$(\mathbf{1}, \mathbf{2}, \mathbf{2})_W$	$(\mathbf{1}, \mathbf{3}, \mathbf{1})_W$	5.84	17.01	0.0587	38.0 ± 1.2

5. Summary

It is becoming apparent that recent LHC searches for supersymmetry have pushed the CMSSM to very high energies where acceptable regions of the parameter space rely on the near degeneracy between the LSP and the NLSP, thus allowing coannihilations to control the relic density. While the stau coannihilation strip is nearly ruled out by LHC searches, possibilities remain for the stop strip and if there are nonuniversal gaugino masses, gluino coannihilation. It is also possible that m_0 is large near the focus point strip so that the LSP is mostly Higgsino-like. There are several variants of the CMSSM which still permit neutralino dark matter. These include models with nonuniversal Higgs scalar masses (NUHM), models where the input universality scale is below the GUT scale (sub-GUT models), or above the GUT scale (super-GUT models) or pure gravity mediated models and minimal anomaly mediated models with either wino or Higgsino dark matter.

While supersymmetry has many motivations beyond dark matter, with the exception of the hierarchy problem, almost all of these motivating factors can be resolved in nonsupersymmetric versions of SO(10) grand unification. The real challenge lies in the detection of dark matter and our ability to discriminate between the various models.

Acknowledgments

I would like to thank my many collaborators involved the work summarized here. This includes J. Ellis, J. Evans, F. Luo, Y. Mambrini, A. Mustafayev, N. Nagata, P. Sandick, J. Zheng, and the entire Mastercode collaboration. This work was supported in part by DOE grant DE–SC0011842 at the University of Minnesota.

References

1. J. R. Ellis, S. Kelley and D. V. Nanopoulos, *Phys. Lett. B* **249**, 441 (1990); J. R. Ellis, S. Kelley and D. V. Nanopoulos, *Phys. Lett. B* **260**, 131 (1991); U. Amaldi, W. de Boer and H. Furstenau, *Phys. Lett. B* **260**, 447 (1991); P. Langacker and M.-X. Luo, *Phys. Rev. D* **44**, 817 (1991); C. Giunti, C. W. Kim and U. W. Lee, *Mod. Phys. Lett. A* **6**, 1745 (1991).
2. L. Maiani, *Proceedings, Gif-sur-Yvette Summer School on Particle Physics* (1979), pp. 1–52; G. 't Hooft *et al.* (eds.), Recent developments in gauge theories, in *Proceedings of the Nato Advanced Study Institute*, Nato Advanced Study Institutes Series: Series B, Physics, Vol. 59, Cargese, France, August 26–September 8, 1979 (Plenum Press, New York, 1980); E. Witten, *Phys. Lett. B* **105**, 267 (1981).
3. J. R. Ellis and D. Ross, *Phys. Lett. B* **506**, 331 (2001), arXiv:hep-ph/0012067.
4. D. Buttazzo, G. Degrassi, P. P. Giardino, G. F. Giudice, F. Sala, A. Salvio and A. Strumia, *J. High Energy Phys.* **1312**, 089 (2013), arXiv:1307.3536 [hep-ph]; G. Degrassi, S. Di Vita, J. Elias-Miro, J. R. Espinosa, G. F. Giudice, G. Isidori and A. Strumia, *J. High Energy Phys.* **1208**, 098 (2012), arXiv:1205.6497 [hep-ph].
5. L. E. Ibanez and G. G. Ross, *Phys. Lett. B* **110**, 215 (1982); K. Inoue, A. Kakuto, H. Komatsu and S. Takeshita, *Prog. Theor. Phys.* **68**, 927 (1982) [Erratum: *ibid.* **70**, 330 (1983)]; L. E. Ibanez, *Phys. Lett. B* **118**, 73 (1982); J. R. Ellis, D. V. Nanopoulos and K. Tamvakis, *Phys. Lett. B* **121**, 123 (1983); J. R. Ellis, J. S. Hagelin, D. V. Nanopoulos and K. Tamvakis, *Phys. Lett. B* **125**, 275 (1983); L. Alvarez-Gaume, J. Polchinski and M. B. Wise, *Nucl. Phys. B* **221**, 495 (1983).
6. H. Goldberg, *Phys. Rev. Lett.* **50**, 1419 (1983); J. Ellis, J. Hagelin, D. Nanopoulos, K. Olive and M. Srednicki, *Nucl. Phys. B* **238**, 453 (1984).
7. R. Barbieri, S. Ferrara and C. A. Savoy, *Phys. Lett. B* **119**, 343 (1982); J. R. Ellis, K. A. Olive, Y. Santoso and V. C. Spanos, *Phys. Lett. B* **573**, 162 (2003), arXiv:hep-ph/0305212; *Phys. Rev. D* **70**, 055005 (2004), arXiv:hep-ph/0405110.
8. M. Drees and M. M. Nojiri, *Phys. Rev. D* **47**, 376 (1993), arXiv:hep-ph/9207234; H. Baer and M. Brhlik, *Phys. Rev. D* **53**, 597 (1996), arXiv:hep-ph/9508321; *Phys. Rev. D* **57**, 567 (1998), arXiv:hep-ph/9706509; H. Baer, M. Brhlik, M. A. Diaz, J. Ferrandis, P. Mercadante, P. Quintana and X. Tata, *Phys. Rev. D* **63**, 015007 (2001), arXiv:hep-ph/0005027; J. R. Ellis, T. Falk, G. Ganis, K. A. Olive and M. Srednicki, *Phys. Lett. B* **510**, 236 (2001), arXiv:hep-ph/0102098.
9. G. L. Kane, C. F. Kolda, L. Roszkowski and J. D. Wells, *Phys. Rev. D* **49**, 6173 (1994), arXiv:hep-ph/9312272; J. R. Ellis, T. Falk, K. A. Olive and M. Schmitt, *Phys. Lett. B* **388**, 97 (1996), arXiv:hep-ph/9607292; *Phys. Lett. B* **413**, 355 (1997), arXiv:hep-ph/9705444; V. D. Barger and C. Kao, *Phys. Rev. D* **57**, 3131 (1998), arXiv:hep-ph/9704403; L. Roszkowski, R. Ruiz de Austri and T. Nihei, *J. High Energy Phys.* **0108**, 024 (2001), arXiv:hep-ph/0106334; A. Djouadi, M. Drees and J. L. Kneur, *J. High Energy Phys.* **0108**, 055 (2001), arXiv:hep-ph/0107316; U. Chattopadhyay, A. Corsetti and P. Nath, *Phys. Rev. D* **66**, 035003 (2002), arXiv:hep-ph/0201001; J. R. Ellis, K. A. Olive and Y. Santoso, *New J. Phys.* **4**, 32 (2002), arXiv:hep-ph/0202110;

H. Baer, C. Balazs, A. Belyaev, J. K. Mizukoshi, X. Tata and Y. Wang, *J. High Energy Phys.* **0207**, 050 (2002), arXiv:hep-ph/0205325; R. Arnowitt and B. Dutta, arXiv:hep-ph/0211417; J. R. Ellis, T. Falk, G. Ganis, K. A. Olive and M. Schmitt, *Phys. Rev. D* **58**, 095002 (1998), arXiv:hep-ph/9801445; J. R. Ellis, T. Falk, G. Ganis and K. A. Olive, *Phys. Rev. D* **62**, 075010 (2000), arXiv:hep-ph/0004169; J. R. Ellis, K. A. Olive, Y. Santoso and V. C. Spanos, *Phys. Lett. B* **565**, 176 (2003), arXiv:hep-ph/0303043; H. Baer and C. Balazs, *J. Cosmol. Astropart. Phys.* **0305**, 006 (2003), arXiv:hep-ph/0303114; A. B. Lahanas and D. V. Nanopoulos, *Phys. Lett. B* **568**, 55 (2003), arXiv:hep-ph/0303130; U. Chattopadhyay, A. Corsetti and P. Nath, *Phys. Rev. D* **68**, 035005 (2003), arXiv:hep-ph/0303201; C. Munoz, *Int. J. Mod. Phys. A* **19**, 3093 (2004), arXiv:hep-ph/0309346; R. Arnowitt, B. Dutta and B. Hu, arXiv: hep-ph/0310103; J. Ellis and K. A. Olive, *Particle Dark Matter*, ed. G. Bertone, pp. 142–163, arXiv:1001.3651 [astro-ph.CO]; J. Ellis and K. A. Olive, *Eur. Phys. J. C* **72**, 2005 (2012), arXiv:1202.3262 [hep-ph]; O. Buchmueller *et al.*, *Eur. Phys. J. C* **74**, 2809 (2014), arXiv:1312.5233 [hep-ph].

10. J. Ellis, F. Luo, K. A. Olive and P. Sandick, *Eur. Phys. J. C* **73**, 2403 (2013), arXiv: 1212.4476 [hep-ph].
11. J. Ellis, J. L. Evans, F. Luo, N. Nagata, K. A. Olive and P. Sandick, *Eur. Phys. J. C* **76**, 8 (2016), arXiv:1509.08838 [hep-ph].
12. O. Buchmueller *et al.*, *Eur. Phys. J. C* **64**, 391 (2009), arXiv:0907.5568 [hep-ph].
13. O. Buchmueller *et al.*, *Eur. Phys. J. C* **74**, 2922 (2014), arXiv:1312.5250 [hep-ph]; O. Buchmueller *et al.*, *Eur. Phys. J. C* **74**, 3212 (2014), arXiv:1408.4060 [hep-ph]; E. A. Bagnaschi *et al.*, *Eur. Phys. J. C* **75**, 500 (2015), arXiv:1508.01173 [hep-ph].
14. CDMS Collab. (Z. Ahmed *et al.*), *Phys. Rev. Lett.* **102**, 011301 (2009), arXiv:0802. 3530 [astro-ph].
15. XENON Collab. (J. Angle *et al.*), *Phys. Rev. Lett.* **100**, 021303 (2008), arXiv:0706. 0039 [astro-ph].
16. XENON100 Collab. (E. Aprile *et al.*), *Phys. Rev. Lett.* **107**, 131302 (2011), arXiv: 1104.2549 [astro-ph.CO].
17. XENON100 Collab. (E. Aprile *et al.*), *Phys. Rev. Lett.* **109**, 181301 (2012), arXiv: 1207.5988 [astro-ph.CO].
18. LUX Collab. (D. S. Akerib *et al.*), *Phys. Rev. Lett.* **112**, 091303 (2014), arXiv:1310. 8214 [astro-ph.CO].
19. D. C. Malling *et al.*, arXiv:1110.0103 [astro-ph.IM].
20. J. Billard, L. Strigari and E. Figueroa-Feliciano, *Phys. Rev. D* **89**, 023524 (2014), arXiv:1307.5458 [hep-ph].
21. P. Cushman *et al.*, arXiv:1310.8327 [hep-ex].
22. WMAP Collab. (G. Hinshaw *et al.*), *Astrophys. J. Suppl.* **208**, 19 (2013), arXiv:1212. 5226 [astro-ph.CO].
23. Planck Collab. (P. A. R. Ade *et al.*), *Astron. Astrophys.* **594**, A13 (2016), arXiv:1502. 01589 [astro-ph.CO].
24. J. Ellis, T. Falk and K. A. Olive, *Phys. Lett. B* **444**, 367 (1998), arXiv:hep-ph/ 9810360; J. Ellis, T. Falk, K. A. Olive and M. Srednicki, *Astropart. Phys.* **13**, 181 (2000) [Erratum: *ibid.* **15**, 413 (2001)], arXiv:hep-ph/9905481.
25. R. Arnowitt, B. Dutta and Y. Santoso, *Nucl. Phys. B* **606**, 59 (2001), arXiv:hep-ph/ 0102181; M. E. Gómez, G. Lazarides and C. Pallis, *Phys. Rev. D* **61**, 123512 (2000), arXiv:hep-ph/9907261; *Phys. Lett. B* **487**, 313 (2000), arXiv:hep-ph/0004028; *Nucl. Phys. B* **638**, 165 (2002), arXiv:hep-ph/0203131; T. Nihei, L. Roszkowski and R. Ruiz de Austri, *J. High Energy Phys.* **0207**, 024 (2002), arXiv:hep-ph/0206266.

26. M. Citron, J. Ellis, F. Luo, J. Marrouche, K. A. Olive and K. J. de Vries, *Phys. Rev. D* **87**, 036012 (2013), arXiv:1212.2886 [hep-ph], and references therein.
27. C. Boehm, A. Djouadi and M. Drees, *Phys. Rev. D* **62**, 035012 (2000), arXiv:hep-ph/9911496; J. R. Ellis, K. A. Olive and Y. Santoso, *Astropart. Phys.* **18**, 395 (2003), arXiv:hep-ph/0112113; J. Edsjo, M. Schelke, P. Ullio and P. Gondolo, *J. Cosmol. Astropart. Phys.* **0304**, 001 (2003), arXiv:hep-ph/0301106; J. L. Diaz-Cruz, J. R. Ellis, K. A. Olive and Y. Santoso, *J. High Energy Phys.* **0705**, 003 (2007), arXiv:hep-ph/0701229; I. Gogoladze, S. Raza and Q. Shafi, *Phys. Lett. B* **706**, 345 (2012), arXiv:1104.3566 [hep-ph]; M. A. Ajaib, T. Li and Q. Shafi, *Phys. Rev. D* **85**, 055021 (2012), arXiv:1111.4467 [hep-ph].
28. J. Ellis, K. A. Olive and J. Zheng, *Eur. Phys. J. C* **74**, 2947 (2014), arXiv:1404.5571 [hep-ph].
29. O. Buchmueller, M. Citron, J. Ellis, S. Guha, J. Marrouche, K. A. Olive, K. de Vries and J. Zheng, *Eur. Phys. J. C* **75**, 469 (2015), arXiv:1505.04702 [hep-ph].
30. J. L. Feng, K. T. Matchev and T. Moroi, *Phys. Rev. Lett.* **84**, 2322 (2000), arXiv:hep-ph/9908309; *Phys. Rev. D* **61**, 075005 (2000), arXiv:hep-ph/9909334; J. L. Feng, K. T. Matchev and F. Wilczek, *Phys. Lett. B* **482**, 388 (2000), arXiv:hep-ph/0004043; H. Baer, T. Krupovnickas, S. Profumo and P. Ullio, *J. High Energy Phys.* **0510**, 020 (2005), arXiv:hep-ph/0507282.
31. J. L. Feng, K. T. Matchev and D. Sanford, *Phys. Rev. D* **85**, 075007 (2012), arXiv:1112.3021 [hep-ph]; P. Draper, J. Feng, P. Kant, S. Profumo and D. Sanford, *Phys. Rev. D* **88**, 015025 (2013), arXiv:1304.1159 [hep-ph].
32. K. Griest and D. Seckel, *Phys. Rev. D* **43**, 3191 (1991).
33. CLEO Collab. (S. Chen *et al.*), *Phys. Rev. Lett.* **87**, 251807 (2001), arXiv:hep-ex/0108032; Belle Collab. (P. Koppenburg *et al.*), *Phys. Rev. Lett.* **93**, 061803 (2004), arXiv:hep-ex/0403004; BaBar Collab. (B. Aubert *et al.*), arXiv:hep-ex/0207076; Heavy Flavor Averaging Group (HFAG) (E. Barberio *et al.*), arXiv:hep-ex/0603003.
34. The Muon g-2 Collab., *Phys. Rev. Lett.* **92**, 161802 (2004), arXiv:hep-ex/0401008; The Muon g-2 Collab. (G. Bennett *et al.*), *Phys. Rev. D* **73**, 072003 (2006), arXiv:hep-ex/0602035.
35. ATLAS Collab. (G. Aad *et al.*), *J. High Energy Phys.* **1409**, 176 (2014), arXiv:1405.7875 [hep-ex]; ATLAS Collab. (G. Aad *et al.*), *J. High Energy Phys.* **1510**, 054 (2015), arXiv:1507.05525 [hep-ex]; CMS Collab. (S. Chatrchyan *et al.*), *J. High Energy Phys.* **1406**, 055 (2014), arXiv:1402.4770 [hep-ex].
36. Super-Kamiokande Collab. (K. Abe *et al.*), *Phys. Rev. D* **90**, 072005 (2014), arXiv:1408.1195 [hep-ex].
37. J. Ellis, J. L. Evans, A. Mustafayev, N. Nagata and K. A. Olive, *Eur. Phys. J. C* **76**, 592 (2016), arXiv:1608.05370 [hep-ph].
38. Super-Kamiokande Collab. (V. Takhistov), arXiv:1605.03235 [hep-ex].
39. M. Ibe, T. Moroi and T. T. Yanagida, *Phys. Lett. B* **644**, 355 (2007), arXiv:hep-ph/0610277; M. Ibe and T. T. Yanagida, *Phys. Lett. B* **709**, 374 (2012), arXiv:1112.2462 [hep-ph]; M. Ibe, S. Matsumoto and T. T. Yanagida, *Phys. Rev. D* **85**, 095011 (2012), arXiv:1202.2253 [hep-ph].
40. M. Dine and D. MacIntire, *Phys. Rev. D* **46**, 2594 (1992), arXiv:hep-ph/9205227; L. Randall and R. Sundrum, *Nucl. Phys. B* **557**, 79 (1999), arXiv:hep-th/9810155; G. F. Giudice, M. A. Luty, H. Murayama and R. Rattazzi, *J. High Energy Phys.* **9812**, 027 (1998), arXiv:hep-ph/9810442; J. A. Bagger, T. Moroi and E. Poppitz, *J. High Energy Phys.* **0004**, 009 (2000), arXiv:hep-th/9911029; P. Binetruy, M. K. Gaillard and B. D. Nelson, *Nucl. Phys. B* **604**, 32 (2001), arXiv:hep-ph/0011081.

41. J. L. Evans, M. Ibe, K. A. Olive and T. T. Yanagida, *Eur. Phys. J. C* **73**, 2468 (2013), arXiv:1302.5346 [hep-ph]; J. L. Evans, K. A. Olive, M. Ibe and T. T. Yanagida, *Eur. Phys. J. C* **73**, 2611 (2013), arXiv:1305.7461 [hep-ph]; J. L. Evans, M. Ibe, K. A. Olive and T. T. Yanagida, *Phys. Rev. D* **91**, 055008 (2015), arXiv:1412.3403 [hep-ph]; J. L. Evans, N. Nagata and K. A. Olive, *Phys. Rev. D* **91**, 055027 (2015), arXiv:1502.00034 [hep-ph].
42. J. D. Wells, arXiv:hep-ph/0306127; N. Arkani-Hamed and S. Dimopoulos, *J. High Energy Phys.* **0506**, 073 (2005), arXiv:hep-th/0405159; G. F. Giudice and A. Romanino, *Nucl. Phys. B* **699**, 65 (2004) [Erratum: *ibid.* **706**, 65 (2005)], arXiv:hep-ph/0406088; N. Arkani-Hamed, S. Dimopoulos, G. F. Giudice and A. Romanino, *Nucl. Phys. B* **709**, 3 (2005), arXiv:hep-ph/0409232; J. D. Wells, *Phys. Rev. D* **71**, 015013 (2005), arXiv:hep-ph/0411041.
43. G. F. Giudice and A. Masiero, *Phys. Lett. B* **206**, 480 (1988).
44. E. Dudas, Y. Mambrini, A. Mustafayev and K. A. Olive, *Eur. Phys. J. C* **72**, 2138 (2012), arXiv:1205.5988 [hep-ph].
45. T. Gherghetta, G. Giudice and J. Wells, *Nucl. Phys. B* **559**, 27 (1999), arXiv:hep-ph/9904378; E. Katz, Y. Shadmi and Y. Shirman, *J. High Energy Phys.* **9908**, 015 (1999), arXiv:hep-ph/9906296; Z. Chacko, M. A. Luty, I. Maksymyk and E. Ponton, *J. High Energy Phys.* **0004**, 001 (2000), arXiv:hep-ph/9905390; J. L. Feng and T. Moroi, *Phys. Rev. D* **61**, 095004 (2000), arXiv:hep-ph/9907319; G. D. Kribs, *Phys. Rev. D* **62**, 015008 (2000), arXiv:hep-ph/9909376; U. Chattopadhyay, D. K. Ghosh and S. Roy, *Phys. Rev. D* **62**, 115001 (2000), arXiv:hep-ph/0006049; I. Jack and D. R. T. Jones, *Phys. Lett. B* **491**, 151 (2000), arXiv:hep-ph/0006116; H. Baer, J. K. Mizukoshi and X. Tata, *Phys. Lett. B* **488**, 367 (2000), arXiv:hep-ph/0007073; A. Datta, A. Kundu and A. Samanta, *Phys. Rev. D* **64**, 095016 (2001), arXiv:hep-ph/0101034; H. Baer, R. Dermisek, S. Rajagopalan and H. Summy, *J. Cosmol. Astropart. Phys.* **1007**, 014 (2010), arXiv:1004.3297 [hep-ph]; A. Arbey, A. Deandrea and A. Tarhini, *J. High Energy Phys.* **1105**, 078 (2011), arXiv:1103.3244 [hep-ph]; B. C. Allanach, T. J. Khoo and K. Sakurai, *J. High Energy Phys.* **1111**, 132 (2011), arXiv:1110.1119 [hep-ph]; A. Arbey, A. Deandrea, F. Mahmoudi and A. Tarhini, *Phys. Rev. D* **87**, 115020 (2013), arXiv:1304.0381 [hep-ph].
46. E. Bagnaschi *et al.*, arXiv:1612.05210 [hep-ph].
47. H. Baer, A. Mustafayev, S. Profumo, A. Belyaev and X. Tata, *Phys. Rev. D* **71**, 095008 (2005), arXiv:hep-ph/0412059; H. Baer, A. Mustafayev, S. Profumo, A. Belyaev and X. Tata, *J. High Energy Phys.* **0507**, 065 (2005), arXiv:hep-ph/0504001.
48. J. R. Ellis, K. A. Olive and P. Sandick, *Phys. Rev. D* **78**, 075012 (2008), arXiv:0805.2343 [hep-ph].
49. J. Ellis, K. Olive and Y. Santoso, *Phys. Lett. B* **539**, 107 (2002), arXiv:hep-ph/0204192; J. R. Ellis, T. Falk, K. A. Olive and Y. Santoso, *Nucl. Phys. B* **652**, 259 (2003), arXiv:hep-ph/0210205.
50. J. R. Ellis, K. A. Olive and P. Sandick, *Phys. Lett. B* **642**, 389 (2006), arXiv:hep-ph/0607002; J. R. Ellis, K. A. Olive and P. Sandick, *J. High Energy Phys.* **0706**, 079 (2007), arXiv:0704.3446 [hep-ph]; J. R. Ellis, K. A. Olive and P. Sandick, *J. High Energy Phys.* **0808**, 013 (2008), arXiv:0801.1651 [hep-ph].
51. L. Calibbi, Y. Mambrini and S. K. Vempati, *J. High Energy Phys.* **0709**, 081 (2007), arXiv:0704.3518 [hep-ph]; L. Calibbi, A. Faccia, A. Masiero and S. K. Vempati, *Phys. Rev. D* **74**, 116002 (2006), arXiv:hep-ph/0605139; E. Carquin, J. Ellis, M. E. Gomez, S. Lola and J. Rodriguez-Quintero, *J. High Energy Phys.* **0905**, 026 (2009), arXiv:0812.4243 [hep-ph]; J. Ellis, A. Mustafayev and K. A. Olive, *Eur. Phys. J. C* **69**, 201 (2010), arXiv:1003.3677 [hep-ph]; J. Ellis, A. Mustafayev and K. A. Olive, *Eur.*

Phys. J. C **69**, 219 (2010), arXiv:1004.5399 [hep-ph]; J. Ellis, A. Mustafayev and K. A. Olive, *Eur. Phys. J. C* **71**, 1689 (2011), arXiv:1103.5140 [hep-ph]; J. Ellis, J. L. Evans, N. Nagata, D. V. Nanopoulos and K. A. Olive, arXiv:1702.00379 [hep-ph].
52. K. A. Olive and M. Srednicki, *Phys. Lett. B* **230**, 78 (1989); K. A. Olive and M. Srednicki, *Nucl. Phys. B* **355**, 208 (1991).
53. H. Georgi, *AIP Conf. Proc.* **23**, 575 (1975); H. Fritzsch and P. Minkowski, *Ann. Phys.* **93**, 193 (1975); M. S. Chanowitz, J. R. Ellis and M. K. Gaillard, *Nucl. Phys. B* **128**, 506 (1977); H. Georgi and D. V. Nanopoulos, *Nucl. Phys. B* **155**, 52 (1979).
54. H. Georgi and D. V. Nanopoulos, *Nucl. Phys. B* **159**, 16 (1979); C. E. Vayonakis, *Phys. Lett. B* **82**, 224 (1979) [*Phys. Lett. B* **83**, 421 (1979)].
55. A. Masiero, *Phys. Lett. B* **93**, 295 (1980); Q. Shafi, M. Sondermann and C. Wetterich, *Phys. Lett. B* **92**, 304 (1980); F. del Aguila and L. E. Ibanez, *Nucl. Phys. B* **177**, 60 (1981); R. N. Mohapatra and G. Senjanovic, *Phys. Rev. D* **27**, 1601 (1983); M. Fukugita and T. Yanagida, *Physics and Astrophysics of Neutrinos*, eds. M. Fukugita and A. Suzuki, Kyoto University, YITP-K-1050, pp. 1–248.
56. Y. Mambrini, K. A. Olive, J. Quevillon and B. Zaldivar, *Phys. Rev. Lett.* **110**, 241306 (2013), arXiv:1302.4438 [hep-ph].
57. Y. Mambrini, N. Nagata, K. A. Olive, J. Quevillon and J. Zheng, *Phys. Rev. D* **91**, 095010 (2015), arXiv:1502.06929 [hep-ph].
58. N. Nagata, K. A. Olive and J. Zheng, *J. High Energy Phys.* **1510**, 193 (2015), arXiv: 1509.00809 [hep-ph].
59. Y. Mambrini, N. Nagata, K. A. Olive and J. Zheng, arXiv:1602.05583 [hep-ph].
60. T. W. B. Kibble, G. Lazarides and Q. Shafi, *Phys. Lett. B* **113**, 237 (1982); L. M. Krauss and F. Wilczek, *Phys. Rev. Lett.* **62**, 1221 (1989); L. E. Ibanez and G. G. Ross, *Phys. Lett. B* **260**, 291 (1991); L. E. Ibanez and G. G. Ross, *Nucl. Phys. B* **368**, 3 (1992); S. P. Martin, *Phys. Rev. D* **46**, 2769 (1992), arXiv:hep-ph/9207218; M. De Montigny and M. Masip, *Phys. Rev. D* **49**, 3734 (1994), arXiv:hep-ph/9309312.
61. M. Kadastik, K. Kannike and M. Raidal, *Phys. Rev. D* **80**, 085020 (2009) [Erratum: *ibid.* **81**, 029903 (2010)], arXiv:0907.1894 [hep-ph]; M. Kadastik, K. Kannike and M. Raidal, *Phys. Rev. D* **81**, 015002 (2010), arXiv:0903.2475 [hep-ph]; M. Frigerio and T. Hambye, *Phys. Rev. D* **81**, 075002 (2010), arXiv:0912.1545 [hep-ph].
62. J. L. Evans, N. Nagata, K. A. Olive and J. Zheng, *J. High Energy Phys.* **1602**, 120 (2016), doi:10.1007/JHEP02(2016)120, arXiv:1512.02184 [hep-ph].
63. P. Minkowski, *Phys. Lett. B* **67**, 421 (1977); M. Gell-Mann, P. Ramond and R. Slansky, *Supergravity*, eds. D. Freedman and P. Van Nieuwenhuizen (North Holland, Amsterdam, 1979), pp. 315–321, ISBN 044485438x; T. Yanagida, *Proc. Workshop on the Unified Theory and the Baryon Number of the Universe*, eds. O. Sawada and S. Sugamoto (1979); R. N. Mohapatra and G. Senjanovic, *Phys. Rev. Lett.* **44**, 912 (1980); J. Schechter and J. W. F. Valle, *Phys. Rev. D* **22**, 2227 (1980); J. Schechter and J. W. F. Valle, *Phys. Rev. D* **25**, 774 (1982).
64. M. Shiozawa, talk presented at *TAUP 2013*, September 8–13, Asilomar, CA, USA.
65. K. S. Babu *et al.*, arXiv:1311.5285 [hep-ph].
66. V. Silveira and A. Zee, *Phys. Lett. B* **161**, 136 (1985); J. McDonald, *Phys. Rev. D* **50**, 3637 (1994), arXiv:hep-ph/0702143; C.P. Burgess, M. Pospelov and T. ter Veldhuis, *Nucl. Phys. B* **619**, 709 (2001), arXiv:hep-ph/0011335; H. Davoudiasl, R. Kitano, T. Li and H. Murayama, *Phys. Lett. B* **609**, 117 (2005), arXiv:hep-ph/0405097.
67. N. G. Deshpande and E. Ma, *Phys. Rev. D* **18**, 2574 (1978); E. Ma, *Phys. Rev. D* **73**, 077301 (2006), arXiv:hep-ph/0601225; R. Barbieri, L. J. Hall and V. S. Rychkov, *Phys. Rev. D* **74**, 015007 (2006), arXiv:hep-ph/0603188; L. Lopez Honorez, E. Nezri, J. F. Oliver and M. H. G. Tytgat, *J. Cosmol. Astropart. Phys.* **0702**, 028 (2007), arXiv:hep-ph/0612275.

68. A. Arhrib, Y. L. S. Tsai, Q. Yuan and T. C. Yuan, *J. Cosmol. Astropart. Phys.* **1406**, 030 (2014), arXiv:1310.0358 [hep-ph]; A. Ilnicka, M. Krawczyk and T. Robens, *Phys. Rev. D* **93**, 055026 (2016), doi:10.1103/PhysRevD.93.055026, arXiv:1508.01671 [hep-ph].

Weighing the Black Hole via Quasi-local Energy

Yuan K. Ha
Department of Physics, Temple University, Philadelphia, PA 19122, USA
yuanha@temple.edu

We set to weigh the black holes at their event horizons in various spacetimes and obtain masses which are substantially higher than their asymptotic values. In each case, the horizon mass of a Schwarzschild, Reissner–Nordström, or Kerr black hole is found to be twice the irreducible mass observed at infinity. The irreducible mass does not contain electrostatic or rotational energy, leading to the inescapable conclusion that particles with electric charges and spins cannot exist inside a black hole. This is proposed as the External Energy Paradigm. A higher mass at the event horizon and its neighborhood is obligatory for the release of gravitational waves in binary black hole merging. We describe how these horizon mass values are obtained in the quasi-local energy approach and applied to the black holes of the first gravitational waves GW150914.

Keywords: Gravitational waves; black holes; quasi-local energy; horizon mass.

1. Black Hole Theorems

Black holes are natural outcome of solutions to Einstein's equation. Since the discovery of the first gravitational waves from binary black hole merging in 2015,[1] black holes are now real astrophysical bodies. They are as legitimate as the elementary particles whose existence is confirmed indirectly. They may be abundant in the Universe and their properties can be investigated from the gravitational waves emitted in binary black hole merging.

A number of important theorems on black holes have been established between 1965 and 2005. They provide the conceptual framework and predict the properties of classical black holes in terms of temperature, entropy, irreversibility, thermodynamics, as well as energy conditions. They are known as

(1) Singularity Theorem (1965),[2]
(2) Area Non-decrease Theorem (1972),[3]
(3) Uniqueness Theorem (1975),[4]
(4) Positive Energy Theorem (1983),[5]
(5) Horizon Mass Theorem (2005).[6]

The first four theorems listed above have been extensively discussed in general relativity for many years and we take for granted that their contents are well known to general relativists. It is the last theorem, the Horizon Mass Theorem, which we shall discuss in this report and apply it to the black holes of the first gravitational waves GW150914.

2. Quasi-local Energy

The Horizon Mass Theorem is the final outcome of the quasi-local energy approach[7] applied to black holes. The quasi-local energy gives the total energy within a spatially bounded region instead of defining locally the energy density for a gravitational system. It is obtained from a Hamiltonian–Jacobi analysis of the Hilbert action in general relativity and it is uniquely suited for investigating the dynamics of the gravitational field.[8] The mass of a black hole can be found anywhere by calculating the total energy contained in a Gaussian surface enclosing the black hole at a given coordinate distance. The usual mass of a black hole is the asymptotic mass seen by an observer at infinity.

A black hole has the strongest gravitational potential energy of any gravitational system. This energy exists outside the black hole. An observer at a distance sees the total of the constituent mass contained at the horizon and the intermediary potential energy. Since gravitational potential energy is always negative and extends throughout all space, the closer an observer gets to the black hole, the less gravitational potential energy the observer will see. Thus, the mass of the black hole increases as the observer gets near the horizon. This is a unique situation for black holes since for any other physical object, the gravitational potential energy is far insignificant compared to its mass and therefore the mass appears to be the same at all distances of observation.

The Brown and York expression for quasi-local energy is given in terms of the total mean curvature of a surface bounding a volume for a gravitational system in four-dimensional spacetime. It is given in the form of an integral,

$$E = \frac{c^4}{8\pi G} \int_{^2B} d^2x \sqrt{\sigma}(k - k^0) \,, \tag{1}$$

where σ is the determinant of the metric defined on the two-dimensional surface ^2B; is the trace of extrinsic curvature of the surface and k^0, the trace of curvature of a reference space. For asymptotically flat reference spacetime, k^0 is zero.

3. Horizon Mass Theorem

The Horizon Mass Theorem can be stated as follows.

Theorem. *For all black holes; neutral, charged or rotating, the horizon mass is always twice the irreducible mass observed at infinity.*

In notation, it has the simple form,

$$M_{\text{horizon}} = 2M_{\text{irr}}\,, \tag{2}$$

where M_{irr} is the irreducible mass. The derivation of this theorem is given fully in Ref. 6. The Horizon Mass Theorem relates the mass of a black hole at the event horizon to its irreducible mass. It is an exact result obtained only with the knowledge of the spacetime metrics of Schwarzschild, Reissner–Nordström, and Kerr without further assumption. It is a new addition to the previous theorems on classical black holes.

In order to understand the Horizon Mass Theorem, it is necessary to introduce the various mass terms involved.

(1) The *asymptotic mass* is the mass of a neutral, charged or rotating black hole including electrostatic and rotational energy. It is the mass observed at infinity used in the various spacetime metrics.
(2) The *horizon mass* is the mass which cannot escape from the horizon of a neutral, charged or rotating black hole. It is the mass of the black hole observed at the horizon.
(3) The *irreducible mass* is the final mass of a charged or rotating black hole when its charge or angular momentum is removed by adding external particles to the black hole. It is the mass observed at infinity.

The Horizon Mass Theorem is remarkable in that the mass contained at the event horizon depends only on the irreducible mass of the black hole. The irreducible mass does not contain electrostatic or rotational energy. This leads to the surprising conclusion that the electrostatic and rotational energy exist only outside the black hole. They are all external quantities. An asymptotic observer investigating a charged or rotating black hole is in fact exploring a Schwarzschild black hole with external energies in between.

4. External Energy Paradigm

There are profound implications which follow from the Horizon Mass Theorem. Since all electric field lines terminate at electric charges and electrostatic energy is external, this indicates that electrical particles cannot exist inside a black hole. They can only stay at the surface. Similarly, since rotational energy is external, any particle with angular momentum also cannot exist inside the black hole and must stay outside, as required by the Horizon Mass Theorem. Together, this implies that all elementary particles possessing charges and spins can only stay outside the horizon. If a black hole is formed from the collapse of a star made of ordinary matter, the result will be a hollow and thin spherical shell with all constituent mass at the horizon. This is a radical view of the black hole and it follows inescapably from the property of the irreducible mass. We are thus led to introduce a new paradigm for black holes to be called the External Energy Paradigm. *All energies of a black hole are external quantities. Matter particles are forbidden inside a black hole and can only stay outside or at the horizon.* These energies include constituent mass, gravitational energy, heat energy, electrostatic energy and rotational energy. It explains naturally why the entropy of a black hole is proportional to the area and not to the volume because matter particles are all at the surface.

5. Schwarzschild Black Hole

The total energy contained in a sphere enclosing the black hole at a coordinate distance r in the quasi-local energy approach is given by the expression[6,7,9]

$$E(r) = \frac{rc^4}{G}\left[1 - \sqrt{1 - \frac{2GM}{rc^2}}\right], \tag{3}$$

where M is the mass of the black hole observed at infinity, c is the speed of light and G is the gravitational constant. At the horizon, the Schwarzschild radius is $r = R_S = 2GM/c^2$. Evaluating the expression in Eq. (3), we find that the metric coefficient $g_{00} = (1 - 2GM/rc^2)^{1/2}$ vanishes identically and the horizon energy is therefore,

$$E(r) = \left(\frac{2GM}{c^2}\right)\frac{c^4}{G} = 2Mc^2\,. \tag{4}$$

The horizon mass of the Schwarzschild black hole is simply twice the asymptotic mass M observed at infinity. The negative gravitational energy outside the black hole has a magnitude as great as the asymptotic mass.

Equation (3) can be used to evaluate the mass seen by an observer at any distance r. We show some particular values in Table 1.

From the listed values, it is seen that 90% of the negative potential energy lies within a distance of two Schwarzschild radii outside the horizon, i.e. $R_S < r < 3R_S$.

Table 1. Mass of black hole observed at a distance r.

Coordinate r in R_S	Mass in $M_\infty = M$
1	2.000
2	1.172
3	1.101
4	1.072
5	1.056
6	1.046
7	1.039
8	1.033
9	1.029
10	1.026
100	1.003
∞	1.000

At a distance of $r = 100R_S$, the mass is seen to be only 0.3% higher than the asymptotic value M. An observer at that location is approaching a near flat spacetime.

6. Reissner–Nordström Black Hole

We investigate next the Reissner–Nordström black hole in the quasi-local energy approach. The total energy of a charged black hole contained within a radius at coordinate r is now given by[6]

$$E(r) = \frac{rc^4}{G}\left[1 - \sqrt{1 - \frac{2GM}{rc^2} + \frac{GQ^2}{r^2c^4}}\right]. \tag{5}$$

Here, M is the mass of the black hole including electrostatic energy observed at infinity and Q is the electric charge. At the horizon radius,

$$r_+ = \frac{GM}{c^2} + \frac{GM}{c^2}\sqrt{1 - \frac{Q^2}{GM^2}}, \tag{6}$$

the metric coefficient g_{00} given by the square root in Eq. (5) also vanishes and the horizon energy becomes

$$E(r_+) = \frac{r_+c^4}{G} = Mc^2 + Mc^2\sqrt{1 - \frac{Q^2}{GM^2}}. \tag{7}$$

For the Reissner–Nordström black hole, the irreducible mass which is obtained when the charge is removed by adding oppositely charged particles has the expression

$$M_{\text{irr}} = \frac{M}{2} + \frac{M}{2}\sqrt{1 - \frac{Q^2}{GM^2}}. \tag{8}$$

Combining Eqs (7) and (8), we find the horizon energy to be exactly twice the irreducible energy

$$E(r_+) = 2M_{\text{irr}}c^2, \tag{9}$$

which depends only on the mass of the black hole when the charge is neutralized.

7. Kerr Black Hole

The rotating black hole is considerably more complicated to handle in the quasi-local energy approach because one is comparing a rotating spacetime with a fixed spacetime. It is therefore not possible to give an exact analytical expression as in the previous two cases. An approximate energy expression[10] is available for a slowly rotating black hole with angular momentum J and angular momentum parameter $\alpha = J/Mc$, where $0 < \alpha \ll 1$,

$$\begin{aligned} E(r) = \frac{rc^4}{G}\left[1 - \sqrt{1 - \frac{2GM}{rc^2} + \frac{\alpha^2}{r^2}}\right] + \frac{\alpha^2c^4}{6rG}\left[2 + \frac{2GM}{rc^2}\right. \\ \left. + \left(1 + \frac{2GM}{rc^2}\right)\sqrt{1 - \frac{2GM}{rc^2} + \frac{\alpha^2}{r^2}}\right] + \cdots. \end{aligned} \tag{10}$$

Again, with the horizon radius of the Kerr black hole

$$r_+ = \frac{GM}{c^2} + \sqrt{\frac{G^2M^2}{c^4} - \frac{J^2}{M^2c^2}} \tag{11}$$

and the definition of the irreducible mass

$$M_{\rm irr}^2 = \frac{M^2}{2} + \frac{M^2}{2}\sqrt{1 - \frac{J^2c^2}{G^2M^4}}, \tag{12}$$

we arrive at a very good approximate relation for the horizon energy

$$E(r_+) \approx 2M_{\rm irr} + O(\alpha^2). \tag{13}$$

For general and fast rotations, the energy can be accurately obtained by numerical evaluation in the teleparallel equivalent of general relativity.[11] The result shows almost perfectly that the horizon mass is twice the irreducible mass. For an exact and impeccable relationship, we have to employ a formula known for the area of a Kerr black hole valid for all rotations,[12] i.e.

$$A = 4\pi(r_+^2 + \alpha^2) = \frac{16\pi G^2 M_{\rm irr}^2}{c^4}. \tag{14}$$

This area is exactly the same as that of a Schwarzschild black hole with an asymptotic mass$M_{\rm irr}$,

$$A = 4\pi R_S^2 = 4\pi\left(\frac{2GM_{\rm irr}}{c^2}\right)^2. \tag{15}$$

As shown earlier, the horizon mass of such a Schwarzschild black hole is twice the irreducible mass. We have therefore established the Horizon Mass Theorem for all black hole cases. The profound consequence of this theorem is that elementary particles with charges or spins cannot exist inside a black hole.

8. Black Holes of GW150914

The discovery of gravitational waves GW150914 by LIGO confirmed the existence of two black holes in a binary system. They merged to form a single black hole with the release of gravitational energy. We realize that the energy of the gravitational waves comes from outside the black holes and not from their interiors. The waves are generated predominantly from near the horizon and they are gravitationally redshifted as they propagate to infinity. The horizon energy therefore becomes important. Without a higher mass at the event horizon and its neighborhood, there can be no gravitational waves emitted in black hole merging.

The two black holes of GW150914 are rotating black holes. For a Kerr black hole, rotation necessarily contributes to the overall mass observed at infinity. To find the irreducible mass of the Kerr black hole, we need to know the dimensionless spin parameter a, which is the ratio of the angular momentum J to the maximum possible angular momentum, i.e.

$$a = \frac{J}{\left(\frac{GM^2}{c}\right)} = \frac{Jc}{GM^2}. \tag{16}$$

The irreducible mass, from Eq. (12), is then given by

$$M_{\mathrm{irr}} = \left[\frac{M^2}{2} + \frac{M^2}{2}\sqrt{1-a^2}\right]^{\frac{1}{2}} \tag{17}$$

and the horizon mass can be found as twice the irreducible mass,

$$M(r_+) = 2M_{\mathrm{irr}} = \left[2M^2 + 2M^2\sqrt{1-a^2}\right]^{\frac{1}{2}} . \tag{18}$$

For the black holes of GW150914,[13] the primary black hole has a mass of 36 M_{Sun} and an average model spin parameter $a = 0.32$. The secondary black hole has a mass of 29 M_{Sun} and average model spin parameter $a = 0.44$. The final black hole has a mass of 62 M_{Sun} and a spin parameter $a = 0.67$. Accordingly, 3 M_{Sun} of energy is released as gravitational waves to infinity as reported by LIGO, i.e.

$$36\, M_{\mathrm{Sun}} + 29\, M_{\mathrm{Sun}} = 62\, M_{\mathrm{Sun}} + 3\, M_{\mathrm{Sun}} . \tag{19}$$

However, this 3 M_{Sun} wave energy has been significantly redshifted and Eq. (19) does not account for the missing energy. Without additional source of energy, the waves cannot propagate away from the deep potential of the black hole. To understand the energy of the waves at the source, we need to know the mass of the black holes at the event horizon. For the primary black hole, the horizon mass is found to be 71 M_{Sun} and for the secondary black hole, a horizon mass of 57 M_{Sun}. The final black hole has a horizon mass of 116 M_{Sun}. In an ideal merging, the energy at the horizon would follow the equation

$$71\, M_{\mathrm{Sun}} + 57\, M_{\mathrm{Sun}} = 116\, M_{\mathrm{Sun}} + (3\, M_{\mathrm{Sun}} + 3\, M_{\mathrm{Sun}}) + 6\, M_{\mathrm{Sun}} . \tag{20}$$

In this account, the energy for the redshift is now available. Analysis of a mass removed from the surface of a black hole shows that the energy required has the same magnitude as the energy of the waves observed at infinity.[14] The total energy required to release 3 M_{Sun} of wave energy to infinity is therefore 3 M_{Sun} + 3 M_{Sun} = 6 M_{Sun}. The remaining 6 M_{Sun} of the energy is for uncertainties in LIGO data. These mass values are additional properties for the binary black hole merger of GW150914.

We may further provide the rotational energy of the black holes by comparing the asymptotic mass with the irreducible mass. The rotational mass of the primary black hole is 36 M_{Sun} − 35.5 M_{Sun} = 0.5 M_{Sun}, while that of the secondary black hole is 29 M_{Sun} − 28.25 M_{Sun} = 0.75 M_{Sun}. The final black hole has a higher rotational mass that is 62 M_{Sun} − 58 M_{Sun} = 4 M_{Sun}. The initial black holes in a binary system generally have different spin orientations. Thus most of the rotational energy of the final black hole comes from the orbiting energy of the binary system.

9. Conclusion

The detection of the first gravitational waves GW150914 shows the need to understand the energy of the black hole at the event horizon. This was first emphasized by the author in 2003 in the paper "The Gravitational Energy of a Black Hole" [14] before the success of LIGO was certain. At the end, there was the remark.

> *Therefore, in detecting any gravitational signals from a black hole collision such as that proposed in the LIGO project, any conclusion about the strength of the signals near its source should be based on the black hole energy formula.*

The quasi-local energy approach and the Horizon Mass Theorem are indispensable tools in the latest development of general relativity.

References

1. B. P. Abbott *et al.*, *Phys. Rev. Lett.* **116**, 061102 (2016).
2. R. Penrose, *Phys. Rev. Lett.* **14**, 57 (1965).
3. S. W. Hawking, *Commun. Math. Phys.* **25**, 152 (1972).
4. D. C. Robinson, *Phys. Rev. Lett.* **34**, 905 (1975).
5. R. Schoen and S. T. Yau, *Commun. Math. Phys.* **65**, 45 (1979).
6. Y. K. Ha, *Int. J. Mod. Phys. D* **14**, 2219 (2005).
7. J. W. Brown and J. W. York, Jr., *Phys. Rev. D* **47**, 1407 (1993).
8. M. T. Wang and S. T. Yau, *Phys. Rev. Lett.* **102**, 021101 (2009).
9. J. W. Maluf, *J. Math. Phys.* **36**, 4242 (1995).
10. E. A. Martinez, *Phys. Rev. D* **50**, 4920 (1994).
11. J. W. Maluf, E. F. Martins and A. Kneip, *J. Math. Phys.* **37**, 6302 (1996).
12. D. Christodoulou and R. Ruffini, *Phys. Rev. D* **4**, 3552 (1971).
13. B. P. Abbott *et al.*, *Phys. Rev. Lett.* **116**, 241102 (2016).
14. Y. K. Ha, *Gen. Relat. Gravit.* **35**, 2045 (2003).

Dark Matter and Baryogenesis from non-Abelian Gauged Lepton Number*

Bartosz Fornal

Department of Physics, University of California, San Diego,
9500 Gilman Drive, La Jolla, CA 92093, USA
bfornal@ucsd.edu

A simple model is constructed based on the gauge symmetry $\mathrm{SU}(3)_c \times \mathrm{SU}(2)_L \times \mathrm{U}(1)_Y \times \mathrm{SU}(2)_\ell$, with only the leptons transforming nontrivially under $\mathrm{SU}(2)_\ell$. The extended symmetry is broken down to the Standard Model gauge group at TeV-scale energies. We show that this model provides a mechanism for baryogenesis via leptogenesis in which the lepton number asymmetry is generated by $\mathrm{SU}(2)_\ell$ instantons. The theory also contains a dark matter candidate — the $\mathrm{SU}(2)_\ell$ partner of the right-handed neutrino.

Keywords: Dark matter; lepton number violation; baryogenesis.

1. Introduction

The Standard Model of elementary particle physics provides an extremely accurate description of Nature at the most fundamental level. Despite its remarkable successes, it explains only 5% of the Universe, while the remaining 95% is attributed to the mysterious dark matter and dark energy. In addition, the Standard Model itself has its own shortcomings: the inability to generate the observable matter–antimatter asymmetry of the Universe, the hierarchy problem, massless neutrinos, unknown origin of flavor, and many more. Although a plethora of models dealing with those issues have been constructed, it is still an open question which of those theories, if any, provides the correct or at least partially correct description of Nature at higher energies. We simply need more experimental data to find this out. In the meantime, further systematizing and rethinking of our model building efforts is definitely required.

The Standard Model is based on the gauge group $\mathrm{SU}(3)_c \times \mathrm{SU}(2)_L \times \mathrm{U}(1)_Y$.[2–6] Apart from this local symmetry, it also has two accidental global

*Based on Ref. 1.

symmetries: baryon number and lepton number. One might wonder whether those are just residual symmetries left over from the breaking of a more fundamental extended gauge symmetry. Efforts of gauging baryon and lepton number were carried in the past,[7–12] but only the models constructed recently[13–17] are experimentally viable. In theories of this type, gauge coupling unification does not occur naturally and so far only partial unification has been achieved.[18–20]

Nevertheless, simple extensions of the Standard Model gauge group provide a good playground for testing various approaches to the dark matter and baryogenesis puzzles. Here, we will discuss one of such extensions containing a dark matter candidate and offering a mechanism for producing a lepton asymmetry, which ultimately can explain the matter–antimatter asymmetry of the Universe.

2. The Model

The theory we propose is based on the gauge group:

$$\mathrm{SU}(3)_c \times \mathrm{SU}(2)_L \times \mathrm{U}(1)_Y \times \mathrm{SU}(2)_\ell \,. \tag{1}$$

2.1. *Fermionic sector*

The Standard Model quarks are singlets under $\mathrm{SU}(2)_\ell$, whereas the left-handed lepton doublet l_L and the right-handed electron e_R are the upper components of $\mathrm{SU}(2)_\ell$ doublets,

$$\hat{l}_L = \begin{pmatrix} l \\ \tilde{l} \end{pmatrix}_L , \qquad \hat{e}_R = \begin{pmatrix} e \\ \tilde{e} \end{pmatrix}_R , \tag{2}$$

where the lower components are the new partner fields $\tilde{l}_L$ and $\tilde{e}_R$, respectively. To cancel the gauge anomalies involving $\mathrm{SU}(2)_\ell$ one requires an extra $\mathrm{SU}(2)_\ell$ doublet of Standard Model singlet fields,

$$\hat{\nu}_R = \begin{pmatrix} \nu \\ \tilde{\nu} \end{pmatrix}_R . \tag{3}$$

The remaining anomalies involving just the Standard Model gauge groups are canceled by introducing new $\mathrm{SU}(2)_\ell$ singlet fields:

$$l'_R \,, \qquad e'_L \,, \qquad \nu'_L \,. \tag{4}$$

The particle content of the model along with the quantum numbers of the fields is shown in Table 1. The Standard Model quarks were not included since they transform trivially under $\mathrm{SU}(2)_\ell$.

2.2. *Higgs and gauge sector*

Although breaking of the extended gauge group down to the Standard Model can be achieved with just one new $\mathrm{SU}(2)_\ell$ doublet Higgs, for reasons discussed later we

Table 1. New field representations in the model.

Field	$SU(2)_\ell$	$SU(2)_L$	$U(1)_Y$
$\hat{l}_L$	2	2	$-\frac{1}{2}$
$\hat{e}_R$	2	1	-1
$\hat{\nu}_R$	2	1	0
l'_R	1	2	$-\frac{1}{2}$
e'_L	1	1	-1
ν'_L	1	1	0
$\hat{\Phi}_{1,2}$	2	1	0

introduce two new Higgs fields $\hat{\Phi}_{1,2}$ and assume that one of the vacuum expectation values (vevs) is much larger than the other, $v_1 \gg v_2$. This can be easily engineered by choosing appropriate values for the parameters in the scalar potential:

$$\begin{aligned} V(\Phi_1, \Phi_2) = &\, m_1^2|\hat{\Phi}_1|^2 + m_2^2|\hat{\Phi}_2|^2 + (m_{12}^2\hat{\Phi}_1^\dagger\hat{\Phi}_2 + \text{h.c.}) \\ &+ \lambda_1|\hat{\Phi}_1|^4 + \lambda_2|\hat{\Phi}_2|^4 + \lambda_3|\hat{\Phi}_1|^2|\hat{\Phi}_2|^2 + \lambda_4|\hat{\Phi}_1^\dagger\hat{\Phi}_2|^2 \\ &+ [\tilde{\lambda}_5\hat{\Phi}_1^\dagger\hat{\Phi}_2|\hat{\Phi}_1|^2 + \tilde{\lambda}_6\hat{\Phi}_1^\dagger\hat{\Phi}_2|\hat{\Phi}_2|^2 + \tilde{\lambda}_7(\hat{\Phi}_1^\dagger\hat{\Phi}_2)^2 + \text{h.c.}] \,, \end{aligned} \tag{5}$$

where we neglected terms involving the Standard Model Higgs field. After $\hat{\Phi}_{1,2}$ develop vevs,

$$\langle\hat{\Phi}_i\rangle = \frac{1}{\sqrt{2}}\begin{pmatrix} 0 \\ v_i \end{pmatrix}, \tag{6}$$

the $SU(3)_c \times SU(2)_L \times U(1)_Y \times SU(2)_\ell$ symmetry is broken down directly to the Standard Model gauge group.

Apart from the new Higgs particles, the theory contains three new vector gauge bosons:

$$Z', \qquad W'_+, \qquad W'_-, \tag{7}$$

which do not mix with the Standard Model electroweak gauge bosons.

2.3. *Particle masses*

The Yukawa part of the Lagrangian is given by:

$$\begin{aligned} \mathcal{L}_Y = &\sum_i [Y_l^{ab}\bar{\hat{l}}_L^a\hat{\Phi}_i l_R^{\prime b} + Y_e^{ab}\bar{\hat{e}}_R^a\hat{\Phi}_i e_L^{\prime b} + Y_\nu^{ab}\bar{\hat{\nu}}_R^a\hat{\Phi}_i\nu_L^{\prime b}] \\ &+ y_e^{ab}\bar{\hat{l}}_L^a H\hat{e}_R^b + y_\nu^{ab}\bar{\hat{l}}_L^a\tilde{H}\hat{\nu}_R^b + y_e^{\prime ab}\bar{l}_R^{\prime a}He'^b{}_L + y_\nu^{\prime ab}\bar{l}_R^{\prime a}\tilde{H}\nu_L^{\prime b} + \text{h.c.} \,, \end{aligned} \tag{8}$$

where $a, b = 1, 2, 3$ are flavor indices. After $SU(2)_\ell$ symmetry breaking, the Yukawa matrices Y_l, Y_e and Y_ν lead to vector-like masses for all new fermions in the theory. The Yukawa matrices y_e and y_ν produce the usual Standard Model lepton masses and, along with y'_e and y'_ν, contribute also to lepton partner masses. The fermionic mass matrix is given by:

$$\frac{1}{\sqrt{2}} \begin{pmatrix} \bar{\tilde{\nu}}_L & \bar{\nu}'_L \end{pmatrix} \begin{pmatrix} Y_l v_\ell & y_\nu v \\ y'^{\dagger}_\nu v & Y_\nu^\dagger v_\ell \end{pmatrix} \begin{pmatrix} \nu'_R \\ \tilde{\nu}_R \end{pmatrix}$$
$$+ \frac{1}{\sqrt{2}} \begin{pmatrix} \bar{\tilde{e}}_L & \bar{e}'_L \end{pmatrix} \begin{pmatrix} Y_l v_\ell & y_e v \\ y'^{\dagger}_e v & Y_e^\dagger v_\ell \end{pmatrix} \begin{pmatrix} e'_R \\ \tilde{e}_R \end{pmatrix} + \text{h.c.}\,, \quad (9)$$

where $v_\ell = \sqrt{v_1^2 + v_2^2}$ and v is the Standard Model Higgs vev. The off-diagonal elements are due to the Yukawa terms involving the Standard Model Higgs and introduce mixing between the electroweak singlets and doublets. We assume $Y_{l,e,\nu} v_\ell \gg y_{e,\nu} v, y'_{e,\nu} v$, which is a phenomenologically natural assumption and frees the model from electroweak precision data constraints.

The mass eigenstates consist of six electrically neutral and six electrically charged states. As shown below, after $SU(2)_\ell$ breaking there remains a residual global $U(1)_\ell$ symmetry which prevents the new particles from decaying to solely Standard Model states. Therefore, if the lightest of the mass eigenstates is electrically neutral, it becomes a natural candidate for dark matter. This implies that the dark matter particle in the model is the $SU(2)_\ell$ partner of the right-handed neutrino, which after electroweak symmetry breaking receives a small admixture from the electroweak doublets:

$$\begin{aligned} \chi_L &= \nu'_L + \epsilon \tilde{\nu}_L \,, \\ \chi_R &= \tilde{\nu}_R + \epsilon \nu'_R \,, \end{aligned} \quad (10)$$

where $\tilde{\nu}_L$ and ν'_R are the upper components of the electroweak doublets $\tilde{l}_L$ and l'_R, respectively, and $\epsilon \sim y_\nu v/(Y_\nu v_\ell) \sim y'_\nu v/(Y_\nu v_\ell) \ll 1$.

The Higgs spectrum of the theory is that of a generic two-Higgs doublet model. There are five physical scalar/pseudoscalar fields remaining after $SU(2)_\ell$ breaking. They are mixtures of the original CP-even and CP-odd components of $\hat{\Phi}_{1,2}$ and their masses depend on the choice of parameter values in the scalar potential (5).

Regarding the gauge sector, since there is no mixing between $SU(2)_\ell$ and the other gauge groups, after $SU(2)_\ell$ breaking the new vector gauge bosons develop equal masses,

$$m_{Z', W'_\pm} \simeq \frac{1}{2} g_\ell v_\ell \,, \quad (11)$$

where g_ℓ is the $SU(2)_\ell$ gauge coupling.

Table 2. Charges under exact and approximate global symmetries.

Field	$U(1)_\ell$	$U(1)_L$	$U(1)_\chi$
$\hat{l}_L$	1	1	0
$\hat{e}_R$	1	1	0
$\hat{\nu}_R$	1	0	1
l'_R	1	1	0
e'_L	1	1	0
ν'_L	1	0	1
$\hat{\Phi}_{1,2}$	0	0	0

2.4. *Global symmetries*

There exist two global symmetries of the Lagrangian. Only one of them, which we denote by $U(1)_\ell$, remains unbroken after $SU(2)_\ell$ breaking. Charges of the fields under this symmetry are provided in Table 2. Using the fact that the Yukawa couplings y_ν are tiny to account for the smallness of the Standard Model neutrino masses, and assuming that y'_ν are small as well, the $U(1)_\ell$ global symmetry is promoted to two global symmetries, $U(1)_L$ and $U(1)_\chi$, which separately survive the breaking of $SU(2)_\ell$. The charges under those global symmetries are also shown in Table 2. Note that the charge under $U(1)_L$ can be interpreted as Standard Model lepton number, whereas the $U(1)_\chi$ charge is the dark matter number.

3. Baryogenesis

We now discuss the details of baryon number asymmetry generation in the model. It relies on the fact that a primordial lepton asymmetry is produced by $SU(2)_\ell$ instantons.[a] The subsequent stages combine key features of several mechanisms: Dirac leptogenesis,[22,23] asymmetric dark matter[24–29] and baryogenesis from an earlier phase transition.[30]

3.1. $SU(2)_\ell$ *instantons*

Because of the non-Abelian nature of $SU(2)_\ell$, the model exhibits non-perturbative dynamics in the form of $SU(2)_\ell$ instantons, which are active only above the $SU(2)_\ell$ breaking scale. The instantons preserve the global $U(1)_\ell$ symmetry, but they do not conserve the global $U(1)_L$ and $U(1)_\chi$ symmetries discussed in the previous section, since those symmetries are both anomalous under $SU(2)_\ell$ interactions. Following the calculation in Ref. 31, we find that the instantons induce the

[a]We note that a similar idea was presented in Ref. 21, which we learned about after the completion of this work.

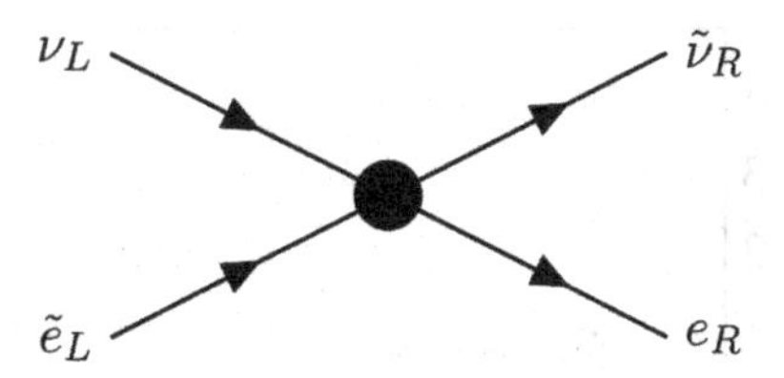

Fig. 1. One of the interactions induced by $\mathrm{SU}(2)_\ell$ instantons for which $\Delta L = -1$ and $\Delta\chi = 1$.

following dimension-six interaction terms:

$$\epsilon_{ij}\big[(l_L^i \cdot \bar{\nu}_R)(l_L^j \cdot \bar{e}_R) - (l_L^i \cdot \bar{\nu}_R)(\tilde{l}_L^j \cdot \bar{\tilde{e}}_R) + (l_L^i \cdot \tilde{l}_L^j)(\bar{\nu}_R \cdot \bar{\tilde{e}}_R) \\ - (l_L^i \cdot \tilde{l}_L^j)(\bar{\tilde{\nu}}_R \cdot \bar{e}_R) + (\tilde{l}_L^i \cdot \bar{\tilde{\nu}}_R)(\tilde{l}_L^j \cdot \bar{\tilde{e}}_R) - (\tilde{l}_L^i \cdot \bar{\tilde{\nu}}_R)(l_L^j \cdot \bar{e}_R)\big], \tag{12}$$

where the dots denote Lorentz contractions and, for simplicity, we assumed just one generation of matter. The generalization to three families is straightforward.

The last term in (12), for example, generates two interaction terms, one of which gives rise to $\nu_L \tilde{e}_L \to \tilde{\nu}_R e_R$ shown in Fig. 1. For this process, as can be read off from Table 2, both the Standard Model lepton number and the dark matter number are violated by one unit: $\Delta L = -1$ and $\Delta\chi = 1$, respectively. Therefore, the first condition for a successful leptogenesis, lepton number violation, is present in the model.

3.2. *CP violation and phase transition*

The remaining Sakharov conditions[32] require sufficient CP violation and out-of-equilibrium dynamics, which in our model can be realized via a first order phase transition. The scalar potential (5) contains four complex parameters: m_{12}^2, $\tilde{\lambda}_5$, $\tilde{\lambda}_6$, and $\tilde{\lambda}_7$. One phase can be rotated away by redefining the phase of $\hat{\Phi}_1^\dagger \hat{\Phi}_2$, leaving three physical phase combinations.[33] It is straightforward to show that for natural values of parameters the amount of CP violation in the model meets the criteria for a successful baryogenesis.[1]

The last condition that needs to be checked is whether the model can actually accommodate a first order phase transition of the Universe. For this purpose we analyze the finite temperature effective potential,[34] under the simplifying assumption $v_1 \gg v_2$:

$$V(u,T) = -\frac{1}{2}m_1^2 u^2 + \frac{1}{4}\lambda_1 u^4 \\ + \frac{1}{64\pi^2}\sum_i n_i \left\{ m_i^4(u)\left[\log\left(\frac{m_i^2(u)}{m_i^2(v_\ell)}\right) - \frac{3}{2}\right] + 2m_i^2(u)\, m_i^2(v_\ell)\right\} \\ + \frac{T^4}{4\pi^2}\sum_i n_i \int_0^\infty dx\, x^2 \left[\log(1 \mp e^{-\sqrt{x^2+m_i^2(u)/T^2}}) - \log(1 \mp e^{-x})\right], \tag{13}$$

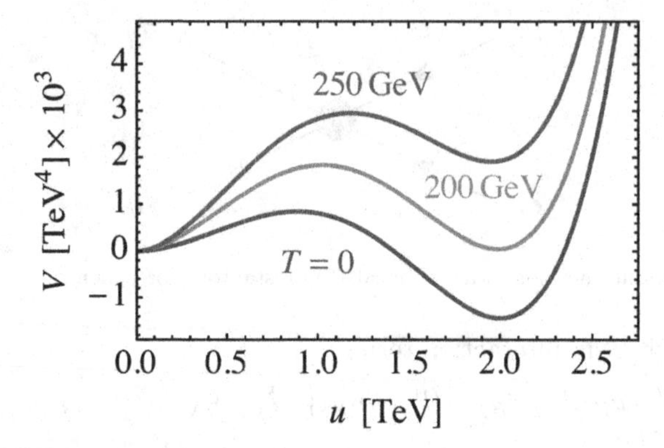

Fig. 2. Plot of the finite temperature effective potential for $v_\ell = 2$ TeV, $\lambda_1 = 2 \times 10^{-3}$ and $g_\ell = 1$.

where the first line is the tree-level Higgs contribution, the second line is the one-loop zero temperature Coleman–Weinberg correction, and the last line is the finite temperature part. The sum is over all particles in the model, with appropriate factors corresponding to the number of degrees of freedom and statistics.

The plot of the effective potential is shown in Fig. 2 for a choice of parameters which set the first order phase transition at the critical temperature $T_c = 200$ GeV. We chose $v_\ell = 2$ TeV, just above the current limit $v_\ell \gtrsim 1.7$ TeV set by the LEP-II experiment,[15] so that the condition $v_\ell(T_c)/T_c \gtrsim 1$ for a strongly first order phase transition is fulfilled.

3.3. *Bubble nucleation and lepton asymmetry*

As the temperature of the Universe decreases and drops to T_c, bubbles of true vacuum start forming and expanding, eventually filling out the entire Universe. A bubble expansion is schematically shown in Fig. 3. Outside the bubble the $SU(2)_\ell$ symmetry is not broken and the $SU(2)_\ell$ instantons remain active. Inside the bubble, on the other hand, $SU(2)_\ell$ is broken and the instanton effects are exponentially

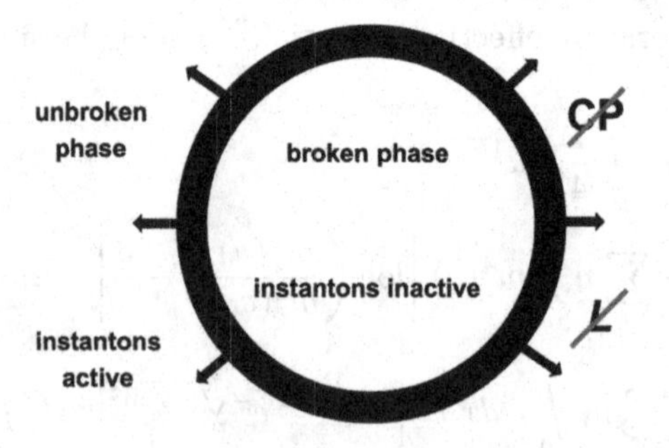

Fig. 3. Expanding bubble of true vacuum.

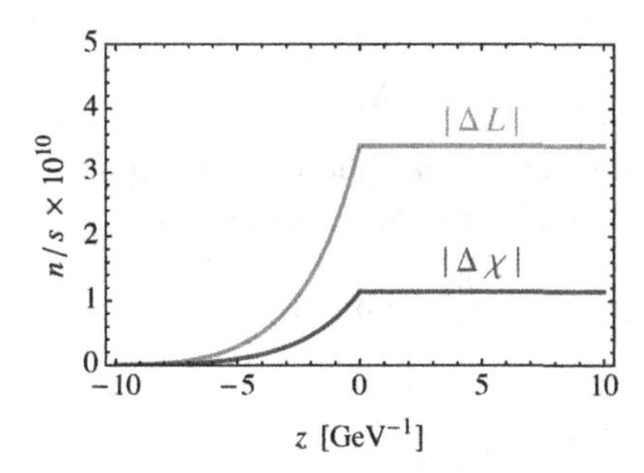

Fig. 4. Standard Model lepton and dark matter particle number densities vs. distance from the bubble wall.

suppressed. As the bubble expands in the presence of CP violation, part of the lepton asymmetry generated by the instantons just outside the bubble becomes trapped inside the bubble. The same is true regarding the dark matter asymmetry.

Although $\mathrm{SU}(2)_\ell$ instantons are not active inside the bubble, one might worry whether the Standard Model lepton and dark matter asymmetries will be washed out by the interactions involving the y_ν and y'_ν Yukawas, since they explicitly violate $\mathrm{U}(1)_L$ and $\mathrm{U}(1)_\chi$, while conserving only their sum $\mathrm{U}(1)_\ell$. This, however, is not an issue, since the small values of y_ν or y'_ν imply that the right-handed neutrinos and their partners reach chemical equillibrium long after the $\mathrm{SU}(2)_\ell$ phase transition. As a result, the Standard Model lepton and dark matter number asymmetries survive until the electroweak phase transition, with just the lepton asymmetry being partially converted into a baryon asymmetry by the electroweak sphalerons, as discussed in the subsequent section.

The process of accumulation of the Standard Model lepton and dark matter asymmetries outside the expanding bubble is described by the diffusion equations:[35,36]

$$\dot{n}_i = D_i \nabla^2 n_i - \sum_j \Gamma_{ij} \frac{n_j}{k_j} + \gamma_i \,, \tag{14}$$

where n_i denotes the number density of a given particle species, D_i is the diffusion constant, Γ_{ij} is the diffusion rate, k_j is the number of degrees of freedom times a factor arising from statistics and γ_i is the CP violating source.[37] In our model, there is a set of 12 diffusion equations and eight constraints coming from the Yukawa and instanton interactions.[1]

The solution to this set of diffusion equations is shown in Fig. 4, which plots the Standard Model lepton and dark matter particle number densities normalized to entropy as a function of the distance from the bubble wall located at $z = 0$, where $z < 0$ corresponds to the outside of the bubble and $z > 0$ describes the inside of the bubble. The ratio of the generated Standard Model lepton and dark matter asymmetries in our model is:

$$\left|\frac{\Delta L}{\Delta \chi}\right| = 3 \tag{15}$$

and is independent of the model parameters. There exists a natural and experimentally allowed choice of parameters for which

$$\frac{n_L}{s} \simeq 3 \times 10^{-10} . \tag{16}$$

3.4. *Baryon asymmetry*

As mentioned earlier, the $\mathrm{SU}(2)_\ell$ instantons become inactive after $\mathrm{SU}(2)_\ell$ breaking and the dark matter number freezes in inside the bubble. This is not exactly the case for the Standard Model lepton asymmetry, since the Standard Model electroweak sphalerons remain active until the electroweak phase transition and convert part of the lepton number to baryon number. The resulting baryon asymmetry generated by the sphalerons is[38]

$$\Delta B = \frac{28}{79} \Delta L , \tag{17}$$

therefore the final baryon asymmetry to entropy ratio is

$$\frac{n_B}{s} \approx 10^{-10} , \tag{18}$$

which agrees with the observed value.

4. Dark Matter

The dark matter candidate in our model (10) is composed mostly of the $\mathrm{SU}(2)_\ell$ doublet partner of the right-handed neutrino. It is therefore predominantly a Standard Model singlet, with only a small admixture of an electroweak doublet picked up by its interactions with the Standard Model Higgs field. As the dark matter and baryon number asymmetries in our model are closely related (approximately equal at present time), the dark matter mass is uniquely determined by the observable ratio of the dark matter and baryonic relic densities. Assuming the dark matter is relativistic at the decoupling temperature, its mass is given by:

$$m_\chi = m_p \frac{\Omega_{\mathrm{DM}}}{\Omega_{\mathrm{B}}} \left|\frac{\Delta B}{\Delta \chi}\right| \simeq \mathrm{GeV} . \tag{19}$$

A dark matter mass of a few GeV is generic in asymmetric dark matter models and it is generally challenging to make its symmetric component efficiently annihilate away. In the current model this issue is circumvented by arranging one of the Higgs components to be lighter than 5 GeV. In such a scenario a successful annihilation can proceed through the channels shown in Fig. 5. A light Higgs component can be realized provided that the quartic terms in the scalar potential are small, which in the case of λ_1 is also needed for a first order phase transition. This

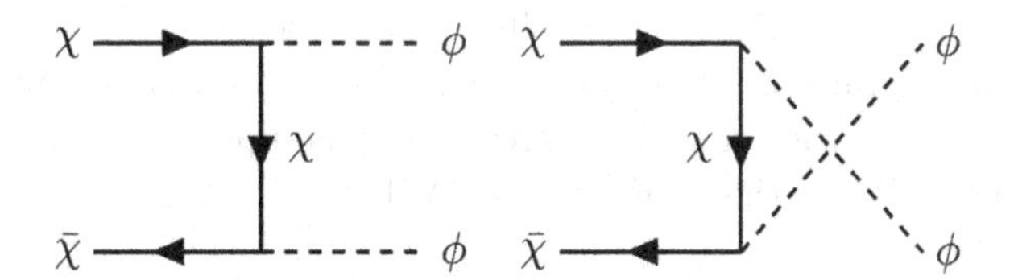

Fig. 5. Dark matter annihilation channels to the pseudoscalar component of $\hat{\Phi}_2$.

mass range for a new scalar/pesudoscalar is not strongly constrained by low energy experiments, but may be accessible in the future.[39]

Since the new Z' gauge bosons do not interact with quarks, there are no tree-level direct detection diagrams involving the Standard Model singlet component of the dark matter. As a result, the direct detection constraint can only come from loop processes, but in this case the GeV-scale dark matter limits coming from the CDMSlite experiment[40] are much less restrictive than the LEP-II constraint of $v_\ell \gtrsim 1.7$ TeV, which we already took into account. Regarding the contribution of the direct detection diagrams involving the electroweak gauge bosons, it can be estimated using the results of Ref. 19 and is fully consistent with experiment in the phenomenologically natural limit $Y_\nu v_\ell \gg y_\nu v, y'_\nu v$ we adopted.

5. Conclusions

Here we discussed a new model extending the Standard Model gauge group with a non-Abelian gauged lepton number $SU(2)_\ell$. The model realizes a mechanism for baryogenesis based on leptogenesis in which the lepton number asymmetry is generated by $SU(2)_\ell$ instantons. It also contains a natural dark matter candidate — the partner of the right-handed neutrino. Despite the theoretical advantages, it is difficult to test this theory experimentally. Since new physics resides in the lepton sector of the model, the best way to probe it would be in a new high energy lepton–lepton collider.

Let us end by saying that there is no reason not to expect the Standard Model gauge symmetry to be enhanced at higher energies. Analyzing other simple theories with extended gauge groups seems like a worthwhile effort, as it may shed more light on the outstanding issues of the Standard Model. However, one should always keep in mind that "Nature will do what Nature does and it's up to experiment to be the final judge".[b]

Acknowledgments

I am grateful to Yuri Shirman, Tim Tait and Jennifer West for a fruitful collaboration. I would also like to thank the organizers of the Conference on Cosmology,

[b]From M. B. Wise, interviewed by A. Ananthaswamy, Hunt is on for quark dark matter, *New Scientist* **3050** (December 5, 2015), p. 10.

Gravitational Waves and Particles in Singapore, especially the chairman, Harald Fritzsch, for the invitation, warm hospitality, and a fantastic scientific and social atmosphere during the conference. This research was supported in part by the DOE grant DE-SC0009919 and the NSF grant PHY-1316792.

References

1. B. Fornal, Y. Shirman, T. M. P. Tait and J. R. West, arXiv:1703.00199, submitted to *Phys. Rev. D.*
2. S. L. Glashow, *Nucl. Phys.* **22**, 579 (1961).
3. S. Weinberg, *Phys. Rev. Lett.* **19**, 1264 (1967).
4. A. Salam, *Conf. Proc.* **C680519**, 367 (1968).
5. H. Fritzsch and M. Gell-Mann, Current algebra: Quarks and what else?, *Proc. of the XVI Int. Conf. on High Energy Physics*, National Accelerator Laboratory, Chicago (1972), pp. 135–165.
6. H. Fritzsch, M. Gell-Mann and H. Leutwyler, *Phys. Lett. B* **47**, 365 (1973).
7. A. Pais, *Phys. Rev. D* **8**, 1844 (1973).
8. Y. Tosa, R. E. Marshak and S. Okubo, *Phys. Rev. D* **27**, 444 (1983).
9. S. Rajpoot, *Int. J. Theor. Phys.* **27**, 689 (1988).
10. R. Foot, G. C. Joshi and H. Lew, *Phys. Rev. D* **40**, 2487 (1989).
11. C. D. Carone and H. Murayama, *Phys. Rev. D* **52**, 484 (1995).
12. H. Georgi and S. L. Glashow, *Phys. Lett. B* **387**, 341 (1996).
13. P. Fileviez Pérez and M. B. Wise, *Phys. Rev. D* **82**, 011901 (2010) [Erratum-*ibid.* **82**, 079901 (2010)].
14. M. Duerr, P. Fileviez Perez and M. B. Wise, *Phys. Rev. Lett.* **110**, 231801 (2013).
15. P. Schwaller, T. M. P. Tait and R. Vega-Morales, *Phys. Rev. D* **88**, 035001 (2013).
16. J. M. Arnold, P. Fileviez Pérez, B. Fornal and S. Spinner, *Phys. Rev. D* **88**, 115009 (2013).
17. P. Fileviez Perez, S. Ohmer and H. H. Patel, *Phys. Lett. B* **735**, 283 (2014).
18. B. Fornal, A. Rajaraman and T. M. P. Tait, *Phys. Rev. D* **92**, 055022 (2015).
19. B. Fornal and T. M. P. Tait, *Phys. Rev. D* **93**, 075010 (2016).
20. P. Fileviez Perez and S. Ohmer, *Phys. Lett. B* **768**, 86 (2017).
21. M. Blennow, B. Dasgupta, E. Fernandez-Martinez and N. Rius, *JHEP* **03**, 014 (2011).
22. K. Dick, M. Lindner, M. Ratz and D. Wright, *Phys. Rev. Lett.* **84**, 4039 (2000).
23. H. Murayama and A. Pierce, *Phys. Rev. Lett.* **89**, 271601 (2002).
24. S. Nussinov, *Phys. Lett. B* **165**, 55 (1985).
25. D. B. Kaplan, *Phys. Rev. Lett.* **68**, 741 (1992).
26. D. Hooper, J. March-Russell and S. M. West, *Phys. Lett. B* **605**, 228 (2005).
27. D. E. Kaplan, M. A. Luty and K. M. Zurek, *Phys. Rev. D* **79**, 115016 (2009).
28. K. Petraki and R. R. Volkas, *Int. J. Mod. Phys. A* **28**, 1330028 (2013).
29. K. M. Zurek, *Phys. Rep.* **537**, 91 (2014).
30. J. Shu, T. M. P. Tait and C. E. M. Wagner, *Phys. Rev. D* **75**, 063510 (2007).
31. D. E. Morrissey, T. M. P. Tait and C. E. M. Wagner, *Phys. Rev. D* **72**, 095003 (2005).
32. A. D. Sakharov, *Pisma Zh. Eksp. Teor. Fiz.* **5**, 32 (1967) [*Usp. Fiz. Nauk* **161**, 61 (1991)].
33. J. F. Gunion and H. E. Haber, *Phys. Rev. D* **72**, 095002 (2005).
34. M. Quiros, Finite temperature field theory and phase transitions, in *Proc. Summer School in High Energy Physics and Cosmology*, Trieste, Italy, June 29–July 17, 1998 (World Scientific, 1999).
35. M. Joyce, T. Prokopec and N. Turok, *Phys. Rev. D* **53**, 2930 (1996).

36. A. G. Cohen, D. B. Kaplan and A. E. Nelson, *Phys. Lett. B* **336**, 41 (1994).
37. A. Riotto, *Phys. Rev. D* **53**, 5834 (1996).
38. J. A. Harvey and M. S. Turner, *Phys. Rev. D* **42**, 3344 (1990).
39. G. Krnjaic, *Phys. Rev. D* **94**, 073009 (2016).
40. R. Agnese *et al.*, *Phys. Rev. Lett.* **116**, 071301 (2016).

Vacuum Dynamics in the Universe versus a Rigid $\Lambda = \text{const.}$

Joan Solà,* Adrià Gómez-Valent† and Javier de Cruz Pérez‡

Departament de Física Quàntica i Astrofísica (FQA)
and Institute of Cosmos Sciences (ICCUB), Universitat de Barcelona,
Av. Diagonal 647, E-08028 Barcelona, Catalonia, Spain
** sola@fqa.ub.edu*
† adriagova@fqa.ub.edu
‡ decruz@fqa.ub.edu

In this year, in which we celebrate 100 years of the cosmological term, Λ, in Einstein's gravitational field equations, we are still facing the crucial question whether Λ is truly a fundamental constant or a mildly evolving dynamical variable. After many theoretical attempts to understand the meaning of Λ, and in view of the enhanced accuracy of the cosmological observations, it seems now mandatory that this issue should be first settled empirically before further theoretical speculations on its ultimate nature. In this review, we summarize the situation of some of these studies. Devoted analyses made recently show that the $\Lambda = \text{const.}$ hypothesis, despite being the simplest, may well not be the most favored one. The overall fit to the cosmological observables SNIa + BAO + $H(z)$ + LSS + BBN + CMB single out the class of "running" vacuum models (RVMs), in which $\Lambda = \Lambda(H)$ is an affine power-law function of the Hubble rate. It turns out that the performance of the RVM as compared to the "concordance" ΛCDM model (with $\Lambda = \text{const.}$) is much better. The evidence in support of the RVM may reach $\sim 4\sigma$ c.l., and is bolstered with Akaike and Bayesian criteria providing strong evidence in favor of the RVM option. We also address the implications of this framework on the tension between the CMB and local measurements of the current Hubble parameter.

Keywords: Cosmology; vacuum energy; dark energy.

1. Introduction

One hundred years ago, in mid-February 1917, the famous seminal paper in which Einstein introduced the cosmological term Λ (actually denoted "λ" in it), as a part of the generally covariant gravitational field equations, was published.[1] Fourteen years later, the idea of Λ as a fundamental piece of these equations was abandoned by Einstein himself;[2] and only one year later the Einstein–de Sitter model, modernly called the cold dark matter model (CDM), was proposed with no further use of the

Λ term for the description of the cosmological evolution.[3] The situation with Λ did not stop here and it took an unexpected new turn when Λ reappeared shortly afterwards in the works of Lemaître,[4] wherein Λ was associated to the concept of vacuum energy density through the expression $\rho_\Lambda = \Lambda/(8\pi G)$ — in which G is Newton's constant. This association is somehow natural if we take into account that the vacuum energy is thought of as being uniformly distributed in all corners of space, thus preserving the Cosmological Principle. The problem (not addressed by Lemaître) is to understand the origin of the vacuum energy in fundamental physics, namely in the quantum theory and more specifically in quantum field theory (QFT). Here is where the cosmological constant problem first pops up. It was first formulated in preliminary form by Zeldovich 50 years ago.[5]

The CC problem[6–8] is the main source of headache for every theoretical cosmologist confronting his/her theories with the measured value of ρ_Λ.[9,10] Furthermore, the purported discovery of the Higgs boson at the LHC has accentuated the CC problem greatly, certainly much more than is usually recognized.[8] Furthermore, owing to the necessary spontaneous symmetry breaking (SSB) of the electroweak (EW) theory, there emerges an induced contribution to ρ_Λ that must also be taken into account. These SSB effects are appallingly much larger (viz. $\sim 10^{56}$) than the tiny value $\rho_\Lambda \sim 10^{-47}$ GeV^4 extracted from observations.[a] So the world-wide celebrated "success" of the Higgs finding in particle physics actually became a cosmological fiasco, since it automatically detonated the "modern CC problem," i.e. the confirmed size of the EW vacuum, which should be deemed as literally "real" (in contrast to other alleged — ultralarge — contributions from QFT) or "unreal" as the Higgs boson itself! One cannot exist without the other. Such uncomfortable situation of the Higgs boson with cosmology might be telling us that the found Higgs boson is not a fundamental particle, as in such a case the EW vacuum could not be counted as a fundamental SSB contribution to the vacuum energy of the universe. I refer the reader to some review papers,[6,7] including,[8,11] for a more detailed presentation of the CC problem. Setting aside the "impossible" task of predicting the Λ value itself — unless it is understood as a "primordial renormalization"[12] — I will focus here on a special class of models in which Λ appears neither as a rigid constant nor as a scalar field (quintessence and the like),[7] but as a "running" quantity in QFT in curved space–time. This is a natural option for an expanding universe. As we will show, such kind of dynamical vacuum models are phenomenologically quite successful; in fact so successful that they are currently challenging the ΛCDM,[13–17] see specially the most recent works,[18–22] in which the most significant signs of vacuum dynamics have been found. Potential time variation of the fundamental constants associated to these models has also been explored in Refs. 23–28, and can be used as complementary experimental signatures for them.

[a]Being ρ_Λ a density, and hence a dimensionful quantity, "tiny" value means only within the particle physics standards, of course.

2. Running Vacuum as the Next-to-minimal Step Beyond the ΛCDM

The "concordance" or ΛCDM model, i.e. the standard model of cosmology is based on the assumption of the existence of dark matter (DM) and the spatial flatness of the Friedmann–Lemaître–Robertson–Walker (FLRW) metric. At the same time the model is based on the existence of a nonvanishing but rigid (and positive) cosmological constant term: $\Lambda = \text{const}$. The model was first proposed as having the minimal ingredients for a possible successful phenomenological description of the data by Peebles in 1984.[29] Nowadays we know it is consistent with a large body of observations, and in particular with the high precision data from the cosmic microwave background (CMB) anisotropies.[9]

The rigid $\Lambda = \text{const.}$ term in the concordance ΛCDM model is the simplest (perhaps too simple) possibility. The running vacuum models (RVMs) (cf. Refs. 8, 11 and references therein) build upon the idea that the cosmological term Λ, and the corresponding vacuum energy density, ρ_Λ, should be time dependent quantities in cosmology. It is difficult to conceive an expanding universe with a strictly constant vacuum energy density that has remained immutable since the origin of time. Rather, a smoothly evolving DE density that inherits its time-dependence from cosmological variables, such as the Hubble rate $H(t)$, or the scale factor $a(t)$, is not only a qualitatively more plausible and intuitive idea, but is also suggested by fundamental physics, in particular by QFT in curved space–time. We denote it in general by $\rho_D = \rho_D(t)$. Despite its time evolution, it may still have the vacuum equation of state (EoS) $w = -1$, or a more general one $w \neq -1$, or even a dynamical effective EoS $w = w(t)$. The main standpoint of the RVM class of dynamical DE models is that ρ_D "runs" because the effective action receives quantum effects from the matter fields. The leading effects may generically be captured from a renormalization group equation (RGE) of the form[8]

$$\frac{d\rho_D}{d\ln\mu^2} = \frac{1}{(4\pi)^2}\sum_i \left[a_i M_i^2 \mu^2 + b_i \mu^4 + c_i \frac{\mu^6}{M_i^2} + \cdots\right]. \tag{1}$$

The running scale μ is typically identified with H or related variables. The RVM ansatz is that $\rho_D = \rho_D(H)$ because μ will be naturally associated to the Hubble parameter at a given epoch $H = H(t)$, and hence ρ_D should evolve with the rate of expansion H. Notice that $\rho_D(H)$ can involve *only* even powers of the Hubble rate H (because of the covariance of the effective action).[8] The coefficients a_i, b_i, $c_i, \ldots$ are dimensionless, and the M_i are the masses of the particles in the loops. Because μ^2 can be in general a linear combination of the homogeneous terms H^2 and $\dot{H}$, it is obvious that upon integration of the above RGE we expect the following general type of (appropriately normalized) RVM density:[8,11,12]

$$\rho_D(H) = \frac{3}{8\pi G}(C_0 + \nu H^2 + \tilde{\nu}\dot{H}) + \mathcal{O}(H^4), \tag{2}$$

where ν and $\tilde{\nu}$ are dimensionless parameters, but C_0 has dimension 2 (energy squared) in natural units. We emphasize that $C_0 \neq 0$ so as to insure a smooth ΛCDM limit when the dimensionless coefficients ν and $\tilde{\nu}$ are set to zero.[b] The interesting possibility that ν and/or $\tilde{\nu}$ are nonvanishing may induce a time evolution of the vacuum energy. These dimensionless coefficients can be computed in QFT from the ratios squared of the masses to the Planck mass,[30] and are therefore small as expected from their interpretation as β-function coefficients of the RGE (1). Since some of the masses inhabit the GUT scale $M_X \sim 10^{16}$ GeV, the values of $\nu, \tilde{\nu}$ need not be very small, typically $\sim 10^{-3}$ at most upon accounting for the large multiplicities that are typical in a GUT — see Ref. 30 for a concrete estimate.

Finally, we note that the $\mathcal{O}(H^4)$-terms in (2) are irrelevant for the study of the current universe, but are essential for the correct account of the inflationary epoch in this context and to explain the graceful exit and entropy problems.[11,12] The RVM (2) is therefore capable of providing a unified dynamical vacuum picture for the entire cosmic evolution.[31–33]

Concerning the parameters ν and $\tilde{\nu}$, they must be determined phenomenologically by confronting the model to the wealth of observations. It is remarkable that the aforementioned theoretical estimate,[30] leading to $\nu, \tilde{\nu} \sim 10^{-3}$, is of the order of magnitude of the phenomenological determination.[13–18] For the current presentation, however, we will assume $\tilde{\nu} = 0$ hereafter and will focus on the implications for the current universe of the canonical RVM:

$$\rho_D(H) = \frac{3}{8\pi G}(C_0 + \nu H^2)\,. \tag{3}$$

We will use the above expression to study the RVM background cosmology and the corresponding perturbations equations. We can compare with the concordance ΛCDM cosmology by just setting $\nu = 0$ in the obtained results. The background cosmological equations for the RVM take on the same form as in the ΛCDM by simply replacing the rigid cosmological term with the dynamical vacuum energy (3). In this way we obtain the generalized Friedmann's and acceleration equations, which read as follows:

$$3H^2 = 8\pi G(\rho_m + \rho_D(H))\,, \quad 2\dot{H} + 3H^2 = -8\pi G(w_m \rho_m + w_D \rho_D(H))\,, \tag{4}$$

where w_m and w_D are the EoS parameters of the matter fluid and of the DE, respectively. It is well known that $w_m = 1/3, 0$ for relativistic and nonrelativistic matter, respectively. Then, for simplicity, we will denote the EoS of the DE component simply as w. The explicit solution will depend of course on whether we assume that the DE is canonical vacuum energy ($w = -1$), in which case ρ_D can

[b]It is important to make clear that models with $C_0 = 0$ (for any ν and $\tilde{\nu}$) are ruled out by the observations, as shown in Refs. 13–15. This conclusion also applies to all DE models of the form $\rho_D \sim aH + bH^2$, with a linear term $\sim H$ admitted only on phenomenological grounds.[13,14] In particular, the model $\rho_D \sim H$ is strongly ruled out, see Ref. 14 (and the discussion in its Appendix).

be properly denoted as ρ_Λ, or dynamical DE with $w \neq -1$. In some cases w can also be a function of time or of some cosmic variable, but we shall not consider this possibility here — see, however, Ref. 13 in which such situation would be mandatory. The solution of the cosmological equations may also depend on whether the gravitational coupling G is constant or also running with the expansion, $G = G(H)$ (as ρ_D itself). And, finally, it may depend on whether we assume that there exists an interaction of the DE with the matter (mainly dark matter, DM). Whatever it be the nature of our assumptions on these important details, they must be of course consistent with the Bianchi identity, which is tantamount to saying with the local covariant conservation laws. In fact, these possibilities have all been carefully studied in the literature and the complete solution of the cosmological equations has been provided in each case.[13–22] In what follows we report only on some of the solutions for the densities of matter and DE in the case when there is an interaction between the DE and matter at fixed G. At the same time we will compare the result when the EoS of the DE is $w \neq -1$ (which means that we will leave this parameter also free in the fit) with the simplest situation $w = 1$ (the strict vacuum case).

3. Canonical RVM with Conserved Baryon and Radiation Densities

The total matter density ρ_m can be split into the contribution from baryons and cold dark matter (DM), namely $\rho_m = \rho_b + \rho_{dm}$. In the following we assume that the DM density is the only one that carries the anomaly, whereas radiation and baryons are self-conserved, so that their energy densities evolve in the standard way $\rho_r(a) = \rho_r^0\, a^{-4}$ and $\rho_b(a) = \rho_b^0\, a^{-3}$. On the other hand the dynamical evolution of the vacuum is given by Eq. (3). Since it is only the DM that exchanges energy with the vacuum, the local conservation law reads as follows:

$$\dot{\rho}_{dm} + 3H\rho_{dm} = Q\,, \qquad \dot{\rho}_\Lambda = -Q\,. \tag{5}$$

The source function Q is a calculable expression from (3) and Friedmann's equation in (4). We find: $Q = -\dot{\rho}_\Lambda = \nu H(3\rho_{dm} + 3\rho_b + 4\rho_r) = \nu H(3\rho_m + 4\rho_r)$. It can be useful to compare the canonical RVM with two alternative dynamical vacuum models (DVMs) with different forms of the interaction sources. Let us therefore list the three DVMs under comparison, which we may denote RVM, Q_{dm} and Q_Λ according to the structure of the interaction source, or also for convenience just I, II and III:

$$\text{Model I }\ (w\text{RVM}):\ Q = \nu\, H(3\rho_m + 4\rho_r) \tag{6}$$

$$\text{Model II }\ (wQ_{dm}):\ Q_{dm} = 3\nu_{dm} H\rho_{dm} \tag{7}$$

$$\text{Model III }\ (wQ_\Lambda):\ Q_\Lambda = 3\nu_\Lambda H\rho_\Lambda\,. \tag{8}$$

Each model has a characteristic (dimensionless) parameter $\nu_i = \nu$, ν_{dm}, ν_Λ as a part of the interaction source, which must be fitted to the observational data. Notice that the three model names are preceded by w to recall that, in the general case, the equation of state (EoS) is very near to the vacuum one ($w = -1 + \epsilon$, with $|\epsilon| \ll 1$). For this reason these dynamical quasi-vacuum models are also denoted as wDVMs. In the particular case $w = -1$ (i.e. $\epsilon = 0$) the wDVMs become just the DVMs. As an example, let us provide the solution of the cosmological equations for the matter and vacuum energy densities in the case of the canonical RVM:

$$\rho_{dm}(a) = \rho_{dm}^0 a^{-3(1-\nu)} + \rho_b^0 (a^{-3(1-\nu)} - a^{-3}) - \frac{4\nu\rho_r^0}{1+3\nu}(a^{-4} - a^{-3(1-\nu)}) \tag{9}$$

and

$$\begin{aligned}\rho_\Lambda(a) &= \rho_\Lambda^0 + \frac{\nu\rho_m^0}{1-\nu}(a^{-3(1-\nu)} - 1) \\ &\quad + \frac{\nu}{1-\nu}\rho_r^0\left(\frac{1-\nu}{1+3\nu}a^{-4} + \frac{4\nu}{1+3\nu}a^{-3(1-\nu)} - 1\right).\end{aligned} \tag{10}$$

For the corresponding expressions of the other models, see Refs. 21 and 22. Models II and III were previously studied in different approximations e.g. in Refs. 34–36. As can be easily checked, for $\nu \to 0$ we recover the corresponding results for the ΛCDM, as it should. The Hubble function can be immediately obtained from these formulas after inserting them in Friedmann's equation, together with the conservation laws for radiation and baryons, $\rho_r(a) = \rho_r^0 a^{-4}$ and $\rho_b(a) = \rho_b^0 a^{-3}$.

4. The XCDM and CPL Parametrizations

In the next section, when we compare the DVMs to the ΛCDM, it is also convenient to fit the data to the simple XCDM parametrization of the dynamical DE.[37] Since both matter and DE are self-conserved (i.e. they are not interacting) in the XCDM, the DE energy density as a function of the scale factor is simply given by $\rho_X(a) = \rho_{X0}\, a^{-3(1+w_0)}$, with $\rho_{X0} = \rho_{\Lambda 0}$, where w_0 is the (constant) equation of state (EoS) parameter of the generic DE entity X in this parametrization. The normalized Hubble function is:

$$E^2(a) = \Omega_m a^{-3} + \Omega_r a^{-4} + \Omega_\Lambda a^{-3(1+w_0)}. \tag{11}$$

For $w_0 = -1$ it boils down to that of the ΛCDM with rigid CC term. Use of the XCDM parametrization becomes useful so as to roughly mimic a (noninteractive) DE scalar field with constant EoS. For $w_0 \gtrsim -1$ the XCDM mimics quintessence, whereas for $w_0 \lesssim -1$ it mimics phantom DE.

A slightly more sophisticated parametrization to the behavior of a noninteractive scalar field playing the role of dynamical DE is furnished by the CPL parametrization,[38] in which one assumes that the generic DE entity X has a slowly varying EoS of the form

$$w_D = w_0 + w_1(1-a) = w_0 + w_1\frac{z}{1+z}. \tag{12}$$

Table 1. Best-fit values for the free parameters of the ΛCDM, XCDM, the three dynamical vacuum models (DVMs) and the three dynamical quasi-vacuum models (wDVMs), including their statistical significance (i.e. the values of the χ^2-test and the difference of the values of the Akaike and Bayesian information criteria, AIC and BIC, with respect to the ΛCDM). For a detailed description of the data and a full list of observational references, see Refs. 19 and 21. The quoted number of degrees of freedom (dof) is equal to the number of data points minus the number of independent fitting parameters (4 for the ΛCDM, 5 for the XCDM and the DVMs, and 6 for the wDVMs). For the CMB data we have used the marginalized mean values and covariance matrix for the parameters of the compressed likelihood for Planck 2015 TT, TE, EE + lowP + lensing as in Ref. 22. Each best-fit value and the associated uncertainties have been obtained by marginalizing over the remaining parameters.

Model	h	ω_b	n_s	Ω_m^0	ν_i	w	$\chi^2_{\rm min}$/dof	ΔAIC	ΔBIC
ΛCDM	0.688 ± 0.004	0.02243 ± 0.00013	0.973 ± 0.004	0.298 ± 0.004	—	—	84.40/85	—	—
XCDM	0.672 ± 0.006	0.02251 ± 0.00013	0.975 ± 0.004	0.311 ± 0.006	—	-0.936 ± 0.023	76.80/84	5.35	3.11
RVM	0.674 ± 0.005	0.02224 ± 0.00014	0.964 ± 0.004	0.304 ± 0.005	0.00158 ± 0.00042	—	68.67/84	13.48	11.24
Q_{dm}	0.675 ± 0.005	0.02222 ± 0.00014	0.964 ± 0.004	0.304 ± 0.005	0.00218 ± 0.00057	—	69.13/84	13.02	10.78
Q_Λ	0.688 ± 0.003	0.02220 ± 0.00015	0.964 ± 0.005	0.299 ± 0.004	0.00673 ± 0.00236	—	76.30/84	5.85	3.61
wRVM	0.671 ± 0.007	0.02228 ± 0.00016	0.966 ± 0.005	0.307 ± 0.007	0.00140 ± 0.00048	-0.979 ± 0.028	68.15/83	11.70	7.27
wQ_{dm}	0.670 ± 0.007	0.02228 ± 0.00016	0.966 ± 0.005	0.308 ± 0.007	0.00189 ± 0.00066	-0.973 ± 0.027	68.22/83	11.63	7.20
wQ_Λ	0.671 ± 0.007	0.02227 ± 0.00016	0.965 ± 0.005	0.313 ± 0.006	0.00708 ± 0.00241	-0.933 ± 0.022	68.24/83	11.61	7.18

The CPL parametrization, in contrast to the XCDM one, gives room for a time evolution of the dark energy EoS owing to the presence of the additional parameter w_1, which satisfies $0 < |w_1| \ll |w_0|$, with $w_0 \gtrsim -1$ or $w_0 \lesssim -1$. The corresponding normalized Hubble function for the CPL can be easily computed:

$$E^2(a) = \Omega_m a^{-3} + \Omega_r a^{-4} + \Omega_\Lambda a^{-3(1+w_0+w_1)} e^{-3w_1(1-a)} \,. \tag{13}$$

Both the XCDM and the CPL parametrizations can be thought of as a kind of baseline frameworks to be referred to in the study of dynamical DE. They can be used as fiducial models to which we can compare other, more sophisticated, models for the dynamical DE, such as the DVMs under study. The XCDM, however, is more appropriate for a fairer comparison with the DVMs, since they have one single vacuum parameter ν_i. For this reason we present the main fitting results with the XCDM, along with the other models, in Table 1. The numerical fitting results for the CPL parametrization are given in Ref. 21. Owing to the presence of an extra parameter the errors in the fitting values of w_0 and w_1 are bigger than the error in the single parameter w_0 of the XCDM parametrization. For this presentation, we limit ourselves to report on the latter, together with the rest of the models.

5. Structure Formation Under Vacuum Dynamics

The DVMs and wDVMs are characterized by a dynamical vacuum/quasi-vacuum energy. Therefore, in order to correctly fit these models to the data, a generalized treatment of the linear structure formation beyond the ΛCDM is of course mandatory. At the (subhorizon) scales under consideration we will neglect the perturbations of the vacuum energy density in front of the perturbations of the matter field. This has been verified for various related cases, see e.g. Refs. 39 and 13. For a recent detailed studied directly involving the DVMs under consideration, see Ref. 41. In the presence of dynamical vacuum energy the matter density contrast $\delta_m = \delta\rho_m/\rho_m$ obeys the following differential equation (cf. Refs. 21 and 22 for details):

$$\delta''_m(a) + \frac{A(a)}{a}\delta'_m(a) + \frac{B(a)}{a^2}\delta_m(a) = 0\,, \tag{14}$$

where the prime denotes differentiation with respect to the scale factor. The functions $A(a)$ and $B(a)$ can be determined after a straightforward application of the general perturbation equations (cf. Ref. 39):

$$A(a) = 3 + \frac{aH'}{H} + \frac{\Psi}{H} - 3r\epsilon\,, \tag{15}$$

$$B(a) = -\frac{4\pi G\rho_m}{H^2} + 2\frac{\Psi}{H} + \frac{a\Psi'}{H} - 15r\epsilon - 9\epsilon^2 r^2 + 3\epsilon(1+r)\frac{\Psi}{H} - 3r\epsilon\frac{aH'}{H}\,. \tag{16}$$

Here $r \equiv \rho_\Lambda/\rho_m$ and $\Psi \equiv Q/\rho_m$. For $\nu_i = 0$ we have $\Psi = 3Hr\epsilon$, and after some calculations one can easily show that (14) can be brought back to the common form for the XCDM and ΛCDM. The (vacuum-matter) interaction source Q for each DVM is given by (6)–(8). For $\rho_\Lambda =$ const. and for the XCDM there is no such an interaction, therefore $Q = 0$, and Eq. (14) reduces to the ΛCDM form:

$$\delta_m''(a) + \left(\frac{3}{a} + \frac{H'(a)}{H(a)}\right)\delta_m'(a) - \frac{4\pi G\rho_m(a)}{H^2}\frac{\delta_m(a)}{a^2} = 0\,. \tag{17}$$

Recalling that $\rho_m(a) = \rho_m^0\, a^{-3}$ in the ΛCDM, the growing mode solution of this equation can be solved by quadrature:[39]

$$\delta_m(a) = \frac{5\Omega_m}{2}\frac{H(a)}{H_0}\int_0^a \frac{d\tilde{a}}{(\tilde{a}\,H(\tilde{a})/H_0)^3}\,. \tag{18}$$

One can easily check that in the matter-dominated epoch ($\Omega_m \simeq 1$, $H^2 \simeq H_0^2 a^{-3}$), the above equation yields $\delta_m \simeq a$, as expected. However, when $\Omega_m < 1$ (owing to $\Omega_\Lambda > 0$) there is an effective suppression of the form $\delta_m \simeq a^s$ with $s < 1$.[40] The last feature is also true in the presence of dynamical vacuum. However, the generalized perturbation equation (14) cannot be solved analytically and one has to proceed numerically. The first thing to do is to fix the initial conditions analytically. This is possible because at high redshift, namely when nonrelativistic matter dominates over radiation and DE, functions (15) and (16) are then approximately constant and Eq. (14) admits power-law solutions $\delta_m(a) = a^s$. The values for the power s can be computed for each model. For example, for the wRVM it can be shown that $s = 1 + (3/5)\nu\left(\frac{1}{w} - 4\right) + \mathcal{O}(\nu^2)$. Notice that $s < 1$ for $\nu > 0$ and w near -1. Using the appropriate initial conditions for each model, the numerical solution of (14) can be obtained — see Refs. 21 and 22 for more details.

Armed with these equations, the linear LSS regime is usually analyzed with the help of the weighted linear growth $f(z)\sigma_8(z)$, where $f(z) = d\ln\delta_m/d\ln a$ is the growth factor and $\sigma_8(z)$ is the rms mass fluctuation on $R_8 = 8h^{-1}$ Mpc scales. It is computed as follows (see e.g. Ref. 21):

$$\sigma_8(z) = \sigma_{8,\Lambda}\frac{\delta_m(z)}{\delta_m^\Lambda(0)}\sqrt{\frac{\int_0^\infty k^{n_s+2}T^2(\vec{p},k)W^2(kR_8)dk}{\int_0^\infty k^{n_{s,\Lambda}+2}T^2(\vec{p}_\Lambda,k)W^2(kR_{8,\Lambda})dk}}\,, \tag{19}$$

where W is a top-hat smoothing function and $T(\vec{p},k)$ the transfer function.[21] The fitting parameters for each model are contained in $\vec{p}$. Following the above mentioned references, we define as fiducial model the ΛCDM at fixed parameter values from the Planck 2015 TT, TE, EE + lowP + lensing data.[9] These fiducial values are collected in $\vec{p}_\Lambda$. The theoretical calculation of $\sigma_8(z)$ and of the product $f(z)\sigma_8(z)$ for each model is essential to compare with the LSS formation data. The calculation is possible after obtaining the fitting results for the parameters, as indicated in Tables 1 and 2 and Fig. 1. The result for $f(z)\sigma_8(z)$ is plotted in Fig. 2, together with the observational points. In Sec. 7 we will further discuss these results. In the next section we discuss some basic facts of the fit analysis.

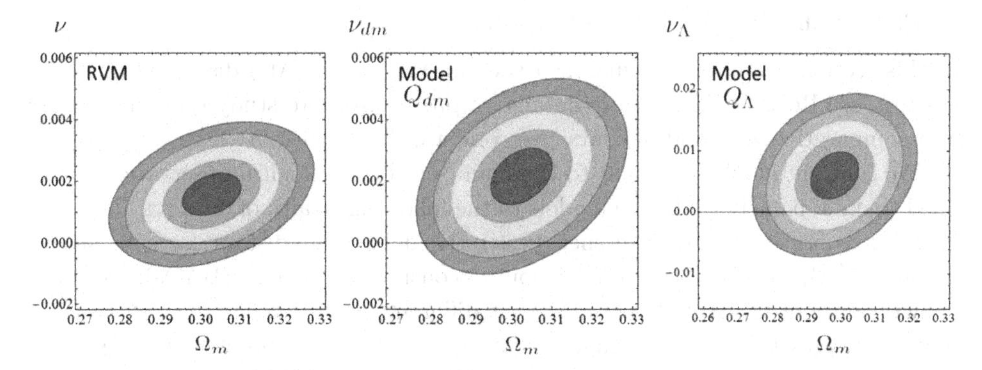

Fig. 1. Likelihood contours in the (Ω_m, ν_i)-plane for the three DVMs I, II and III (we are restricting here to the vacuum case $w = -1$ in all cases) defined in (6)–(8). Shown are the regions corresponding to $-2 \ln \mathcal{L}/\mathcal{L}_{\max} = 2.30, 6.18, 11.81, 19.33, 27.65$ (corresponding to 1σ, 2σ, 3σ, 4σ and 5σ c.l.) after marginalizing over the rest of the fitting parameters indicated in Table 1. The elliptical shapes have been obtained applying the standard Fisher approach. We estimate that for the RVM, 94.80% (resp. 89.16%) of the 4σ (resp. 5σ) area is in the $\nu > 0$ region. The ΛCDM ($\nu_i = 0$) appears disfavored at $\sim 4\sigma$ c.l. in the RVM and Q_{dm}, and at $\sim 2.5\sigma$ c.l. for Q_Λ. For more details, see Ref. 21.

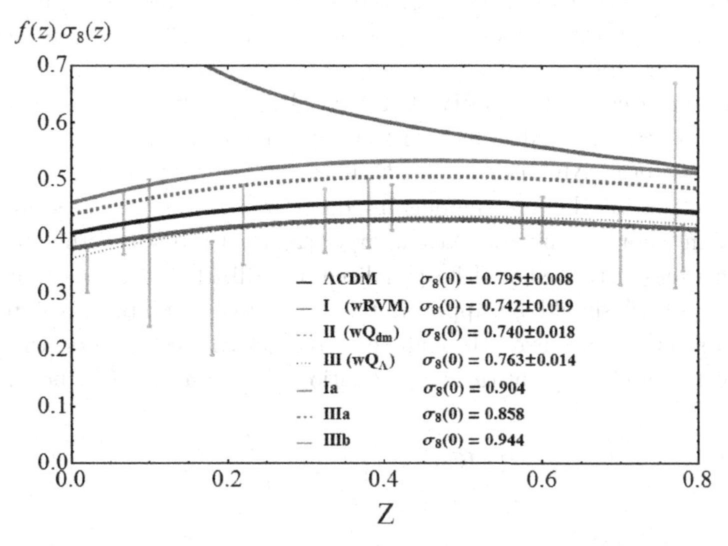

Fig. 2. The large scale structure (LSS) formation data $(f(z)\sigma_8(z))$ and the theoretical predictions for Models I, II and III in the case $w \neq -1$ (i.e. the wDVMs). The computed values of $\sigma_8(0)$ for each model are also indicated. The curves for the cases Ia and IIIa correspond to special scenarios for Models I and III, in which only the BAO and CMB data are used (not the LSS). Despite the agreement of the CMB measurement $H_0^{\rm Planck}$ with the Riess *et al.* local value $H_0^{\rm Riess}$ can be better for these special scenarios, the price to enforce such "agreement" is that the concordance with the LSS data is now spoiled (the curves for Ia and IIIa are higher). Case IIIb is our theoretical calculation of the impact on the LSS data for the scenario proposed in Ref. 62, aimed at optimally relaxing the tension with $H_0^{\rm Riess}$. Unfortunately it is in severe disagreement with the LSS data. The last three scenarios lead to phantom-like DE and are all disfavored (at different degrees) by the LSS data.[22]

6. Fitting the DVMs to Observations

In this section, we put the dynamical vacuum models (DVMs) discussed above to the test, see Refs. 18–22 for more details. It proves useful to study them altogether in a comparative way, and of course we compare them to the ΛCDM.

We confront all these models to the main set of cosmological observations compiled to date, namely we fit the models to the following wealth of data (cf. Refs. 19–21 for a complete list of references): (i) the data from distant type Ia supernovae (SNIa); (ii) the data on baryonic acoustic oscillations (BAOs); (iii) the known values of the Hubble parameter at different redshift points, $H(z_i)$; (iv) the large scale structure (LSS) formation data encoded in the weighted linear growth rate $f(z_i)\sigma_8(z_i)$; (v) the CMB distance priors from WMAP9, Planck 2013 and Planck 2015. Thus, we use the essential observational data represented by the cosmological observables SNIa + BAO + $H(z)$ + LSS + BBN + CMB. For the analysis we have defined a joint likelihood function $\mathcal{L}$ from the product of the likelihoods for all the data sets discussed above. For Gaussian errors, the total χ^2 to be minimized reads:

$$\chi^2_{\rm tot} = \chi^2_{\rm SNIa} + \chi^2_{\rm BAO} + \chi^2_{H} + \chi^2_{f\sigma_8} + \chi^2_{\rm CMB}\,. \tag{20}$$

Each one of these terms is defined in the standard way, including the covariance matrices for each sector.[c] The fitting results for models (6)–(8) are displayed in terms of contour plots in Fig. 1.

For the numerical fitting details, cf. Table 1. In such table we assess also the comparison of the various models in terms of the time-honored Akaike and Bayesian information criteria, AIC and BIC.[42,43] These information criteria are extremely useful for comparing different models in competition. The reason is obvious: the models having more parameters have a larger capability to adjust the observations, and for this reason they should be penalized accordingly. It means that the minimum value of χ^2 should be appropriately corrected so as to take into account this feature. This is achieved through the AIC and BIC estimators, which can be thought of as a modern quantitative formulation of Occam's razor. They are defined as follows:[42–45]

$$21{\rm AIC} = \chi^2_{\rm min} + \frac{2nN}{N-n-1} \tag{21}$$

and

$$22{\rm BIC} = \chi^2_{\rm min} + n\,\ln N\,. \tag{22}$$

Here n is the number of independent fitting parameters and N the number of data points. The larger are the differences ΔAIC (ΔBIC) with respect to the model that carries smaller value of AIC (BIC) — the DVMs here — the higher is the evidence against the model with larger value of AIC (BIC) — the ΛCDM. For ΔAIC and/or ΔBIC in the range 6–10 we can speak of "strong evidence" against the ΛCDM,

[c]See the details in Appendix B of Ref. 21.

and hence in favor of the DVMs. Above 10, we are entitled to claim "very strong evidence"[42–45] in favor of the DVMs.

From Table 1 we can read off the results: for Models I and II we find ΔAIC > 10 and ΔBIC > 10 simultaneously. The results from both are consistent with the fact that these models yield a nonvanishing value for ν_i with the largest significance ($\sim 4\sigma$). It means that the DVMs I and II fit better the overall data than the ΛCDM at such confidence level. Model III (i.e. Q_Λ) also fits better the observations than the ΛCDM, but with a lesser c.l. Indeed, in this case ΔAIC > 5 and ΔBIC > 3, and the parameter ν_Λ is nonvanishing at near 3σ. Thus, the three DVMs are definitely more favored than the ΛCDM, and the most conspicuous one is the RVM, Eq. (3).

We conclude that the wealth of cosmological data at our disposal currently suggests that the hypothesis $\Lambda =$ const. despite being the simplest may well not be the most favored one. The absence of vacuum dynamics is excluded at nearly 4σ c.l. as compared to the best DVM considered here (the RVM). The strength of this statement is riveted with the firm verdict of Akaike and Bayesian criteria. Overall we have collected a fairly strong statistical support of the conclusion that the SNIa+ BAO + $H(z)$ + LSS + CMB cosmological data do favor a mild dynamical vacuum evolution.

7. Dynamical Vacuum and the H_0 Tension. A Case Study

The framework outlined in the previous sections suggest that owing to a possible small interaction with matter, the vacuum energy density might be evolving with the cosmic expansion. This opens new vistas for a possible understanding of the well known discrepancy between the CMB measurements of H_0,[9,10] suggesting a value below 70 km/s/Mpc, and the local determinations emphasizing a higher range clearly above 70 km/s/Mpc.[46]

7.1. *A little of history*

Such tension is reminiscent of the prediction by the famous astronomer A. Sandage in the sixties, who asserted[47] that the main task of future observational cosmology would be the search for two parameters: the Hubble constant H_0 and the deceleration parameter q_0. The first of these parameters defines the most important distance (and time) scale in cosmology prior to any other cosmological quantity. Sandage's last published value with Tammann (in 2010) is 62.3 km/s/Mpc[48] — a value that was slightly revised in Ref. 49 as $H_0 = 64.1 \pm 2.4$ km/s/Mpc after due account of the high-weight TRGB (tip of the red-giant branch) calibration of SNIa.

As for the deceleration parameter, q_0, its measurement is equivalent to determining Λ in the context of the concordance model. As indicated in the introduction, on fundamental grounds understanding the value of Λ may not just be a matter of observation; in actual fact it embodies one of the most important and unsolved conundrums of theoretical physics and cosmology of all times: the cosmological constant problem.[6,d] It is our contention that a better understanding of H_0 is related to a deeper knowledge of the nature of Λ, in particular whether the rigid option $\Lambda = $ const. could be superseded by the more flexible notion of dynamical vacuum energy. Obviously this has an implication on the value of H_0, as shown in Table 1.

The actual value of H_0 has a long and tortuous history, and the tension among the different measurements is inherent to it. Let us only recall that after Baade's revision (by a factor of one half[50]) of the exceedingly large value ~ 500 km/s/Mpc originally estimated by Hubble (implying a universe of barely two billion years old only), the Hubble parameter was further lowered to 75 km/s/Mpc and finally was captured in the range $H_0 = 55 \pm 5$ km/s/Mpc, where it remained for 20 years (until 1995), mainly under the influence of Sandage's devoted observations.[51] See nevertheless[52] for an alternative historical point of view, in which a higher range of values is advocated. Subsequently, with the first historical observations of the accelerated expansion of the universe, suggesting a positive value of Λ,[53,54] the typical range for H_0 moved upwards to ~ 65 km/s/Mpc. In the meantime, a wealth of observational values of H_0 have piled up in the literature using different methods (see e.g. the median statistical analysis of > 550 measurements considered in Refs. 55 and 56).

7.2. *The documented tension*

As mentioned above, two kinds of *precision* (few percent level) measurements of H_0 have generated considerable perplexity in the recent literature, specifically between the latest Planck values ($H_0^{\rm Planck}$) obtained from the CMB anisotropies, and the local HST measurement (based on distance estimates from Cepheids). The latter, obtained by Riess *et al.* is[46]

$$H_0^{\rm Riess} = 73.24 \pm 1.74 \text{ km/s/Mpc} \tag{23}$$

whereas the CMB value, depending on the kind of data used, reads[9]

$$H_0^{\rm Planck\ 2015} = 67.51 \pm 0.64 \text{ km/s/Mpc (TT, TE, EE + lowP + lensing data)}, \tag{24}$$

and[10]

$$H_0^{\rm Planck\ 2016} = 66.93 \pm 0.62 \text{ km/s/Mpc (TT, TE, EE + SIMlow data)}. \tag{25}$$

[d]There is a famous saying by Allan Sandage: "... it is not a matter of debate, it is a matter of observation,"[49] which we may as well apply here. The CC problem is a matter of debate, no doubt about it, but once more we should concur with him that it is above all a matter of observation; for observing if Λ is a rigid parameter or a dynamical variable can also greatly help in the way we should finally tackle the CC problem! It may actually lead to the clue for its understanding. Is this not, after all, how real physics works? It is time for less formal theory and more observations!

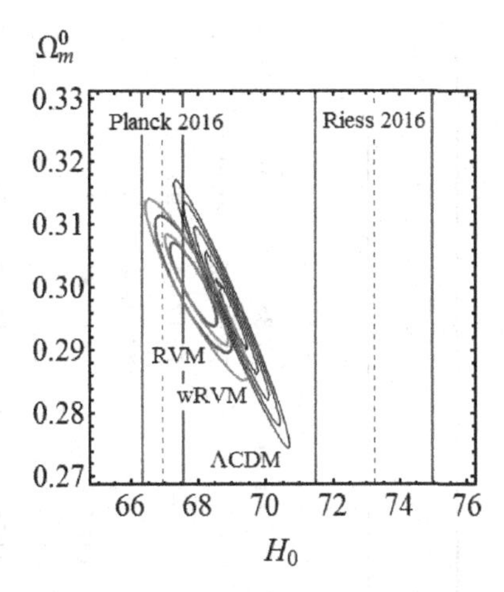

Fig. 3. (Color online) Contour plots in the (H_0, Ω_m^0)-plane for the RVM (blue) and wRVM (orange) up to 2σ, together with those for the ΛCDM (black) up to 5σ, corresponding to the situation when the local H_0 value of Riess *et al.*[46] is included as a data point in the fit (cf. Table 2).[22]

In both cases there is a tension above 3σ c.l. (viz. 3.1σ and 3.4σ, respectively) with respect to the local measurement.[e] We will refer the Planck measurement collectively as $H_0^{\rm Planck}$ since the tension with the local measurement (23) is essentially the same. This situation, and in general a certain level of tension with some independent observations in intermediate cosmological scales, has stimulated a number of discussions and possible solutions in the literature, see e.g. Refs. 57–62.

7.3. *Refitting the overall data in the presence of* $H_0^{\rm Riess}$

We wish to reexamine here the $H_0^{\rm Riess} - H_0^{\rm Planck}$ tension, but not as an isolated conflict between two particular sources of observations, but rather in light of the overall fit to the current cosmological SNIa + BAO + $H(z)$ + LSS + CMB data. In other words, it is worthwhile to reconsider the fitting results of Table 1 when we introduce $H_0^{\rm Riess}$ as an explicit data point in the fit. The result is recorded in Table 2.

In Fig. 2 we plot $f(z)\sigma_8(z)$ for the various Models I, II and III for the case with constant $w \neq -1$ using the fitted values of Table 2 and some other special situations described in the caption. The case $w = -1$ is not plotted in Fig. 2 because it is visually undistinguishable from the case w near -1. The numerical differences in the fitting results, however, are not negligible as can be see on comparing Tables 1 and 2.

Finally, in Fig. 3 we display the contour plots in the (H_0, Ω_m^0)-plane for the RVM (blue) and wRVM (orange) up to 2σ, together with those for the ΛCDM (black)

[e]It is suggestive to see that the late local measurements of H_0 obtained by Sandage and Tammann,[48] based on Cepheids and SNIa, are closer to the current CMB measurements than to the recent Riess *et al.* local measurements based on similar techniques. It makes one think.

Table 2. Best-fit values for the free parameters of the ΛCDM, XCDM, the three dynamical vacuum models (DVMs) and the three dynamical quasi-vacuum models (wDVMs), including their statistical significance (the values of the χ^2-test and the difference of the Akaike and Bayesian information criteria AIC and BIC with respect to the ΛCDM), i.e. the same as in Table 1 but now adding the H_0 local measurement $H_0^{\rm Riess}$, as indicated in Eq. (23).

Model	h	ω_b	n_s	Ω_m^0	ν_i	w	$\chi^2_{\rm min}/{\rm dof}$	ΔAIC	ΔBIC
ΛCDM	0.690 ± 0.003	0.02247 ± 0.00013	0.974 ± 0.003	0.296 ± 0.004	—	—	90.59/86	—	—
XCDM	0.680 ± 0.006	0.02252 ± 0.00013	0.975 ± 0.004	0.304 ± 0.006	—	-0.960 ± 0.023	87.38/85	0.97	−1.29
RVM	0.679 ± 0.005	0.02232 ± 0.00014	0.967 ± 0.004	0.300 ± 0.004	0.00133 ± 0.00040	—	78.96/85	9.39	7.13
Q_{dm}	0.679 ± 0.005	0.02230 ± 0.00014	0.966 ± 0.004	0.300 ± 0.004	0.00185 ± 0.00057	—	79.17/85	9.18	6.92
Q_Λ	0.690 ± 0.003	0.02224 ± 0.00016	0.965 ± 0.005	0.297 ± 0.004	0.00669 ± 0.00234	—	82.48/85	5.87	3.61
wRVM	0.680 ± 0.007	0.02230 ± 0.00015	0.966 ± 0.005	0.300 ± 0.006	0.00138 ± 0.00048	-1.005 ± 0.028	78.93/84	7.11	2.66
wQ_{dm}	0.679 ± 0.007	0.02230 ± 0.00016	0.966 ± 0.005	0.300 ± 0.006	0.00184 ± 0.00066	-0.999 ± 0.028	79.17/84	6.88	2.42
wQ_Λ	0.679 ± 0.006	0.02227 ± 0.00016	0.966 ± 0.005	0.306 ± 0.006	0.00689 ± 0.00237	-0.958 ± 0.022	78.98/84	7.07	2.61

up to 5σ, corresponding to the situation when the local H_0 value of Riess *et al.*[46] is included as a data point in the fit (cf. Table 2).[22] One can see that when all data sources SNIa + BAO + $H(z)$ + LSS + BBN + CMB are used, the price for reaching the vicinity of $H_0^{\rm Riess}$ is a too small value of Ω_m^0 around 0.27 and requires extended contours beyond 5σ c.l. The figure also shows that both the RVM and wRVM intersect much better (already at 1σ) the $H_0^{\rm Planck}$ range than the ΛCDM. The latter requires also 5σ contours to reach $H_0^{\rm Planck}$, and Ω_m^0 near 0.32. In other words, when the local value $H_0^{\rm Riess}$ enters the fit, the ΛCDM is in a rather uncomfortable position, as it is almost far-equidistant from both the $H_0^{\rm Planck}$ and $H_0^{\rm Riess}$ domains! In short, our analysis of the overall SNIa + BAO + $H(z)$ + LSS + BBN + CMB observations shows that the ΛCDM model cannot comfortably account neither for the Planck nor for the Riess *et al.* measurements of the Hubble parameter. However, when we reanalyze the same set of data within the RVM we find that the Planck range of values of $H_0^{\rm Planck}$ is clearly favored. It is interesting to remark at this point that in two recent observational works, in which the corresponding authors perform a reanalysis of the local determination of H_0 within the ΛCDM, they find values that are in between the Planck and Riess et al ranges (and with larger errors), see Refs. 63 and 64. Their results are more compatible with our own determination of H_0 in Table 1 for the ΛCDM. The fact that the RVM introduces a certain degree of vacuum dynamics, allows this model to favor lower values of H_0, what eases the RVM prediction to overlap more easily with the Planck range. This holds good even in the presence of the Riess *et al.* data point (cf. Fig. 3), which tends to drag the curves away from the Planck range, i.e. towards the right part of Fig. 3.

8. Conclusions

In this work, we have reviewed the status of the dynamical vacuum models (DVMs) in their ability to compete with the ΛCDM model (namely the standard or concordance model of cosmology) to fit the overall SNIa+BAO+$H(z)$+LSS+BBN+CMB cosmological observations. We find that the current cosmological data disfavors the ΛCDM, and hence the Λ = const. hypothesis, in a very significant way. The best fit value to the overall data is provided by the running vacuum model (RVM), at a significance level of roughly $\sim 4\sigma$ as compared to the ΛCDM.

We have also used these models to reanalyze the tension between the Riess *et al.* local measurement $H_0^{\rm Riess}$ and the value obtained in the CMB measurements from the Planck satellite, $H_0^{\rm Planck}$, which is 3σ smaller. We find that the fit quality to the overall SNIa+BAO+$H(z)$+LSS+CMB cosmological data increases to the maximum level only when the local $H_0^{\rm Riess}$ measurement is not taken into account. In other words, the CMB determination of H_0 is clearly preferred. We demonstrate that not only the CMB and BAO, but also the LSS data, are essential to grant these results. We have also comparatively considered the performance of the wDVMs (i.e. the dynamical quasi-vacuum models with $w \neq 1$), and we have found that they are also able to improve the ΛCDM fit, although to a lesser extent than the best DVMs.

However, the extra degree of freedom associated to the free parameter w in these models can be used to try to enforce a minimal $H_0^{\rm Riess} - H_0^{\rm Planck}$ tension. What we find[22] is that if the LSS data are not considered in the fit analysis, the tension can indeed be diminished, but only at the expense of a phantom-like dynamical behavior of the DE, namely w turns out to satisfy $w \lesssim -1$. But the main problem is that this results in a serious disagreement with the structure formation data. Such disagreement disappears when the LSS data are restored, and in fact a good fit quality to the overall observations (better than the ΛCDM at 3σ c.l.) can be achieved, with no trace of phantom dynamical DE energy. In the absence of the local $H_0^{\rm Riess}$ measurement, the fit quality further increases in favor of the main dynamical vacuum models up to $\sim 4\sigma$ c.l. In general the vacuum dynamics tends to favor the CMB determination of H_0 against the local measurement $H_0^{\rm Riess}$, but this measurement can still be accommodated in the fit without seriously spoiling the capacity of the RVM to improve the ΛCDM fit. These results are bolstered by outstanding marks of the information criteria (yielding ΔAIC > 10 and ΔBIC > 10) in favor of the main DVMs and against the concordance model. To summarize, we claim that significant signs of dynamical vacuum energy are sitting in the current cosmological data, which the concordance ΛCDM model is unable to accommodate.

Acknowledgments

J. Solà is thankful to Prof. Harald Fritzsch for the kind invitation to this stimulating conference on Cosmology, Gravitational Waves and Particles. J. Solà would also like to thank Prof. K. K. Phua for inviting him to present this contribution in the review section of IJMPA. We have been supported by MINECO FPA2016-76005-C2-1-P, Consolider CSD2007-00042, 2014-SGR-104 (Generalitat de Catalunya) and MDM-2014-0369 (ICCUB). J. Solà is also particularly grateful for the support by the Institute for Advanced Study of the Nanyang Technological University in Singapore.

References

1. A. Einstein, Kosmologische Betrachtungen zur allgemeinen Relativitätstheorie, *Sitzungsber. Königl. Preuss. Akad. Wiss. Phys.-Math. Klasse* **VI**, 142 (1917).
2. A. Einstein, Zum kosmologischen Problem der allgemeinen Relativitätstheorie, *Sitzungsber. Königl. Preuss. Akad. Wiss. Phys.-Math. Klasse* **XII**, 235 (1931).
3. A. Einstein and W. de Sitter, On the relation between the expansion and the mean density of the universe, *Proc. Nat. Acad. Sci.* **18**, 213–214 (1932).
4. G. Lemaître, Evolution of the expanding universe, *Proc. Nat. Acad. Sci.* **20**, 12–17 (1934).
5. Y. B. Zeldovich, Cosmological constant and elementary particles, *JETP Lett.* **6**, 316 (1967); *Pisma Zh. Eksp. Teor. Fiz.* **6**, 883 (1967); Cosmological constant and the theory of elementary particles, *Sov. Phys. Usp.* **11**, 381 (1968); republished in *Gen. Rel. Grav.* **40**, 1557 (2008) (edited by V. Sahni and A. Krasinski).
6. S. Weinberg, *Rev. Mod. Phys.* **61**, 1 (1989).
7. V. Sahni and A. Starobinsky, *Int. J. Mod. Phys. A* **9**, 373 (2000); T. Padmanabhan, *Phys. Rep.* **380**, 235 (2003); P. J. E. Peebles and B. Ratra, *Rev. Mod. Phys.* **75**, 559 (2003).

8. J. Solà, Cosmological constant and vacuum energy: Old and new ideas, *J. Phys. Conf. Ser.* **453**, 012015 (2013), arXiv:1306.1527.
9. Planck Collab. (P. A. R. Ade *et al.*), Planck 2015 results. XIII, *Astron. Astrophys.* **594**, A13 (2016).
10. Planck Collab. (N. Aghanim *et al.*), Planck 2016 intermediate results. XLVI, *Astron. Astrophys.* **596**, A107 (2016).
11. J. Solà and A. Gómez-Valent, *Int. J. Mod. Phys. D* **24**, 1541003 (2015).
12. J. Solà, *Int. J. Mod. Phys. D* **24**, 1544027 (2015).
13. A. Gómez-Valent, E. Karimkhani and J. Solà, *JCAP* **12**, 048 (2015).
14. A. Gómez-Valent and J. Solà, *Mon. Not. R. Astron. Soc.* **448**, 2810 (2015).
15. A. Gómez-Valent, J. Solà and S. Basilakos, *JCAP* **01**, 004 (2015).
16. S. Basilakos, M. Plionis and J. Solà, *Phys. Rev. D* **80**, 3511 (2009).
17. J. Grande, J. Solà, S. Basilakos and M. Plionis, *JCAP* **1108**, 007 (2011).
18. J. Solà, A. Gómez-Valent and J. de Cruz Pérez, *Astrophys. J.* **811**, L14 (2015).
19. J. Solà, A. Gómez-Valent and J. de Cruz Pérez, *Astrophys. J.* **836**, 43 (2017).
20. J. Solà, A. Gómez-Valent and J. de Cruz Pérez, *Mod. Phys. Lett. A* **32**, 1750054.
21. J. Solà, J. de Cruz Pérez and A. Gómez-Valent, Towards the firsts compelling signs of vacuum dynamics in modern cosmological observations, arXiv:1703.08218.
22. J. Solà, A. Gómez-Valent and J. de Cruz Pérez, The H_0 tension in light of vacuum dynamics in the Universe, arXiv:1705.06723.
23. H. Fritzsch and J. Solà, *Class. Quantum Grav.* **29**, 215002 (2012).
24. J. Solà, *Mod. Phys. Lett. A* **30**, 1502004 (2015).
25. H. Fritzsch and J. Solà, *Mod. Phys. Lett. A* **30**, 1540034 (2015).
26. J. Solà (ed.), Fundamental constants in physics and their time variation, *Mod. Phys. Lett. A* **30** (2015), Special Issue.
27. J. Solà, *Int. J. Mod. Phys. A* **29**, 1444016 (2014).
28. H. Fritzsch, R. C. Nunes and J. Solà, *Eur. Phys. J. C* **77**, 193 (2017).
29. P. J. E. Peebles, *Astrophys. J.* **284**, 439 (1984).
30. J. Solà, *J. Phys. A* **41**, 164066 (2008).
31. J. A. S. Lima, S. Basilakos and J. Solà, *Mon. Not. R. Astron. Soc.* **431**, 923 (2013).
32. J. A. S. Lima, S. Basilakos and J. Solà, *Gen. Rel. Grav.* **47**, 40 (2015).
33. J. A. S. Lima, S. Basilakos and J. Solà, *Eur. Phys. J. C* **76**, 228 (2016).
34. V. Salvatelli *et al.*, *Phys. Rev. Lett.* **113**, 181301 (2014).
35. Y. H. Li, J. F. Zhang and X. Zhang, *Phys. Rev. D* **93**, 023002 (2016).
36. R. Murgia, S. Gariazzo and N. Fornengo, *JCAP* **1604**, 014 (2016).
37. S. M. Turner and M. White, *Phys. Rev. D* **56**, R4439 (1997).
38. M. Chevallier and D. Polarski, *Int. J. Mod. Phys. D* **10**, 213 (2001); E. V. Linder, *Phys. Rev. Lett.* **90**, 091301 (2003); *Phys. Rev. D* **70**, 023511 (2004).
39. J. Grande, A. Pelinson and J. Solà, *Phys. Rev. D* **79**, 043006 (2009).
40. J. Grande, R. Opher, A. Pelinson and J. Solà, *J. Cosmol. Astropart. Phys.* **0712**, 007 (2007).
41. A. Gómez-Valent and J. Solà, in preparation.
42. H. Akaike, *IEEE Trans. Autom. Control* **19**, 716 (1974).
43. G. Schwarz, *Annals of Statistics* **6**, 461 (1978).
44. R. E. Kass and A. Raftery, *J. Amer. Statist. Assoc.* **90**, 773 (1995).
45. K. P. Burnham and D. R. Anderson, *Model Selection and Multimodel Inference* (Springer, New York, 2002).
46. A. G. Riess *et al.*, *Astrophys. J.* **826**, 56 (2016).
47. A. Sandage, The ability of the 200-inch telescope to discriminate between selected world models, *Astrophys. J.* **133**, 355 (1961).

48. G. A. Tammann and A. Sandage, The hubble constant and HST, in *The Impact of HST on European Astronomy, Astrophysics and Space Science Proceedings*, ed. F. Macchetto (Springer, Dordrecht, 2010), p. 289.
49. G. A. Tammann and B. Reindl, *IAU Symp.* **289**, 13 (2013); *Astron. Astrophys.* **549**, A136 (2013).
50. W. Baade, *Astrophys. J.* **100**, 137 (1944).
51. G. A. Tammann, *Publ. Astron. Soc. Pac.* **108**, 1083 (1996).
52. S. van den Bergh, *Publ. Astron. Soc. Pac.* **108**, 1091 (1996).
53. A. G. Riess *et al.*, *Astron. J.* **116**, 1009 (1998).
54. S. Perlmutter *et al.*, *Astrophys. J.* **517**, 565 (1999).
55. G. Chen and B. Ratra, *Publ. Astron. Soc. Pac.* **123**, 1127 (2011).
56. S. Bethapudi and S. Desai, *Eur. Phys. J. Plus* **132**, 78 (2017).
57. E. D. Valentino, A. Melchiorri and J. Silk, *Phys. Lett. B* **761**, 242 (2016).
58. J. L. Bernal, L. Verde and A. G. Riess, *JCAP* **1610**, 019 (2016).
59. A. Shafieloo and D. K. Hazra, *JCAP* **1704**, 012 (2017).
60. W. Cardona, M. Kunz and V. Pettorino, *JCAP* **1703**, 056 (2017).
61. E. D. Valentino, A. Melchiorri, E. V. Linder and J. Silk, Constraining dark energy dynamics in extended parameter space, arXiv:1704.00762.
62. E. D. Valentino, A. Melchiorri and O. Mena, Can interacting dark energy solve the H_0 tension?, arXiv:1704.08342.
63. B. R. Zhang *et al.*, A blinded determination of H_0 from low-redshift Type Ia supernovae, calibrated by Cepheid variables, arXiv:1706.07573.
64. Y. Wang, L. Xu and G.-B. Zhao, A measurement of the Hubble constant using galaxy redshift surveys, arXiv:1706.09149.

Dark Matter and Excited Weak Bosons

Harald Fritzsch
Department für Physik, Ludwig-Maximilians-Universität, München, Germany
fritzsch@mpp.mpg.de

The weak bosons are not elementary gauge bosons, but bound states of two fermions. Here the excitations of the weak bosons and the new fermions are discussed — they might provide the dark matter in the universe.

The weak bosons might not be elementary gauge bosons, but bound states of two fermions, analogous to the ρ-mesons in QCD. The weak bosons are the ground states. The scalar boson with a mass of 125 GeV, discovered at the LHC,[1,2] would be an excitation of the Z-boson.

A theory, in which the weak bosons are bound states, describes also the leptons and quarks as bound states. Such theories were studied in the past by many theorists.[3–11]

The weak bosons are bound states of a fermion and its antiparticle, which are denoted as "haplons" (see also Refs. 7 and 9). Their dynamics is described by a confining gauge theory, denoted as "quantum haplodynamics" (QHD).

The QHD mass scale is given by a mass parameter Λ_h, which determines the size of the weak bosons. The QHD mass scale is about 0.2 TeV.[12–14] The haplons are massless and interact with each other through the exchange of massless gauge bosons.

Two types of haplons are needed as constituents of the weak bosons, denoted by α and β. The charges of the weak bosons do not fix the charges of the haplons. They could be, for example, $(+1/2)$ and $(-1/2)$, but they might also be the same as the charges of the quarks: $(+2/3)$ and $(-1/3)$. We shall assume, that this is the case.

The three weak bosons have the following internal structure:

$$W^+ = (\bar{\beta}\alpha)\,, \qquad W^- = (\bar{\alpha}\beta)\,, \qquad W^3 = \frac{1}{\sqrt{2}}(\bar{\alpha}\alpha - \bar{\beta}\beta)\,. \tag{1}$$

The weak bosons consist of pairs of haplons, which are in an s-wave. The spins of the two haplons are aligned, as the spins of the quarks in a ρ-meson. The first excited states are those, in which the two haplons are in a p-wave. The weak isospin and the angular momentum of these states are described by the two numbers (I, J).

There are three SU(2) singlets: $S(0)$, $S(1)$ and $S(2)$ as well as three SU(2) triplets: $T(0)$, $T(1)$ and $T(2)$. These bosons are analogous to the mesons in strong interaction physics, in which the quarks are in a p-wave. The scalar meson σ, the vector meson $h_1(1170)$ and the tensor meson $f_2(1270)$ are the QCD-analogs of the singlet states $S(0)$, $S(1)$ and $S(2)$.

The isospin triplet mesons, the scalar meson $a_0(980)$, the vector meson $b_1(1235)$ and the tensor meson $a_2(1320)$, correspond to the bosons $T(0)$, $T(1)$ and $T(2)$.

We assume that the boson $S(0)$ is the particle, discovered at CERN — thus the mass of $S(0)$ is about 125 GeV. In analogy to QCD we expect that the masses of the other p-wave states are between 0.26 TeV and 0.41 TeV. Here are the expected masses of the bosons $S(1)$ and $S(2)$:

$$\begin{aligned} &S(1)\colon 0.32 \text{ TeV} \pm 0.06 \text{ TeV}\,, \\ &S(2)\colon 0.34 \text{ TeV} \pm 0.06 \text{ TeV}\,. \end{aligned} \tag{2}$$

The masses of the SU(2)-triplet bosons T are slightly larger than the masses of the S-bosons:

$$\begin{aligned} &T(0)\colon 0.25 \text{ TeV} \pm 0.05 \text{ TeV}\,, \\ &T(1)\colon 0.33 \text{ TeV} \pm 0.05 \text{ TeV}\,, \\ &T(2)\colon 0.36 \text{ TeV} \pm 0.06 \text{ TeV}\,. \end{aligned} \tag{3}$$

The $S(0)$-boson will decay into two charged weak bosons, into two Z-bosons, into a photon and a Z-boson, into two photons, into a lepton and an anti-lepton or into a quark and an anti-quark. Details were discussed in Ref. 5. The expected decay rates are similar to the decay rates, expected for a Higgs-particle.

According to the theory a Higgs particle decays into two fermions with the probability of about 66.7%, into two gluons with the probability of about 8.6% and into two gauge bosons with the probability of about 24.7%. In our model there is no decay of the $S(0)$-boson into two gluons. Decays into pairs of leptons and quarks are possible, but cannot be calculated.

The bosons $S(1)$ and $S(2)$ and the nine T-bosons will decay mainly into two or three weak bosons or photons and into a pair of leptons or quarks. The bosons $S(1)$ and $T(1)$ are vector bosons, thus they cannot decay into two photons. We shall consider in detail the decays of the charged boson $T(0,+)$ and of the neutral boson $T(0,0)$, taking into account the available phase space.

If we assume that the mass of $T(0,0)$ is 0.25 TeV and that there is no decay into two leptons or quarks, we find the following branching ratios:

$$
\begin{aligned}
&\mathrm{BR}(T(0,0) \Rightarrow W^+ + W^-) \approx 0.37\,, \\
&\mathrm{BR}(T(0,0) \Rightarrow Z + Z) \approx 0.09\,, \\
&\mathrm{BR}(T(0,0) \Rightarrow Z + \gamma) \approx 0.08\,, \\
&\mathrm{BR}(T(0,0) \Rightarrow \gamma + \gamma) \approx 0.02\,, \\
&\mathrm{BR}(T(0,0) \Rightarrow W^+ + W^- + \gamma) \approx 0.08\,, \\
&\mathrm{BR}(T(0,0) \Rightarrow W^+ + W^- + Z) \approx 0.21\,, \\
&\mathrm{BR}(T(0,0) \Rightarrow Z + Z + Z) \approx 0.05\,, \\
&\mathrm{BR}(T(0,0) \Rightarrow Z + Z + \gamma) \approx 0.07\,, \\
&\mathrm{BR}(T(0,0) \Rightarrow Z + \gamma + \gamma) \approx 0.03\,.
\end{aligned} \tag{4}
$$

In a similar way we calculate the decay rates for the decay of the charged boson $T(0,+)$. If the mass of this boson is 0.25 TeV, we obtain for the branching ratios:

$$
\begin{aligned}
&\mathrm{BR}(T(0,+) \Rightarrow W^+ + Z) \approx 0.40\,, \\
&\mathrm{BR}(T(0,+) \Rightarrow W^+\gamma) \approx 0.15\,, \\
&\mathrm{BR}(T(0,+) \Rightarrow W^+ + W^+ + W^-) \approx 0.30\,, \\
&\mathrm{BR}(T(0,+) \Rightarrow W^+ + Z + Z) \approx 0.08\,, \\
&\mathrm{BR}(T(0,+) \Rightarrow W^+ + Z + \gamma) \approx 0.06\,, \\
&\mathrm{BR}(T(0,+) \Rightarrow W^+ + \gamma + \gamma) \approx 0.01\,.
\end{aligned} \tag{5}
$$

The excited weak bosons can be observed in the LHC experiments. If the $T(0,0)$ boson has a mass of 0.25 TeV, the expected rate for the decay into two photons should be about 0.08 times the rate of the decay of the $S(0)$ boson into two photons. The reduction is due to the higher mass of the $T(0,0)$ boson (see also Ref. 15).

The $T(0,0)$ boson might also be observed in the decay into two charged weak bosons (branching ratio about 37%), in the decay into two Z-bosons (branching ratio about 9%) or in the decay into a Z-boson and a photon (branching ratio about 8%).

The charged bosons $T(0,+)$ and $T(0,-)$ can be observed by the decay into a charged weak boson and a photon (branching ratio about 16%) or into a charged weak boson and a Z-boson (branching ratio about 40%). Decays into three particles, e.g. the decay into a charged weak boson and two Z-bosons, have small branching ratios and very difficult to observe.

If our model is correct, the Large Hadron Collider should soon discover the new boson $T(0,0)$, e.g. by observing the decay of this particle into two photons, and the new charged bosons $T(0,+)$ and $T(0,-)$, e.g. by observing the decay into a charged weak boson and a photon.

We also expect that there are fermions, composed of three haplons, in particular the two ground states:

$$D^+ = (\alpha\alpha\beta)\,, \qquad D^0 = (\alpha\beta\beta)\,. \tag{6}$$

These fermions are the QHD-analogies of the proton and neutron in QCD. Since the haplons are massless, the mass of the charged D-fermion is slightly larger than the mass of the neutral D-fermion. Thus it would decay into the neutral fermion, emitting a virtual weak boson, decaying into a muon or a positron and a neutrino.

The neutral D-fermion has haplon number 3 and would be stable. The mass of this particle is expected to be in the range 0.5–1 TeV. The mass difference between the two D-fermions should be about 1 GeV.

Shortly after the Big Bang the D-fermions and their antiparticles would be produced. Since there is a small asymmetry between the matter and the antimatter in the universe, due to the violation of the CP-symmetry, there would be more D-fermions than anti-D-fermions. These would annihilate with the D-fermions, and finally the universe would contain, besides protons and neutrons, a gas of stable neutral D-fermions, providing the dark matter in our universe.

The average density of dark matter in our galaxy is about 0.4 GeV/cm^3. If we assume, for example, that the mass of a D-fermion is 0.5 TeV, there should be 780 D-fermions/m^3. The average energy of the D-fermions should be about the same as the energy of the photons in the cosmic background radiation (0.24 meV). Thus the D-fermions would have a velocity of about 7 m/sec.

A neutral D-fermion can emit a virtual Z-boson, which interacts with an atomic nucleus. In a collision of a neutral D-fermion with a nucleus one can observe in specific experiments the sudden change of the momentum of the nucleus. After the collision the nucleus would have a velocity of about 10 m/sec. Thus the neutral D-fermions can be observed indirectly, e.g. at the Gran Sasso Laboratory. The present experiments give a limit on the mass for such a dark matter particle — it must be larger than 400 GeV.

The D-fermions can be produced in pairs at the LHC. The production of a D-fermion and its antiparticle could be observed, since a large amount of the energy would be missing. A charged D-fermion would decay weakly into a neutral D-fermion by emitting e.g. a muon. Thus one should observe two myons, accompanied by missing energy.

References

1. ATLAS Collab., *Phys. Lett. B* **716**, 1 (2012).
2. CMS Collab., *Phys. Lett. B* **716**, 30 (2012).
3. I. A. D'Souza and C. S. Kalman, *Preons* (World Scientific, 1992).
4. J. C. Pati and A. Salam, *Phys. Rev. D* **10**, 275 (1974).
5. H. Harari and N. Seiberg, *Phys. Lett. B* **98**, 269 (1981).
6. L. F. Abbott and E. Farhi, *Phys. Lett. B* **101**, 69 (1961).
7. H. Fritzsch and G. Mandelbaum, *Phys. Lett. B* **102**, 319 (1981).

8. R. Barbieri, R. Mohapatra and A. Masiero, *Phys. Lett. B* **105**, 369 (1981).
9. H. Fritzsch, D. Schildknecht and R. Koegerler, *Phys. Lett. B* **114**, 157 (1982).
10. T. Kugo, S. Uehara and T. Yanagida, *Phys. Lett. B* **147**, 32 (1984).
11. J. J. Dugne, S. Fredriksson and J. Hanson, *EPL* **60**, 188 (2002).
12. H. Fritzsch, *Phys. Lett. B* **712**, 231 (2012).
13. H. Fritzsch, *Mod. Phys. Lett. A* **26**, 2305 (2011).
14. H. Fritzsch, *Mod. Phys. Lett. A* **31**, 1630019 (2016).
15. J. Baglio and A. Djouadi, *JHEP* **03**, 055 (2011).

The Matter-antimatter Asymmetry Problem

B. A. Robson
Department of Theoretical Physics, Research School of Physics and Engineering, The Australian National University, Canberra ACT 2601, Australia
brian.robson@anu.edu.au

The matter-antimatter asymmetry problem, corresponding to the virtual nonexistence of antimatter in the universe, is one of the greatest mysteries of cosmology. According to the prevailing cosmological model, the universe was created in the so-called 'Big Bang' from pure energy and it is generally considered that the Big Bang and its aftermath produced equal numbers of particles and antiparticles, although the universe today appears to consist almost entirely of matter rather than antimatter. This constitutes the matter-antimatter asymmetry problem: where have all the antiparticles gone? Within the framework of the Generation Model (GM) of particle physics, it is demonstrated that the asymmetry problem may be understood in terms of the composite leptons and quarks of the GM. It is concluded that there is essentially no matter-antimatter asymmetry in the present universe and that the observed hydrogen-antihydrogen asymmetry may be understood in terms of statistical fluctuations associated with the complex many-body processes involved in the formation of either a hydrogen atom or an antihydrogen atom.

1. Introduction

The subject of my talk today is "The matter-antimatter asymmetry problem", corresponding to the virtual nonexistence of antimatter in the universe. Let me outline my talk. First, I shall introduce the matter-antimatter asymmetry problem. Currently there is no acceptable understanding of this problem. Second, I shall briefly introduce the Standard Model[1] (SM) of particle physics, indicating its shortcomings and the need for an improved model such as the Generation Model[2] (GM) in which the elementary particles of the SM have a substructure. Finally, it will be shown that the substructure of the leptons and quarks leads to an understanding of the matter-antimatter problem.

According to the prevailing cosmological model,[3] the universe was created in the so-called 'Big Bang' from pure energy, and is currently composed of about 5% ordinary matter, 27% dark matter and 68% dark energy. It is generally considered that the Big Bang and its aftermath produced equal numbers of particles and antiparticles, although the universe today appears to consist almost entirely of matter (particles) rather than antimatter (antiparticles). The 5% ordinary matter of the universe in the SM, prior to the nucleosynthesis, i.e., the fusion into heavier elementsm is estimated[4] to consist of 92% hydrogen atoms and 8% helium atoms.

This constitutes the matter-antimatter asymmetry problem: where have all the antiparticles gone? Currently there is no acceptable understanding of this asymmetry problem within the framework of the SM. An understanding of this problem requires both knowledge of the physical nature of the Big Bang and a precise definition of *matter*. Unfortunately, knowledge of the physical nature of the Big Bang is far from complete and matter has not been defined precisely within the framework of the SM.

The prevailing model of the Big Bang is based upon the theory of General Relativity:[5] extrapolation of the expansion of the universe backwards in time yields an infinite density and temperature at a finite time in the past (approximately 13.8 billion years ago). Thus the 'birth' of the universe seems to be associated with a 'singularity', which not only signals a breakdown of the theory of General Relativity, but also all the laws of physics. This is a serious impediment to understanding the matter-antimatter asymmetry problem. Consequently, this problem will be discussed in terms of the observed nature of the universe, ignoring the singularity.

A consistent definition of the terms *matter* and *antimatter* is the following: *matter* is built of elementary *matter* particles and *antimatter* is built of elementary *antimatter* antiparticles.

In the SM, the elementary *matter* particles are assumed to be the leptons and quarks so that electrons, neutrons and protons are all *matter*. In the GM (see Sect. 3), the elementary *matter* particles are 'rishons' and the elementary *antimatter* particles are 'antirishons' so that electrons, neutrons and protons are not all *matter*.

2. Standard Model of Particle Physics

The Standard Model[1] of particle physics provides an excellent account of all the experimental data involving the interactions of leptons and the multitude of hadrons (baryons and mesons) with each other and the decay modes of the unstable leptons and hadrons. This is achieved by assuming twelve elementary particles, six leptons and six quarks, and three fundamental interactions, the electromagnetic, strong and weak interactions. It has been very successful. However, most physicists consider that the SM is *incomplete*.

This is because the SM provides no understanding of several empirical observations. For example, it does not explain the occurence of three generations of the elementary particles.[6] The first generation comprising the up and down quarks, the electron and its neutrino; the second generation comprising the charmed and strange quarks, the muon and its neutrino and the third generation comprising the top and bottom quarks, the tau and its neutrino. Each generation behaves similarly except for mass. Second the SM does not provide a unified description of the origin of mass nor describe the mass hierarchy of leptons and quarks. It also fails to describe the nature of gravity, dark matter, dark energy and the matter-antimatter asymmetry problem.

In 2001, when I started to construct an improved model, I considered that the basic problem with the SM was its classification of its elementary particles in terms of additive quantum numbers. The SM employed a nonunified and complicated scheme of additive quantum numbers, some of which were not conserved in weak interaction processes. Moreover, the SM failed to provide any physical basis for its classification scheme.

Table 1. SM additive quantum numbers for leptons and quarks.

particle	Q	L	L_μ	L_τ	A	S	C	B	T
e^-	-1	1	0	0	0	0	0	0	0
ν_e	0	1	0	0	0	0	0	0	0
μ^-	-1	1	1	0	0	0	0	0	0
ν_μ	0	1	1	0	0	0	0	0	0
τ^-	-1	1	0	1	0	0	0	0	0
ν_τ	0	1	0	1	0	0	0	0	0
u	$+\frac{2}{3}$	0	0	0	$\frac{1}{3}$	0	0	0	0
d	$-\frac{1}{3}$	0	0	0	$\frac{1}{3}$	0	0	0	0
c	$+\frac{2}{3}$	0	0	0	$\frac{1}{3}$	0	1	0	0
s	$-\frac{1}{3}$	0	0	0	$\frac{1}{3}$	-1	0	0	0
t	$+\frac{2}{3}$	0	0	0	$\frac{1}{3}$	0	0	0	1
b	$-\frac{1}{3}$	0	0	0	$\frac{1}{3}$	0	0	-1	0

Table 1 shows the additive quantum numbers allotted to classify the six leptons and the six quarks that constitute the elementary matter particles of the SM. For the leptons we have: charge Q, lepton number L, muon lepton number L_μ and tau lepton number L_τ. For quarks we have: charge Q, baryon number A, strangeness S, charm C, bottomness B and topness T. Antiparticles have minus the quantum numbers of the corresponding particle. It should be noted that except for charge, leptons and quarks have different kinds of quantum numbers so that this classification is *nonunified.* Each of the additive quantum numbers is conserved in any interaction, except for S, C, B and T, which can undergo a change of one unit in weak interactions.

3. The Generation Model of Particle Physics

The Generation Model,[2] which I have been developing since 2001, based upon a simpler classification scheme, provides agreement with the SM for all the experimental data described by the SM. However, the GM also provides new physical insights into many of the empirical observations for which the SM fails to provide any understanding. It gives new paradigms for both mass and gravity. In particular, for today's talk, it allows the development of a composite model of the leptons and quarks.

Table 2. GM additive quantum numbers for leptons and quarks.

particle	Q	p	g	particle	Q	p	g
ν_e	0	-1	0	u	$+\frac{2}{3}$	$\frac{1}{3}$	0
e^-	-1	-1	0	d	$-\frac{1}{3}$	$\frac{1}{3}$	0
ν_μ	0	-1	± 1	c	$+\frac{2}{3}$	$\frac{1}{3}$	± 1
μ^-	-1	-1	± 1	s	$-\frac{1}{3}$	$\frac{1}{3}$	± 1
ν_τ	0	-1	$0, \pm 2$	t	$+\frac{2}{3}$	$\frac{1}{3}$	$0, \pm 2$
τ^-	-1	-1	$0, \pm 2$	b	$-\frac{1}{3}$	$\frac{1}{3}$	$0, \pm 2$

Table 2 shows the additive quantum numbers allotted to both leptons and quarks in the GM. This is a much simpler and unified classification scheme involving only three additive quantum numbers: charge Q, particle number p and generation quantum number g. All three quantum numbers are conserved in all interactions. In particular this classification scheme allows the development of a composite model of leptons and quarks, which I considered a necessary condition for an improved model.

Let me now turn to another topic - whether leptons and quarks are composite or not. There actually exists considerable indirect evidence that leptons and quarks are *composites.* First, the electrical charges of the electron and proton are opposite in sign but are exactly equal in magnitude so that atoms with the same number of electrons and protons are neutral. Consequently, in a proton consisting of quarks, the electrical charges of the quarks are intimately related to that of the electron: in fact the up quark has charge $Q = +\frac{2}{3}$ and the down quark has charge $Q = -\frac{1}{3}$, if the electron has charge $Q = -1$. These relations are readily comprehensible if leptons and quarks are composed of the same kinds of particles.

Second, all leptons and quarks have mass. Now it is known that most of the mass of a nucleon arises from the internal energy of its constituents, quarks and gluons, which are massless or have small mass. This suggests that leptons and quarks may also be composite. Third, the six leptons and the six quarks can be grouped into three generations or families. Each generation contains particles, which have similar properties. The existence of three repeating patterns, which is like a miniature Mendeleev table of the elements, also suggests strongly that the members of each generation are composites.

4. Composite Generation Model (CGM)

In 2005 I began construction of a GM in which leptons and quarks are composite particles. This composite GM was based on the unified classification scheme and also on early 1979 composite models of Harari[7] and Shupe.[8] The current composite GM (CGM) was proposed[9] in 2011 and is described[2] in detail in Chapter 1 of the book *Particle Physics* published by InTech in 2012. Unfortunately, today I have only the time to indicate some of the features of the CGM that are relevant for today's talk.

Both the models of Harari and Shupe are very similar and treat leptons and quarks as composites of two kinds of spin-1/2 particles, which Harari named 'rishons' from the Hebrew word for primary. The CGM adopts this name for the constituents of both leptons and quarks and for consistency the same three additive quantum numbers are assigned to the constituents as were previously allotted in the GM to leptons and quarks (see Table 2).

In the Harari-Shupe Model (HSM), two elementary spin-1/2 rishons: (i) a T-rishon with $Q = +\frac{1}{3}$ and (ii) a V-rishon with $Q = 0$ and their corresponding antiparticles (denoted in the usual way by a bar over the defining particle symbol): a $\bar{T}$-antirishon with $Q = -\frac{1}{3}$ and a $\bar{V}$-antirishon with $Q = 0$ are used to construct the leptons and quarks. In the HSM, each spin-1/2 lepton or quark is composed of three rishons or three antirishons. The HSM provided arguably the most economical and impressive description of the first generation of leptons and quarks. However, it did not provide a satisfactory understanding of the second and third generations.

Table 3. HSM of first generation of leptons and quarks.

particle	*structure*	Q
e^+	TTT	$+1$
u	TTV	$+\frac{2}{3}$
$\bar{d}$	TVV	$+\frac{1}{3}$
ν_e	VVV	0
$\bar{\nu}_e$	$\bar{V}\bar{V}\bar{V}$	0
d	$\bar{T}\bar{V}\bar{V}$	$-\frac{1}{3}$
$\bar{u}$	$\bar{T}\bar{T}\bar{V}$	$-\frac{2}{3}$
e^-	$\bar{T}\bar{T}\bar{T}$	-1

Table 3 shows the proposed HSM structures of the first generation of leptons and quarks. Basically the HSM describes only the charge structure of the first generation of leptons and quarks and does not provide a satisfactory understanding of the second and third generations.

The CGM is a major extension of the Harari-Shupe model. First, it introduced a third kind of rishon, which is required to describe the higher generations of leptons and quarks. Second, the rishons are allotted the same three additive quantum numbers as the leptons and quarks. Third, the rishons are each assumed to carry a single color charge, red, green or blue, analogous to the quarks in the SM. Table 4 shows the quantum numbers allotted to the three kinds of rishons. For each rishon additive quantum number N, the corresponding antirishon has the additive quantum number $-N$. The CGM accounts for the conservation of the three additive quantum numbers, Q, p and g as simply the conservation of each of the three kinds of rishons.

Table 4. CGM additive quantum numbers for rishons.

rishon	Q	p	g
T	$+\frac{1}{3}$	$+\frac{1}{3}$	0
V	0	$+\frac{1}{3}$	0
U	0	$+\frac{1}{3}$	-1

In the CGM, the substructure of the leptons and quarks is described in terms of massless rishons and/or antirishons. Each rishon carries a color charge, red, green or blue, while each antirishon carries an anticolor charge, antired, antigreen or antiblue. The constituents of leptons and quarks are bound together by a strong color-type interaction, corresponding to a local gauged $SU(3)$ symmetry (analogous to QCD in the SM) mediated by massless hypergluons (analogous to gluons in the SM).

Table 5. CGM of first generation of leptons and quarks.

particle	*structure*	Q	p	g
e^+	TTT	$+1$	$+1$	0
u	$TT\bar{V}$	$+\frac{2}{3}$	$+\frac{1}{3}$	0
$\bar{d}$	$T\bar{V}\bar{V}$	$+\frac{1}{3}$	$-\frac{1}{3}$	0
ν_e	$\bar{V}\bar{V}\bar{V}$	0	-1	0
$\bar{\nu}_e$	VVV	0	$+1$	0
d	$\bar{T}VV$	$-\frac{1}{3}$	$+\frac{1}{3}$	0
$\bar{u}$	$\bar{T}\bar{T}V$	$-\frac{2}{3}$	$-\frac{1}{3}$	0
e^-	$\bar{T}\bar{T}\bar{T}$	-1	-1	0

Table 5 displays both the structures and their additive quantum numbers of the first generation of composite leptons and quarks in the CGM. The additive quantum numbers allotted to each particle are determined from those of the constituent rishons and antirishons. The u-quark has $p = +\frac{1}{3}$ since it contains two T-rishons and one $\bar{V}$-antirishon. It is essential that the u-quark should contain a $\bar{V}$-antirishon rather than a V-rishon as in the HSM, since its particle number is required to agree with its baryon number $A = +\frac{1}{3}$. It should be noted that the leptons are composed of three rishons, while the quarks are composed of one rishon and one rishon-antirishon pair. One should also note that the quantum number p essentially defines the particle or antiparticle nature of each composite particle. In particular, the quarks and antiquarks contain both rishons and antirishons, while the leptons and antileptons of the SM are composed of three antirishons and three rishons, respectively. Thus in the GM the electron is composed of three antiparticles not three particles.

The CGM is a viable alternative to the SM. The essential difference between the CGM and the SM is that in the CGM leptons and quarks are composite particles rather than elementary particles as in the SM. The constituents (massless rishons or antirishons) are bound together by strong color interactions mediated by massless hypergluons acting between color charges of the constituents.

These strong color interactions are analogous to those of the SM, mediated by massless gluons, acting between quarks or antiquarks - described by a theory called quantum chromodynamics (QCD).[10] In the CGM the strong color interaction has been taken down one layer of complexity to describe the composite nature of leptons and quarks.

5. Matter-antimatter Asymmetry Problem

The solution of the matter-antimatter asymmetry problem involves the particle number additive quantum number p of the GM. In particular the values of p corresponding to the electron and the quarks comprising the proton. These are the constituents of the hydrogen atom. In the GM the proton contains two weak eigenstate up quarks and one weak eigenstate down quark.[2] These three quarks each have $p = +\frac{1}{3}$ so that the proton has $p = +1$. The values of $p = +\frac{1}{3}$ of the quarks, correspond to the values of their baryon number in the SM, while the value of $p = -1$ of the electron, corresponds to minus the value of the lepton number of the electron in the SM. In the GM, the electron consists entirely of antirishons, i.e., antiparticles, while in the SM it is assumed to be a particle. The electron has $p = -1$ so that the hydrogen atom has $p = 0$. The hydrogen atom in the GM consists of an equal number of rishons and antirishons so that there is no asymmetry of matter and antimatter there.

In the GM the neutron consists of three weak eigenstate quarks, one up quark and two down quarks, so that the neutron also has particle number $p = +1$. Consequently, a helium atom, consisting of two protons, two neutrons and two electrons has particle number $p = +2$: the helium atom in the GM consists of six more rishons than antirishons, i.e., more matter than antimatter. In the GM it is assumed that during the formation of helium in the aftermath of the Big Bang that an equivalent surplus of antimatter was formed as neutrinos, which have $p = -1$, so that overall equal numbers of rishons and antirishons prevailed. This assumption is a consequence of the conservation of p in all interactions.

Thus the ordinary matter present in the universe, prior to the fusion process into heavier elements, has essentially particle number $p = 0$. Since the additive quantum number p is conserved in all interactions, this implies that the overall particle number of the universe will remain essentially as $p = 0$, i.e., symmetric in particle and antiparticle matter, and the universe contains equal numbers of both rishons and antirishons.

To summarize: the ordinary matter present in the universe has an overall particle number of $p = 0$, so that it contains equal numbers of both rishons and antirishons. This implies that the original antimatter created in the Big Bang is now contained within the stable composite leptons, i.e., electrons and neutrinos, and the stable composite quarks, i.e., the weak eigenstate up and down quarks, which comprise the protons and neutrons. The hydrogen, helium and heavier atoms all consist of electrons, protons and neutrons. This explains where all the antiparticles have gone. However, it does not explain why the universe consists primarily of hydrogen atoms and not antihydrogen atoms. It is suggested that the hydrogen-antihydrogen asymmetry may be understood as follows.

In the GM, antihydrogen consists of the same rishons and antirishons as does hydrogen, although the rishons and antirishons are differently arranged in the two systems. This implies that both hydrogen atoms and antihydrogen atoms should be formed during the aftermath of the Big Bang with about the same probability. In fact, estimates from the cosmic microwave background data suggest that for every billion hydrogen-antihydrogen pairs there was just one extra hydrogen atom. It is suggested that this extremely small difference, one extra hydrogen atom in 10^9 hydrogen-antihydrogen pairs, may arise from statistical fluctuations associated with the complex many-body processes involved in the formation of either a hydrogen atom or an antihydrogen atom. The uniformity of the universe,[11] in particular, the lack of antihydrogen throughout the universe, indicates that the above statistical fluctuations took place prior to the 'inflationary period'[12,13] associated with the Big Bang scenario.

6. Conclusion

Within the framework of the GM of particle physics, it has been demonstrated that the matter-antimatter asymmetry problem may be understood in terms of the particle additive quantum number (p) and the composite nature of the leptons and quarks of the GM. The ordinary matter present in the universe has an overall particle number $p = 0$, so that it contains the same number of particles (rishons) as antiparticles (antirishons).

This implies that the original antimatter created in the Big Bang is now contained within the stable composite leptons, the electrons and neutrinos, and the stable composite quarks, the weak eigenstate up and down quarks that comprise the protons and neutrons. The hydrogen, helium and heavier atoms all consist of electrons, protons and neutrons.

Thus there is no matter-antimatter asymmetry in the present universe. However, there does exist a hydrogen-antihydrogen asymmetry: the present universe consists predominently of hydrogen atoms and virtually no antihydrogen atoms. In the SM this is tantamount to the matter-antimatter asymmetry, since both protons and electrons are assumed to be matter. In the GM this is not the case, since both hydrogen and antihydrogen atoms contain the same number of rishons as the number of antirishons.

Thus there are two main conclusions: (1) there is no matter-antimatter asymmetry in the present universe and (2) it is suggested that the observed small hydrogen-antihydrogen asymmetry in the present universe may be understood in terms of statistical fluctuations associated with the complex many-body processes involved in the formation of either a hydrogen atom or an antihydrogen atom.

Finally it should be noted that if the Big Bang produced equal numbers of particles and antiparticles so that the initial state of the universe had particle number $p = 0$, then the GM *predicts* that the present state of the universe should also have $p = 0$, since particle number p is conserved in all interactions.

References

1. K. Gottfried and V. F. Weisskopf, *Concepts of Particle Physics* Vol. 1 (Oxford University Press, New York, 1984).
2. B. A. Robson, The Generation Model of Particle Physics in *Particle Physics*, Ed. E. Kennedy (InTech Open Access Publisher, Rijeka, Croatia, 2012).
3. P. A. R. Ade et al., (Planck Collaboration) *Astronomy and Astrophysics* **571** (2014) Article Number: A1.
4. R. Morris, *Cosmic Questions*, (John Wiley and Sons, New York 1993).
5. A. Einstein, *Ann. Physik* **49** (1916) 769.
6. M. Veltman, *Facts and Mysteries in Elementary Physics* (World Scientific, Singapore, 2003).
7. H. Harari, *Phys. Lett. B* **86** (1979) 83.
8. M. A Shupe, *Phys. Lett. B* **86** (1979) 87.
9. B. A. Robson, *Int. J. Mod. Phys. E* **20** (2011) 733.
10. F. Halzen and A. D. Martin, *Quarks and Leptons: An Introductory Course in Modern Particle Physics*, (John Wiley and Sons, New York 1984).
11. D. Lincoln, *Understanding the Universe from Quarks to the Cosmos*, rev. edn. (World Scientific, Singapore, 2012).
12. A. H. Guth, *Phys. Rev. D* **23** (1981) 347.
13. A. D. Linde, *Phys. Lett. B* **108** (1982) 389.

Neutrinos and Cosmological Matter–antimatter Asymmetry: A Minimal Seesaw with Frampton–Glashow–Yanagida Ansatz

Jue Zhang

Center for High Energy Physics, Peking University, Beijing 100871, China
juezhang87@pku.edu.cn

Shun Zhou

Institute of High Energy Physics, Chinese Academy of Sciences, Beijing 100049, China
School of Physical Sciences,
University of Chinese Academy of Sciences, Beijing 100049, China
Center for High Energy Physics, Peking University, Beijing 100871, China
zhoush@ihep.ac.cn

In light of the latest neutrino data, we revisit a minimal seesaw model with the Frampton–Glashow–Yanagida ansatz. Renormalization-group running effects on neutrino masses and flavor mixing parameters are discussed and found to essentially have no impact on testing such a minimal scenario in low-energy neutrino experiments. However, since renormalization-group running can modify neutrino mixing parameters at high energies, it does affect the leptogenesis mechanism, which is responsible for the observed matter–antimatter asymmetry in our Universe. Furthermore, to ease the conflict between the naturalness argument and the successful leptogenesis, a special regime for resonant leptogenesis is also emphasized.

Keywords: Neutrino masses; matter–antimatter asymmetry; seesaw mechanism.

1. Introduction

Awarding Nobel Prize in physics to Takaaki Kajita and Arthur B. McDonald in 2015 was undoubtedly an exciting event in the field of neutrino physics. In retrospect, the Super-Kamiokande experiment led by Takaaki Kajita discovered atmospheric neutrino oscillations for the first time in 1998, while the SNO collaboration directed by Arthur B. McDonald finally settled down the issue of solar neutrinos in 2002, and the KamLAND experiment singled out the LMA-MSW solution to solar neutrino problem in 2003. Including the recent discovery of the smallest mixing angle from reactor neutrino experiments Daya Bay, RENO and Double Chooz, we now have a complete picture of three-flavor neutrino oscillations, which indisputably shows that neutrinos have masses.

The standard description of neutrino oscillations involves three mixing angles θ_{ij} for $ij = 12, 13, 23$, one Dirac CP-violating phase δ and two mass-squared differences Δm^2_{21} and Δm^2_{31}, where $\Delta m^2_{ij} \equiv m^2_i - m^2_j$ with m_i for $i = 1, 2, 3$ being three neutrino masses. These mixing parameters reside in the leptonic flavor mixing matrix, or the so-called PMNS matrix, the analog of CKM matrix for the quark flavor mixing. Up to now, all these three mixing angles and the absolute values of two mass-squared differences have been quite accurately measured, while the octant of θ_{23} and the sign of Δm^2_{31} are still unresolved. The latter ambiguity on the sign of Δm^2_{31} is often rephrased as the problem of neutrino mass ordering, i.e. Normal Ordering (NO) with $m_1 < m_2 < m_3$ or Inverted Ordering (IO) with $m_3 < m_1 < m_2$. The latest global-fit results on the neutrino mixing parameters can be found in Ref. 1. An interesting comparison between this latest version with the previous version in two years back[2] shows that the favored mass ordering was shifted from IO to NO, although the statistical significance is still quite low. Moreover, the hint for a maximal Dirac CP-violating phase, i.e. $\delta \sim -90°$, in the previous version is sustained till the latest one. It is worth pointing out that the CP-conserving cases of $\delta_{\rm CP} = 0$ or π have recently been excluded by T2K at the 90% confidence level.[3] Upcoming neutrino experiments will further pursue these issues of discriminating neutrino mass orderings and measuring the Dirac CP-violating phase. For instance, the JUNO experiment is currently under construction and will determine the neutrino mass ordering at $(3 \sim 4)\sigma$ level after a 6-year data taking.[4]

In addition to the neutrino mass ordering and CP violation in the lepton sector, there remain other mysteries in neutrino physics, such as the nature of neutrinos (Dirac or Majorana), the absolute scale of neutrino masses, the possible existence of extra neutrino species, the origin of neutrino masses and flavor mixing. In this talk,[a] we will mainly focus on the theoretical aspects and try to understand neutrino masses and leptonic flavor mixing pattern via a minimal scenario, namely, minimal seesaw with Frampton–Glashow–Yanagida (FGY) ansatz.[5] Moreover, the issue of cosmological matter–antimatter asymmetry will also be addressed in this minimal scenario via the leptogenesis mechanism.[6]

2. Minimal Seesaw with Frampton–Glashow–Yanagida Ansatz

There exist many proposals trying to accommodate tiny neutrino masses; some are seesaw-like mechanisms at the tree level, while others generate neutrino masses at the loop level.[7] Here we choose to work within a tree-level framework, namely, the so-called type-I seesaw mechanism,[8–12] in which right-handed neutrinos $N_{\rm R}$ that are singlets under the $\mathrm{SU(2)_L \times U(1)_Y}$ gauge group of the Standard Model (SM) are introduced. The most general gauge-invariant Lagrangian relevant for lepton

[a]This article is based on a talk given at the Conference on Cosmology, Gravitational Waves and Particles, 6–10 February 2017, NTU, Singapore.

masses and flavor mixing can be written as

$$-\mathcal{L}_{\rm m} = \overline{\ell_{\rm L}} Y_l H E_{\rm R} + \overline{\ell_{\rm L}} Y_\nu \tilde{H} N_{\rm R} + \frac{1}{2}\overline{N^{\rm c}_{\rm R}} M_{\rm R} N_{\rm R} + {\rm h.c.}\,, \tag{1}$$

where $\ell_{\rm L}$ and $\tilde{H} \equiv i\sigma_2 H^*$ denote the left-handed lepton and Higgs doublets, respectively, while $E_{\rm R}$ the right-handed charged-lepton singlets. In addition, Y_l and Y_ν stand respectively for the Yukawa coupling matrices of charged leptons and neutrinos, and $M_{\rm R}$ is the Majorana mass matrix for right-handed neutrino singlets. After the Higgs field acquires its vacuum expectation value $\langle H \rangle = v \approx 174$ GeV and the gauge symmetry is spontaneously broken down, the charged-lepton mass matrix is given by $M_l = Y_l v$, while the Dirac neutrino mass matrix is $M_{\rm D} = Y_\nu v$. In this canonical seesaw model, the lightness of ordinary neutrinos can be ascribed to the heaviness of right-handed Majorana neutrinos. Moreover, the mismatch between the diagonalizations of M_l and M_ν leads to lepton flavor mixing.

Type-I seesaw mechanism is theoretically attractive in several aspects. On the one hand, the introduction of heavy right-handed neutrinos fits well with the grand unified theories, where the SM matter fields together with these right-handed neutrinos neatly form a **16**-dimensional spinor representation under the unified gauge group SO(10). On the other hand, one can implement the leptogenesis mechanism to explain the observed baryon number asymmetry in our Universe. In the very early Universe, the temperature is high enough to thermally produce heavy Majorana neutrinos $N_{\rm R}$. As the Universe cools down, the out-of-equilibrium and CP-violating decays of $N_{\rm R}$ generate lepton number asymmetries, which will further be converted into the baryon asymmetry via nonperturbative sphaleron processes.[13,14] Excellent reviews on leptogenesis can be found in Refs. 15–17.

In spite of the above appealing features, the flavor structure of type-I seesaw mechanism in its original form with *three* generations of right-handed neutrinos is hard to probe experimentally, due to the existence of too many free parameters. In the basis where Y_l is diagonal, the number of free parameter in Y_ν and $M_{\rm R}$ is eighteen, which shall be compared with the nine independent observables in low-energy neutrino experiments. Therefore, one is forced to consider some more minimal cases of type-I seesaw mechanism so as to directly confront theory with data.

While one can reduce the number of free parameters by imposing some family symmetries in Yukawa and mass matrices, another straightforward way can be just cutting the number of generations of right-handed neutrinos from *three* to *two*,[b] resulting in the minimal seesaw scenario.[5,18–20] Consequently, the neutrino Yukawa matrix becomes a 3×2 matrix with rank two, and one of three light neutrinos is massless, a prediction still in compatible with current neutrino data. This practical

[b]Alternatively, one can interpret the minimal seesaw scenario as if the third generation of right-handed neutrino is so heavy or couples so weakly with SM fields that they become irrelevant in the discussion of neutrino masses.

approach of reducing the number of free parameters follows closely the spirit of Occam's razor.[21,22]

A parameter count in this minimal seesaw scenario reveals that the remaining number of free parameters is eleven, still too large to confront with data. In 2002 Frampton, Glashow and Yanagida proposed a further minimal scenario by imposing two texture zeros in the neutrino Yukawa matrix Y_ν.[5] Finally, we obtain a theory with only five parameters, which can all be determined from five currently well-measured observables, i.e. three mixing angles and two mass-squared differences.

To systematically study this FGY ansatz in minimal seesaw scenario, one can categorize the patterns of neutrino Yukawa matrix Y_ν as follows:

- **Case A** — Two texture zeros are located in the same row, namely, $(Y_\nu)_{\alpha i} = (Y_\nu)_{\alpha j} = 0$ with $i \neq j$. There are only three patterns:

$$\mathbf{A}_1 : \begin{pmatrix} \mathbf{0} & \mathbf{0} \\ \times & \times \\ \times & \times \end{pmatrix}, \quad \mathbf{A}_2 : \begin{pmatrix} \times & \times \\ \mathbf{0} & \mathbf{0} \\ \times & \times \end{pmatrix}, \quad \mathbf{A}_3 : \begin{pmatrix} \times & \times \\ \times & \times \\ \mathbf{0} & \mathbf{0} \end{pmatrix}, \tag{2}$$

  where the cross "×" denotes a nonzero matrix element.
- **Case B** — Two texture zeros are located in different columns and rows, namely, $(Y_\nu)_{\alpha i} = (Y_\nu)_{\beta j} = 0$ with $\alpha \neq \beta$ and $i \neq j$. There are six patterns:

$$\begin{aligned} &\mathbf{B}_1 : \begin{pmatrix} \mathbf{0} & \times \\ \times & \mathbf{0} \\ \times & \times \end{pmatrix}, \quad \mathbf{B}_2 : \begin{pmatrix} \mathbf{0} & \times \\ \times & \times \\ \times & \mathbf{0} \end{pmatrix}, \quad \mathbf{B}_3 : \begin{pmatrix} \times & \times \\ \mathbf{0} & \times \\ \times & \mathbf{0} \end{pmatrix}, \\ &\mathbf{B}_4 : \begin{pmatrix} \times & \mathbf{0} \\ \mathbf{0} & \times \\ \times & \times \end{pmatrix}, \quad \mathbf{B}_5 : \begin{pmatrix} \times & \mathbf{0} \\ \times & \times \\ \mathbf{0} & \times \end{pmatrix}, \quad \mathbf{B}_6 : \begin{pmatrix} \times & \times \\ \times & \mathbf{0} \\ \mathbf{0} & \times \end{pmatrix}, \end{aligned} \tag{3}$$

  where $\mathbf{B}_{4,5,6}$ are derived from $\mathbf{B}_{1,2,3}$ by exchanging two columns.
- **Case C** — Two texture zeros are located in the same column, namely, $(Y_\nu)_{\alpha i} = (Y_\nu)_{\beta i} = 0$ with $\alpha \neq \beta$. There are six patterns:

$$\begin{aligned} &\mathbf{C}_1 : \begin{pmatrix} \mathbf{0} & \times \\ \mathbf{0} & \times \\ \times & \times \end{pmatrix}, \quad \mathbf{C}_2 : \begin{pmatrix} \mathbf{0} & \times \\ \times & \times \\ \mathbf{0} & \times \end{pmatrix}, \quad \mathbf{C}_3 : \begin{pmatrix} \times & \times \\ \mathbf{0} & \times \\ \mathbf{0} & \times \end{pmatrix}, \\ &\mathbf{C}_4 : \begin{pmatrix} \times & \mathbf{0} \\ \times & \mathbf{0} \\ \times & \times \end{pmatrix}, \quad \mathbf{C}_5 : \begin{pmatrix} \times & \mathbf{0} \\ \times & \times \\ \times & \mathbf{0} \end{pmatrix}, \quad \mathbf{C}_6 : \begin{pmatrix} \times & \times \\ \times & \mathbf{0} \\ \times & \mathbf{0} \end{pmatrix}, \end{aligned} \tag{4}$$

  where $\mathbf{C}_{4,5,6}$ can be obtained from $\mathbf{C}_{1,2,3}$ by exchanging two columns.

It is worth pointing out that the patterns in each class can be related by the elementary transformations, i.e. the 3×3 elementary matrices $\mathcal{P}_{ij}$ (for $ij = 12, 23, 13$) and the 2×2 elementary matrix $\mathcal{Q}$. The action of $\mathcal{P}_{ij}$ from left (or right) induces an exchange between ith and jth rows (or columns), and likewise for $\mathcal{Q}$. With the help of $\mathcal{P}_{ij}$ and $\mathcal{Q}$, one can change the positions of texture zeros. For instance, we have $Y_\nu(\mathbf{A}_2) = \mathcal{P}_{12} Y_\nu(\mathbf{A}_1)$ and $Y_\nu(\mathbf{A}_3) = \mathcal{P}_{13} Y_\nu(\mathbf{A}_1)$.

Below the seesaw scale, one can integrate out heavy Majorana neutrinos and obtain the unique Weinberg operator $\mathcal{O}_5 = (\kappa/2)\,(\overline{\ell_{\rm L}}\tilde{H})\cdot(\tilde{H}^{\rm T}\ell^{\rm c}_{\rm L})$ of dimension five[23] with $\kappa = -Y_\nu \hat{M}_{\rm R}^{-1} Y_\nu^{\rm T}$. After the spontaneous gauge symmetry breaking, neutrinos acquire tiny Majorana masses from the Weinberg operator and their mass matrix is $M_\nu = \kappa v^2$. Given Y_ν in Eqs. (2)–(4), we are ready to check if κ inherits some texture zeros from Y_ν. Since all the patterns in each class are related by $\mathcal{P}_{ij}$ and $\mathcal{Q}$ matrices, it is sufficient to consider the first pattern and perform the corresponding elementary transformations to derive the results for the others. More explicitly, we have $\kappa(M_1)$ at the seesaw scale

$$\kappa_{\mathbf{A}_1}: \begin{pmatrix} \mathbf{0} & \mathbf{0} & \mathbf{0} \\ \mathbf{0} & \times & \times \\ \mathbf{0} & \times & \times \end{pmatrix}, \quad \kappa_{\mathbf{B}_1}: \begin{pmatrix} \times & \mathbf{0} & \times \\ \mathbf{0} & \times & \times \\ \times & \times & \times \end{pmatrix}, \quad \kappa_{\mathbf{C}_1}: \begin{pmatrix} \times & \times & \times \\ \times & \times & \times \\ \times & \times & \times \end{pmatrix}, \tag{5}$$

where one can observe that the patterns $\mathbf{C}_i$ (for $i = 1, 2, \ldots, 6$) do not lead to any texture zeros in κ. For **Case A** in Eq. (2), it is easy to derive $\kappa_{\mathbf{A}_j} = \mathcal{P}_{1j}\kappa_{\mathbf{A}_1}\mathcal{P}_{1j}$ for $i = 2, 3$, so κ in this case has a nonzero 2×2 block submatrix. For **Case B** in Eq. (3), we arrive at the following identities

$$\kappa_{\mathbf{B}_2} = \mathcal{P}_{23}\kappa_{\mathbf{B}_1}\mathcal{P}_{23}\,, \quad \kappa_{\mathbf{B}_3} = \mathcal{P}_{12}\mathcal{P}_{23}\kappa_{\mathbf{B}_1}\mathcal{P}_{23}\mathcal{P}_{12}\,, \quad \kappa_{\mathbf{B}_{i+3}} = \kappa_{\mathbf{B}_i}\,, \tag{6}$$

where the last identity indicates that one texture zero is located in the same position in κ for $\mathbf{B}_{i+3}$ and $\mathbf{B}_i$ for $i = 1, 2, 3$.

It should be noted that the above obtained κ matrices are those at the seesaw scale. To confront with the data at low energies, it is necessary to take into account the renormalization-group (RG) running effects.

3. Renormalization-Group Running Effects

At the one-loop level, the RG running effects of neutrino masses and flavor mixing parameters can be studied by solving the RG equation of κ (Refs. 24–26)

$$16\pi^2 \frac{{\rm d}\kappa}{{\rm d}t} = \alpha_\kappa \kappa + C_\kappa \left[\left(Y_l Y_l^\dagger\right)\kappa + \kappa\left(Y_l Y_l^\dagger\right)^{\rm T} \right], \tag{7}$$

with $t \equiv \ln(\mu/\Lambda_{\rm EW})$. In the SM, the relevant coefficients in Eq. (7) are $C_\kappa = -3/2$ and $\alpha_\kappa \approx -3g_2^2 + 6y_t^2 + \lambda$, where g_2 stands for the ${\rm SU}(2)_{\rm L}$ gauge coupling, y_t the top-quark Yukawa coupling, and λ the Higgs self-coupling constant. If the dimension-five Weinberg operator is derived in the minimal supersymmetric standard model (MSSM), we have $M_\nu = \kappa(v \sin\beta)^2$ with $\tan\beta$ being the ratio of vacuum expectation values of two MSSM Higgs doublets. In this framework, the RG equation of κ is still given by Eq. (7) but with $C_\kappa = 1$ and $\alpha_\kappa \approx -6g_1^2/5 - 6g_2^2 + 6y_t^2$. Note that only the top-quark Yukawa coupling is retained in α_κ, as the Yukawa couplings of other fermions are much smaller and have safely been neglected.

Working in the basis where the charged-lepton Yukawa coupling matrix $Y_l = \mathrm{diag}\{y_e, y_\mu, y_\tau\}$ is diagonal, we can solve Eq. (7) and obtain

$$\kappa(\Lambda_{\rm EW}) = I_0 \begin{pmatrix} I_e & 0 & 0 \\ 0 & I_\mu & 0 \\ 0 & 0 & I_\tau \end{pmatrix} \kappa(M_1) \begin{pmatrix} I_e & 0 & 0 \\ 0 & I_\mu & 0 \\ 0 & 0 & I_\tau \end{pmatrix} , \tag{8}$$

where the evolution functions read

$$I_0 = \exp\left[-\frac{1}{16\pi^2} \int_0^{\ln(M_1/\Lambda_{\rm EW})} \alpha_\kappa(t) {\rm d}t \right] , \tag{9}$$

$$I_\alpha = \exp\left[-\frac{C_\kappa}{16\pi^2} \int_0^{\ln(M_1/\Lambda_{\rm EW})} y_\alpha^2(t) {\rm d}t \right] , \tag{10}$$

for $\alpha = e,\ \mu,\ \tau$. From Eq. (8), it is now evident how the low-energy observables residing in $M_\nu = \kappa(\Lambda_{\rm EW}) v^2$ are related to the model parameters in $\kappa(M_1)$ at a high-energy scale. In the following, we show that it is already possible to exclude most patterns in Eqs. (2)–(4) based on the solution in Eq. (8).

(1) $\kappa(\Lambda_{\rm EW})$ in **Case A** inherits the same structure of $\kappa(M_1)$, leading to just one nontrivial mixing angle, which has already been excluded by current neutrino oscillation data.

(2) For **Case B**, there is only one texture zero in the off-diagonal position of $\kappa(\Lambda_{\rm EW})$, namely,

$$(M_\nu)_{\alpha\beta} = \sum_i m_i U_{\alpha i} U_{\beta i} = 0 , \tag{11}$$

for $(\alpha, \beta) = (e, \mu)$, (e, τ) and (μ, τ), where $U_{\alpha i}$ is the matrix element in the PMNS matrix. When the RG running effects are considered, Eq. (8) indicates that the texture zero remains in the effective neutrino mass matrix M_ν. The constraints on neutrino masses and mixing matrix elements in Eq. (11) can be expressed as

$$\begin{aligned} U_{\alpha 2} U_{\beta 2} m_2 + U_{\alpha 3} U_{\beta 3} m_3 &= 0 \quad \text{for NO} , \\ U_{\alpha 1} U_{\beta 1} m_1 + U_{\alpha 2} U_{\beta 2} m_2 &= 0 \quad \text{for IO} , \end{aligned} \tag{12}$$

which have been investigated in Ref. 21, where the latest neutrino oscillation data are implemented but the RG running effects are entirely ignored. In the NO case, it has been found that all the patterns in Eq. (3) are ruled out mainly due to the observed θ_{13}. In the IO case, $(M_\nu)_{\mu\tau} = 0$ is shown to be strongly disfavored, so the patterns $\mathbf{B}_3$ and $\mathbf{B}_6$ are excluded. Hence, according to Ref. 21, only $\mathbf{B}_{1,2}$ and $\mathbf{B}_{4,5}$ in the IO case are compatible with the latest neutrino oscillation data.

(3) Lastly, for **Case C** although an investigation on texture zeros similar to the above two cases is no longer applicable, one can verify that the following relation

$$\frac{(M_\nu)_{ee}}{(M_\nu)_{\mu e}} = \frac{(M_\nu)_{e\mu}}{(M_\nu)_{\mu\mu}} = \frac{(M_\nu)_{e\tau}}{(M_\nu)_{\mu\tau}} \tag{13}$$

holds both for $\mu = \Lambda_{\rm EW}$ and for $\mu = M_1$. Therefore, it is adequate to inspect if the relationship in Eq. (13) is satisfied by current neutrino oscillation data. It was shown in Ref. 21 that all the patterns in Eq. (4) are excluded regardless of the type of neutrino mass ordering.

In summary, we have proved that texture zeros or proportionality relations in $\kappa(M_1)$ are not spoiled by the RG running effects, so they also exist in $\kappa(\Lambda_{\rm EW})$ at the low-energy scale. Consequently, neutrino oscillation data can be directly implemented to rule out most patterns of Y_ν with two texture zeros. It turns out that only $\mathbf{B}_{1,2}$ and $\mathbf{B}_{4,5}$ in Eq. (3) in the case of IO are consistent with experimental data, which generalizes the conclusions reached in Ref. 21 to the situation including radiative corrections.

RG running effects, however, do play an important role when considering physics at high-energy scales, such as the high-scale leptogenesis mechanism. This arises from the fact that the model parameters at high energy scale can be significantly modified because of RG running. We next take **Pattern $\mathbf{B}_1$** to illustrate this.

First of all, we introduce the Casas–Ibarra parametrization[27] for the neutrino Dirac mass matrix $M_{\rm D}$,

$$\begin{aligned} M_{\rm D} &= U\sqrt{\hat{M}_\nu}O\sqrt{\hat{M}_{\rm R}} \\ &= U\begin{pmatrix} \sqrt{m_1} & 0 & 0 \\ 0 & \sqrt{m_2} & 0 \\ 0 & 0 & 0 \end{pmatrix}\begin{pmatrix} \cos z & -\sin z \\ \sin z & \cos z \\ 0 & 0 \end{pmatrix}\begin{pmatrix} \sqrt{M_1} & 0 \\ 0 & \sqrt{M_2} \end{pmatrix}, \end{aligned} \tag{14}$$

where U is the PMNS matrix, and O is a 3×2 orthogonal matrix with z being a complex parameter, satisfying $O^{\rm T}O = OO^{\rm T} = \mathbf{1}$. Thanks to the texture zeros in $M_{\rm D}$, the complex parameter z can be determined by

$$\tan z = -\frac{U_{e1}}{U_{e2}}\sqrt{\frac{m_1}{m_2}} = -\frac{\sqrt{\zeta}}{\tan\theta_{12}}e^{-{\rm i}\sigma}, \tag{15}$$

where $\zeta = \sqrt{1 - \Delta m^2_{21}/|\Delta m^2_{32}|}$ and σ is the only Majorana phase in the PMNS matrix (taken to be in the (22) element of the Majorana phase matrix in our convention). From the previous discussion we know that only IO is allowed in this case, therefore $m_3 = 0$ and $\zeta = m_1/m_2$. Moreover, because of only five free parameters in the model, we can first express them in terms of the well-measured mixing angles and mass-squared differences, and then yield predictions on those which have not

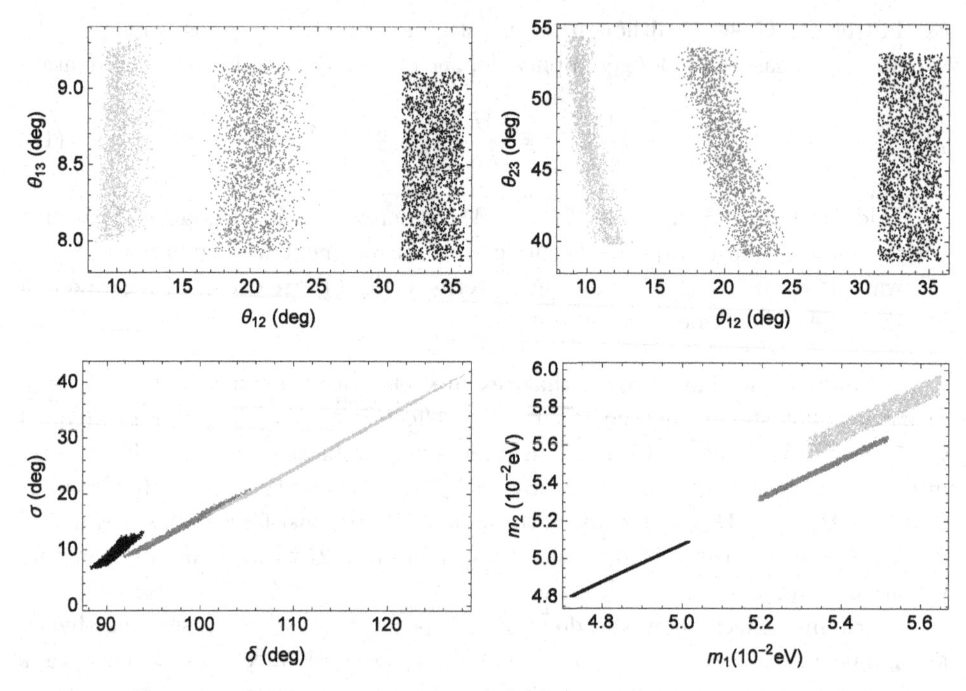

Fig. 1. Illustration for the RG running effects on neutrino mixing angles $\{\theta_{12}, \theta_{13}, \theta_{23}\}$, leptonic CP-violating phases $\{\delta, \sigma\}$ and neutrino masses $\{m_1, m_2\}$ for **Pattern B**$_1$ in the MSSM, where the black points denote the parameters at $M_Z = 91.2$ GeV, while the dark- and light-gray points represent the parameters at the seesaw scale $\Lambda_{\rm SS} = 10^{13}$ GeV for $\tan\beta = 30$ and $\tan\beta = 50$, respectively. Note that δ and σ also have another branch of solutions with their signs inverted simultaneously, and the mass scale of sparticles is taken to be $M_{\rm SUSY} = 1$ TeV.

been properly measured. For instance, we have

$$\cos\delta = \frac{s_{12}^2 c_{12}^2 c_{23}^2 (1-\zeta^2) + s_{23}^2 s_{13}^2 \left(s_{12}^4 - \zeta^2 c_{12}^4\right)}{2 s_{12} c_{12} s_{23} c_{23} s_{13} \left(s_{12}^2 + \zeta^2 c_{12}^2\right)} \,, \tag{16}$$

$$\cos 2\sigma = \frac{s_{12}^2 c_{12}^2 c_{23}^2 (1+\zeta^2) - s_{23}^2 s_{13}^2 \left(s_{12}^4 + \zeta^2 c_{12}^4\right)}{2\zeta s_{12}^2 c_{12}^2 \left(c_{23}^2 + s_{23}^2 s_{13}^2\right)} \,, \tag{17}$$

where $s_{ij} \equiv \sin\theta_{ij}$ and $c_{ij} \equiv \cos\theta_{ij}$. Thus, the whole RG running effects are encoded in the running of three mixing angles and two masses.

As is known that RG running effects are insignificant in the SM, in Fig. 1 we shown the allowed regions of three neutrino mixing angles $\{\theta_{12}, \theta_{13}, \theta_{23}\}$, two leptonic CP-violating phases $\{\delta, \sigma\}$ and two nonzero neutrino masses $\{m_1, m_2\}$ in the case of MSSM with $\tan\beta = 30$ and $\tan\beta = 50$. The allowed parameter space at the low-energy scale is denoted by black points, and one can observe that δ and σ are restricted to a small area around $\delta = 90°$ and $\sigma = 10°$. At the high-energy seesaw scale $\Lambda_{\rm SS} = 10^{13}$ GeV, the parameter space in the MSSM with $\tan\beta = 30$ and $\tan\beta = 50$ has been represented by dark- and light-gray points, respectively.

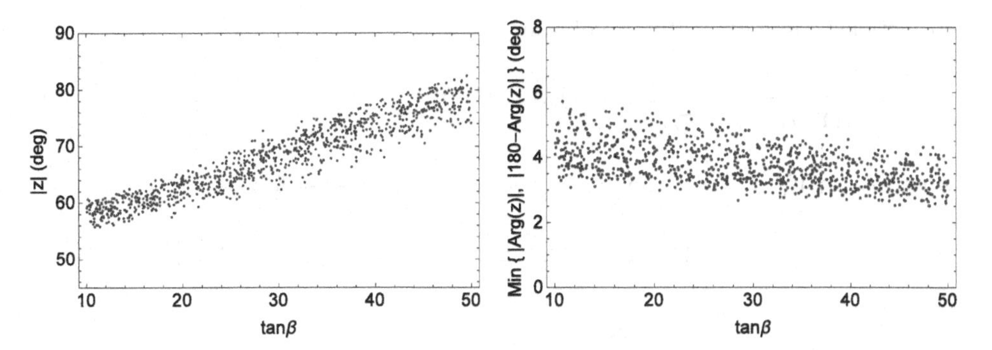

Fig. 2. The absolute value $|z|$ and the phase $\arg z$ of the complex parameter z are given in units of degrees in the left and right panels, respectively. For a given value of $\tan\beta$, $|z|$ and $\arg z$ are calculated by varying the low-energy parameters in their 3σ ranges and the high-energy scales from 10^8 GeV to 10^{13} GeV. Note that z is almost real in all cases as indicated in the right panel.

One can see that the RG running effects on θ_{13} and θ_{23} are insignificant, whereas the running effects on θ_{12}, δ and σ are indeed remarkable.

Furthermore, for the later convenience in discussing leptogenesis we also show the RG running on the complex parameter z in Fig. 2. One can see that a small phase of z is obtained in all cases, implying the suppression of CP violation at the high-energy scale. The latter observation becomes clearer when we calculate the CP asymmetries in the decays of heavy Majorana neutrinos.

4. Accommodate Naturalness with Resonant Leptogenesis

One salient feature of the canonical seesaw model is to simultaneously explain tiny neutrino masses and the observed baryon number asymmetry in our Universe, which is usually measured by the baryon-to-photon density ratio[28]

$$\eta^0_{\rm B} \equiv \frac{n_{\rm B}}{n_\gamma} = (6.065 \pm 0.090) \times 10^{-10} \, , \tag{18}$$

where $n_{\rm B}$ and n_γ stand for today's baryon and photon number density, respectively. In thermal leptogenesis, we first assume that in the very early Universe heavy Majorana neutrinos $N_{{\rm R}i}$ can be produced in thermal equilibrium. As the Universe cools down, the CP-violating decays of $N_{i{\rm R}}$ go out of thermal equilibrium. The CP asymmetries in the decays of $N_{i{\rm R}}$ into leptons of different flavors are defined as[15–17]

$$\varepsilon_{i\alpha} \equiv \frac{\Gamma(N_{i{\rm R}} \to l_\alpha H) - \Gamma(N_{i{\rm R}} \to \overline{l}_\alpha \bar{H})}{\Gamma(N_{i{\rm R}} \to l_\alpha H) + \Gamma(N_{i{\rm R}} \to \overline{l}_\alpha \bar{H})} \, , \tag{19}$$

where $\Gamma(N_{i{\rm R}} \to l_\alpha H)$ and $\Gamma(N_{i{\rm R}} \to \overline{l}_\alpha \bar{H})$ for $\alpha = e, \, \mu, \, \tau$ denote the decay rates of $N_{i{\rm R}}$ into leptons l_α and anti-leptons $\overline{l}_\alpha$, respectively. It is the interference between the tree-level and one-loop decay amplitudes that gives rise to CP asymmetries, which receive both contributions from the one-loop self-energy and vertex corrections. More explicitly, we obtain

$$\varepsilon_{i\alpha} = \frac{1}{8\pi (Y_\nu^\dagger Y_\nu)_{ii}} \mathrm{Im} \sum_{k\neq i} (Y_\nu^*)_{\alpha i} (Y_\nu)_{\alpha k} \left[(Y_\nu^\dagger Y_\nu)_{ik} f(x_{ki}) + (Y_\nu^\dagger Y_\nu)^*_{ik} g(x_{ki}) \right] , \quad (20)$$

where $x_{ki} \equiv M_k^2/M_i^2$ and the loop functions are defined as follows

$$\begin{aligned} f(x_{ki}) &= \sqrt{x_{ki}} \left[\frac{1 - x_{ki}}{(1-x_{ki})^2 + r_{ki}^2} + 1 - (1 + x_{ki}) \ln \frac{1 + x_{ki}}{x_{ki}} \right] , \\ g(x_{ki}) &= \frac{1 - x_{ki}}{(1-x_{ki})^2 + r_{ki}^2} . \end{aligned} \quad (21)$$

If the mass spectrum of heavy Majorana neutrinos is strongly hierarchical, r_{ki} can be neglected in the denominators in Eq. (21). For instance, for **Pattern B$_1$** with $(Y_\nu)_{e1} = (Y_\nu)_{\mu 2} = 0$, we obtain $\varepsilon_{1e} = \varepsilon_{1\mu} = 0$, and

$$\varepsilon_{1\tau} = \varepsilon_1 \approx -\frac{3}{16\pi} \frac{M_1}{v^2} \frac{\Delta m_{21}^2 \,\mathrm{Im}\left[c_z^2\right]}{m_1 |c_z|^2 + m_2 |s_z|^2} . \quad (22)$$

r_{ki}, however, serves as an important regulator to avoid any singularity in the limit of mass degeneracy $M_k^2 = M_i^2$ or equivalently $x_{ki} = 1$. Although in this resonant regime the true form of r_{ki} is still controversial at present,[29] we have verified that it has little effect in our discussion.

The produced lepton-number asymmetries in the $N_{i\mathrm{R}}$ decays will partly be washed out by the inverse decays and lepton-number-violating scattering. To describe these washout effects, we introduce the decay parameters $K_i \equiv \Gamma_i / H(M_i)$, where $\Gamma_i = (Y_\nu^\dagger Y_\nu)_{ii} M_i / 8\pi$ is the total decay width of $N_{i\mathrm{R}}$ and $H(M_i)$ is the Hubble parameter at temperature $T = M_i$. The lepton number asymmetries will be converted into the baryon number asymmetry through the $(B+L)$-violating and $(B-L)$-conserving sphaleron processes.[13,14] The final baryon number asymmetry is then given by[15,16]

$$\eta_{\mathrm{B}} \approx -0.96 \times 10^{-2} \sum_i \sum_\alpha \varepsilon_{i\alpha} \kappa_{i\alpha} , \quad (23)$$

where the efficiency factors $\kappa_{i\alpha}$ can be determined by solving the Boltzmann equations of heavy Majorana neutrino and lepton number densities. Note that although particles are doubled in the MSSM compared to the SM, the finally obtained baryon number asymmetry is not much changed.[17]

In the vanilla scenario of leptogenesis, the mass spectrum of heavy Majorana neutrinos is taken to be hierarchical, and only the lightest Majorana neutrino N_1 and the one-flavor approximation are considered. This is actually done for the FGY model in the previous papers,[5,18,19,21] where a narrow mass range of the lightest heavy Majorana neutrino $M_1 \sim 5 \times 10^{13}$ GeV has been found in the IO case. However, we now know that lepton flavor effects do play a nontrivial role in leptogenesis, and also given the ignorance of high energy physics the hierarchy of heavy Majorana neutrinos may not be as hierarchical as one usually assumes. More careful treatment involving lepton flavor effects can be found in Ref. 30. We next pay more attention to the latter issue of non-hierarchical right-handed neutrinos.

In fact, the above obtained high-mass scale of heavy Majorana neutrinos in the hierarchical limit also causes the so-called naturalness problem or the fine-tuning problem for the light Higgs boson mass,[31–35] and the gravitino overproduction problem if the model is supersymmetrized.[36] In Ref. 35, a detailed analysis of the naturalness problem in the type-I seesaw model yields an upper bound on the heavy Majorana neutrino masses, namely, $M_1 < 4\times10^7$ GeV and $M_2 < 7\times10^7$ GeV. Therefore, in view of naturalness it is also necessary to go beyond the hierarchical limit and consider both mild mass hierarchy and a nearly-degenerate mass spectrum.

To quantify the level of mass degeneracy, we introduce a dimensionless parameter

$$\Delta \equiv \frac{M_2 - M_1}{M_2} , \tag{24}$$

which approaches to zero (one) when $M_1 = M_2$ $(M_2 \gg M_1)$. Because of a mild hierarchy between M_1 and M_2, both $N_{1\mathrm{R}}$ and $N_{2\mathrm{R}}$ participate in the production and washout processes of lepton number asymmetries. The evolution of these asymmetries therefore involves solving the Boltzmann equations with both $N_{1\mathrm{R}}$ and $N_{2\mathrm{R}}$. More details on setting up the Boltzmann equations can be found in Ref. 30. Given the initial conditions of the thermal abundance of $n_{N_{i\mathrm{R}}}$ and vanishing $B-L$ asymmetries, in Fig. 3 we present the allowed parameter space for M_1 and Δ in the case of **Pattern B**$_1$. The black solid curve represents a contour of $\eta_\mathrm{B} = 6.065 \times 10^{-10}$, for which the observational uncertainty is so small that it will be hidden by the line width in the figure. The mass regions, which are represented by the shading areas, are characterized by the charged-lepton flavor effects.

In the mild hierarchy case, we observe from Fig. 3 that M_1 still sits around 5×10^{13} GeV. Therefore, including the contributions from N_2 cannot significantly enhance the amount of CP asymmetry, and one then still needs to raise the mass scale of M_1 so as to reach the required value of η_B. In the nearly-degenerate case, we see that a mass degeneracy at the level of $\Delta = 10^{-7}$ is required to meet the naturalness bound $M_1 < 4 \times 10^7$ GeV and account for the baryon number asymmetry via resonant leptogenesis.[37–39] Although it seems unnatural to require such a high mass degeneracy, it can actually be achieved by implementing a flavor symmetry and its soft breaking at a superhigh-energy scale,[38] or by the RG running effects.[40,41] As one can see, there is a kink around $M_1 = 10^{12}$ GeV. The reason is simply that we use different Boltzmann equations for the two cases of below and above 10^{12} GeV. The kink should disappear if the fully quantum Boltzmann equations with coherent flavor effects are used.[29]

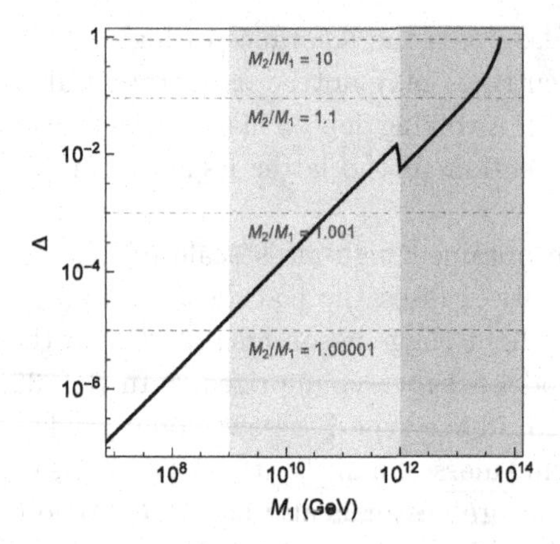

Fig. 3. Illustration for the dependence of baryon number asymmetry on the lightest heavy Majorana neutrino mass M_1 and the mass degeneracy parameter Δ. The black and solid curve corresponds to the allowed regions of model parameters, for which the observed baryon number asymmetry $\eta_{\rm B} \approx 6.065 \times 10^{-10}$ can be naturally explained. The dashed lines indicate a few typical values of the mass ratio M_2/M_1.

5. Summary

Neutrino physics has witnessed a swift development in the last two decades: from an initial discovery of neutrino oscillation phenomena in 1998 to the current prospect for entering a precision era. To fully utilize those low-energy neutrino data, one sets out to consider some minimal scenarios that can potentially account for the mixing and mass patterns of neutrinos. In this talk we focus on one of such minimal scenarios, i.e. the minimal seesaw with Frampton–Glashow–Yanagida ansatz, in which all five free model parameters can be determined from current neutrino data. Renormalization-group running effects are firstly considered, and our investigation generalizes the conclusion that was made in the literature when no RG running effects are taken into account. Moreover, we also discuss the cosmological implication of this minimal scenario by invoking the leptogenesis mechanism to explain the observed matter–antimatter asymmetry in our Universe. To ease the conflict between the naturalness argument and the successful leptogenesis, the special regime of resonant leptogenesis is emphasized. Upcoming neutrino experiments will further test this minimal scenario, and if it could indeed stand the test of time, we should go further to identify the underlying symmetries and dynamics for both neutrino masses and cosmological matter–antimatter asymmetry.

Acknowledgments

One of us (S. Zhou) is indebted to the organizers for a kind invitation to the conference. This work was supported in part by the National Recruitment Program for Young Professionals and by the CAS Center for Excellence in Particle Physics (CCEPP).

References

1. I. Esteban, M. C. Gonzalez-Garcia, M. Maltoni, I. Martinez-Soler and T. Schwetz, *J. High Energy Phys.* **1701**, 087 (2017), arXiv:1611.01514.
2. M. C. Gonzalez-Garcia, M. Maltoni and T. Schwetz, *J. High Energy Phys.* **1411**, 052 (2014), arXiv:1409.5439.
3. T2K Collab. (K. Abe *et al.*), *Phys. Rev. Lett.* **118**, 151801 (2017), arXiv:1701.00432.
4. Y. F. Li, J. Cao, Y. Wang and L. Zhan, *Phys. Rev. D* **88**, 013008 (2013), arXiv: 1303.6733.
5. P. H. Frampton, S. L. Glashow and T. Yanagida, *Phys. Lett. B* **548**, 119 (2002), arXiv:hep-ph/0208157.
6. M. Fukugita and T. Yanagida, *Phys. Lett. B* **174**, 45 (1986).
7. T. Ohlsson and S. Zhou, *Nature Commun.* **5**, 5153 (2014), arXiv:1311.3846.
8. P. Minkowski, *Phys. Lett. B* **67**, 421 (1977).
9. T. Yanagida, *Proceedings of the Workshop on Unified Theory and the Baryon Number of the Universe*, eds. O. Sawada and A. Sugamoto (KEK, Tsukuba, 1979), p. 95.
10. M. Gell-Mann, P. Ramond and R. Slansky, *Supergravity*, eds. P. van Nieuwenhuizen and D. Z. Freeman (North-Holland, Amsterdam, 1979), p. 315.
11. S. L. Glashow, *Quarks and Leptons*, eds. M. Levy *et al.* (Plenum, New York, 1980), p. 707.
12. R. N. Mohapatra and G. Senjanovic, *Phys. Rev. Lett.* **44**, 912 (1980).
13. N. S. Manton, *Phys. Rev. D* **28**, 2019 (1983).
14. F. R. Klinkhamer and N. S. Manton, *Phys. Rev. D* **30**, 2212 (1984).
15. W. Buchmuller, P. Di Bari and M. Plumacher, *Ann. Phys.* **315**, 305 (2005), arXiv:hep-ph/0401240.
16. W. Buchmuller, R. D. Peccei and T. Yanagida, *Annu. Rev. Nucl. Part. Sci.* **55**, 311 (2005), arXiv:hep-ph/0502169.
17. S. Davidson, E. Nardi and Y. Nir, *Phys. Rep.* **466**, 105 (2008), arXiv:0802.2962.
18. W. L. Guo and Z. Z. Xing, *Phys. Lett. B* **583**, 163 (2004), arXiv:hep-ph/0310326.
19. J. W. Mei and Z. Z. Xing, *Phys. Rev. D* **69**, 073003 (2004), arXiv:hep-ph/0312167.
20. W. L. Guo, Z. Z. Xing and S. Zhou, *Int. J. Mod. Phys. E* **16**, 1 (2007), arXiv: hep-ph/0612033.
21. K. Harigaya, M. Ibe and T. T. Yanagida, *Phys. Rev. D* **86**, 013002 (2012), arXiv: 1205.2198.
22. T. Ohlsson, *Phys. Rev. D* **86**, 097301 (2012).
23. S. Weinberg, *Phys. Rev. Lett.* **43**, 1566 (1979).
24. P. H. Chankowski and Z. Pluciennik, *Phys. Lett. B* **316**, 312 (1993), arXiv:hep-ph/ 9306333.
25. K. S. Babu, C. N. Leung and J. T. Pantaleone, *Phys. Lett. B* **319**, 191 (1993), arXiv: hep-ph/9309223.
26. S. Antusch, M. Drees, J. Kersten, M. Lindner and M. Ratz, *Phys. Lett. B* **519**, 238 (2001), arXiv:hep-ph/0108005.
27. J. A. Casas and A. Ibarra, *Nucl. Phys. B* **618**, 171 (2001), arXiv:hep-ph/0103065.
28. Planck Collab. (P. A. R. Ade *et al.*), *Astron. Astrophys.* **571**, A16 (2014), arXiv: 1303.5076.
29. P. S. Bhupal Dev, P. Millington, A. Pilaftsis and D. Teresi, *Nucl. Phys. B* **886**, 569 (2014), arXiv:1404.1003.
30. J. Zhang and S. Zhou, *J. High Energy Phys.* **1509**, 065 (2015), arXiv:1505.04858 [hep-ph].

31. F. Vissani, *Phys. Rev. D* **57**, 7027 (1998), arXiv:hep-ph/9709409.
32. A. Abada, C. Biggio, F. Bonnet, M. B. Gavela and T. Hambye, *J. High Energy Phys.* **0712**, 061 (2007), arXiv:0707.4058.
33. Z. Z. Xing, *Prog. Theor. Phys. Suppl.* **180**, 112 (2009), arXiv:0905.3903.
34. M. Farina, D. Pappadopulo and A. Strumia, *J. High Energy Phys.* **1308**, 022 (2013), arXiv:1303.7244.
35. J. D. Clarke, R. Foot and R. R. Volkas, *Phys. Rev. D* **91**, 073009 (2015), arXiv: 1502.01352.
36. G. F. Giudice, A. Notari, M. Raidal, A. Riotto and A. Strumia, *Nucl. Phys. B* **685**, 89 (2004), arXiv:hep-ph/0310123.
37. A. Pilaftsis, *Phys. Rev. D* **56**, 5431 (1997), arXiv:hep-ph/9707235.
38. A. Pilaftsis and T. E. J. Underwood, *Nucl. Phys. B* **692**, 303 (2004), arXiv:hep-ph/ 0309342.
39. Z. Z. Xing and S. Zhou, *Phys. Lett. B* **653**, 278 (2007), arXiv:hep-ph/0607302.
40. R. Gonzalez Felipe, F. R. Joaquim and B. M. Nobre, *Phys. Rev. D* **70**, 085009 (2004), arXiv:hep-ph/0311029.
41. G. C. Branco, R. Gonzalez Felipe, F. R. Joaquim and B. M. Nobre, *Phys. Lett. B* **633**, 336 (2006), arXiv:hep-ph/0507092.

The Apparent Likeness of Gravitational and Chargelike Gauges

Peter Minkowski

Albert Einstein Center for Fundamental Physics — ITP,
University of Bern, Switzerland

Gauging orientation on a differentiable manifold (classical configurations)

We consider parallel transport of a (contravariant) vector v^{ϱ}

$$\begin{aligned} &\delta_{\|} v^{\varrho} = -dx^{\kappa} (\Gamma_{\kappa})^{\varrho}{}_{\sigma}(x) v^{\sigma} \\ &(\Gamma^{(1)} = dx^{\kappa} \Gamma_{\kappa})^{\varrho}{}_{\sigma} : \qquad \textbf{matrix valued 1-form} \end{aligned} \tag{1}$$

and along a curve C from x to y, giving rise to the (curve associated) parallel transport matrix, denoted* $T(y \overset{C}{\leftarrow} x)$

$$\begin{aligned} &\left\{ T(y \overset{C}{\leftarrow} x) = P \exp - \int_x^y \Gamma^{(1)} \right\}^{\varrho}_{\sigma} \\ &v_{\|}(y \overset{C}{\leftarrow} x) = T(y \overset{C}{\leftarrow} x) v \\ &\qquad \textbf{matrix notation} \end{aligned} \tag{2}$$

In Eq. (2) P denotes ordering **from left (further along) to right** along the path C. Now we imagine the same parallel transport done using other local coordinates

$$x'^{\varrho} = x'^{\varrho}(x) \to \{ M^{\varrho}{}_{\sigma} = \partial_{\sigma} x'^{\varrho} \}(x) \tag{3}$$

*Some still original works go back to Wolfgang Pauli [1] and Élie Cartan [2] – [3].

Eq. (2) takes the (trans-) form

$$\left\{ T'(y' \overset{C}{\leftarrow} x') = P \exp - \int_{x'}^{y'} \Gamma'^{(1)} \right\}_{\sigma}^{\varrho}$$

$$v'_{\|}(y' \overset{C}{\leftarrow} x') = T'(y' \overset{C}{\leftarrow} x')v' \tag{4}$$

$$v'_{\|} = M(y)v_{\|}, \quad v' = M(x)v$$

and substituting one system relative to the other

$$M(y)T(y \overset{C}{\leftarrow} x)v = T'(y' \overset{C}{\leftarrow} x')M(x)v \tag{5}$$

$$T'(y' \overset{C}{\leftarrow} x') = M(y)T(y \overset{C}{\leftarrow} x)(M(x))^{-1} \tag{6}$$

In Eqs. (2)–(6)

$$\{M(z)|\forall z\} \tag{7}$$

forms the family of local transformations, gauging orientation.†

The role of the entire set of parallel transport matrices $T(y \overset{C}{\leftarrow} x)$ is clear and perfectly covariant, while the local connection $\Gamma^{(1)}$ transforms inhomogeneously.

The parallel transported vectors along the path C, using a path parameter s

$$C : \{z(s)|z(1) = y; z(0) = x\} \tag{8}$$

satisfy the differential equation ($\dot{}\ = d/ds$)

$$\dot{v}(s) = -\dot{z}^k(s)\Gamma_k(s)v(s)$$

$$v(s) = T(z(s) \overset{C}{\leftarrow} x)v \tag{9}$$

†They form the group $GL(d, R)$, where d is the (real) dimension of the manifold.

Comparing with the coordinate transformed equation and using $M(s) = M[z(s)]$

$$\begin{aligned}
\dot{v}(s) &= -\dot{z}^k(s)\Gamma_k(s)v(s) \\
\dot{v}'(s) &= -\dot{z}'^k(s)\Gamma'_k(s)v'(s) \\
\begin{pmatrix} v'(s) \\ \dot{z}'(s) \end{pmatrix} &= M(s)\begin{pmatrix} v(s) \\ \dot{z}(s) \end{pmatrix}
\end{aligned} \tag{10}$$

The second relation in Eq. (10) thus becomes

$$M\dot{v} + \dot{M}v = -\dot{z}^k M^r{}_k \Gamma'_r M v \tag{11}$$

and substituting the first on

$$\begin{aligned}
\dot{z}^k\Gamma_k v &= \dot{z}^k M^r{}_k M^{-1}\Gamma'_r M v + M^{-1}\dot{M}v \\
\dot{M} &= \dot{z}^k \partial_{zk} M \qquad \rightarrow \\
M^r{}_k \Gamma'_r &= M\Gamma_k M^{-1} + M\partial_{zk}M^{-1} \\
\Gamma'_r &= \{M\Gamma_k M^{-1} + M\partial_{zk}M^{-1}\}(M^{-1})^k_r
\end{aligned} \tag{12}$$

From Eq. (12) the transformation of the one-form $\Gamma^{(1)} = dx^\kappa \Gamma_\kappa$ (Eq. (1)) follows

$$\begin{aligned}
\Gamma'^{(1)} &= dx'^\kappa \Gamma'_\kappa \\
\Gamma'^{(1)} &= M\Gamma^{(1)}M^{-1} + MdM^{-1} \\
dF &= dx^\kappa \partial_{x\kappa} F; \quad F : \textbf{matrix valued}
\end{aligned} \tag{13}$$

Torsion ... and ..., is it relevant?

It shall remain relevant, until proven otherwise.

We proceed noting the one special feature of the connection transformation (Eq. (12)), written in full, upon using

$$\begin{aligned}
MdM^{-1} &= -(dM)M^{-1} \\
\Gamma'_{r'}{}^{u'}{}_{t'} &= \left[M^{u'}_u \Gamma_r{}^u{}_t (M^{-1})^r{}_{r'}(M^{-1})^t{}_{t'} + I'_{r'}{}^{u'}{}_{t'}\right] \\
I'_{r'}{}^{u'}{}_{t'} &= -(\partial_r M^{u'}_u)(M^{-1})^r{}_{r'}(M^{-1})^u{}_{t'} \\
\partial_r M^{u'}_u &= \partial_r \partial_u x'^{u'}(x) = \partial_u M^{u'}_r
\end{aligned} \tag{14}$$

It follows that the inhomogeneous orientation gauging part is symmetric

$$\begin{aligned} \Gamma'_{r'}{}^{u'}{}_{t'} &= [M^{u'}_{u}\Gamma_{r}{}^{u}{}_{t}(M^{-1})^{r}{}_{r'}(M^{-1})^{t}{}_{t'} + I'_{r'}{}^{u'}{}_{t'}] \\ I'_{r'}{}^{u'}{}_{t'} &= I'_{t'}{}^{u'}{}_{r'} \end{aligned} \tag{15}$$

Three things emerge:

(a) the antisymmetric part of the connection defines a 3-tensor $T_{[r}{}^{u}{}_{t]}$: torsion

$$T_{[r}{}^{u}{}_{t]} = \frac{1}{2}(\Gamma_{r}{}^{u}{}_{t} - \Gamma_{t}{}^{u}{}_{r}) \tag{16}$$

(b) it does *not* follow, that the symmetric part derives from a metric.

(c) a symmetric metric yields a symmetric Riemannian (minimal) connection

$$\begin{aligned} \overset{o}{\Gamma}{}_{\{r}{}^{u}{}_{t\}} &= \frac{1}{2}g^{uv}[\partial_{r}g_{vt} + \partial_{t}g_{vr} - \partial_{v}g_{rt}] \\ \gamma_{\{r}{}^{u}{}_{t\}} &= \frac{1}{2}(\Gamma_{r}{}^{u}{}_{t} + \Gamma_{t}{}^{u}{}_{r}) - \overset{o}{\Gamma}{}_{\{r}{}^{u}{}_{t\}} \end{aligned} \tag{17}$$

$\overset{o}{\Gamma}{}_{\{r}{}^{u}{}_{t\}}$ defined in Eq. (17) — if not vanishing — defines a symmetric 3-tensor, in addition to torsion.

Notwithstanding the eventual presence of 3-tensors $T_{[r}{}^{u}{}_{t]}$ and $\gamma_{[r}{}^{u}{}_{t]}$ the general 1-form, defined in Eq. (1) with transformation properties given in Eq. (13) (repeated below for clarity)

$$\begin{aligned} &(\Gamma^{(1)} = dx^{\kappa}\Gamma_{\kappa})^{\varrho}{}_{\sigma} : \textbf{matrix valued 1-form} \\ &\Gamma'^{(1)} = dx'^{\kappa}\Gamma'_{\kappa} \\ &\Gamma'^{(1)} = M\Gamma^{(1)}M^{-1} + MdM^{-1} \end{aligned} \tag{18}$$

generate a **matrix valued 1-form** curvature 2-form, a 4-tensor

$$\begin{aligned} R^{(2)} &= d\Gamma^{(1)} + (\Gamma^{(1)})^{2} \\ &\to \frac{1}{2}(R_{[\sigma\tau]})^{u}{}_{v}dx^{\sigma} \wedge dx^{\tau} \end{aligned} \tag{19}$$

as follows from the transformation properties (Eq. (18))‡

$$R'^{(2)} = MR^{(2)}M^{-1} \tag{20}$$

$$R'^{(2)} = \begin{bmatrix} M(d\Gamma^{(1)})M^{-1} & 1 \\ +(dM)M^{-1}M\Gamma^{(1)}M^{-1} & 2 \\ -(M\Gamma^{(1)}M^{-1})MdM^{-1} & 3 \\ +(dM)dM^{-1} & 4 \\ +(M\Gamma^{(1)}M^{-1})MdM^{-1} & 5 \\ +M(dM^{-1})M\Gamma^{(1)}M^{-1} & 6 \\ +(MdM^{-1})(MdM^{-1}) & 7 \\ +M(\Gamma^{(1)})^2M^{-1} & 8 \end{bmatrix} \tag{21}$$ §

References

[1] Wolfgang Pauli, 'Relativitätstheorie', new edition of the original edition, "Encyclopädie der Wissenschaften", Vol. V, Art. 19, 1921, Paolo Boringhieri, ed., 1953.

[2] Élie Cartan, 'Sur les variétés à connexion projective', Bull. Soc. Math. de France, 52, 2 (1923) 205.

[3] Élie Cartan, 'Sur une classe remarquable d'espaces de Riemann', Bull. Soc. Math. de France, 54 (1926) 214, and 55 (1927) 114.

[4] P. Debye and P. Scherrer, 'Kristallpulver', Gött. Nachr. (1916) 1, 'Flüssigkeiten', Gött. Nachr. (1916) 16 and Phys. Z. 17 (1916) 277.

[5] P. Minkowski, 'On the hypothesis that curvature freezes a set of space-like variables beyond the observed four at energies much below the Planck mass', Bern University preprint 17/1977, October 1977, unpublished; see URL — http://www.mink.itp.unibe.ch/.

[6] L. Smolin, 'Towards a theory of space-time structure at very short distances', HUTP-79/A010, April 1979, Nucl. Phys. B 160 (1979) 253.

[7] S. Kobayashi and K. Nomizu, 'Foundations of differential geometry', Interscience Tracts in Pure and Applied Mathematics, No. 15, Vols. I, II, L. Bers, R. Courant and J.J. Stoker eds., John Wiley and Sons, New York 1963, 1969.

[8] Charles Ehresmann — oeuvres complètes et commentées, 'Topologie algébrique et géometrie différentielle', parties I-1 et I-2, suppléments No. 1 et 2 au Vol. XXIV (1983) des 'Cahiers de topologie et géometrie différentielle', A.C. Ehresmann ed., Imprimerie Evrard, Amiens 1984.

[9] H. Weyl, 'Raum Zeit Materie', Julius Springer Verlag, Berlin 1923.

‡ ... well known yet remarkable ...

§ The rows 2–7 cancel.

Scale Hierarchies and String Cosmology

I. Antoniadis
LPTHE, UMR CNRS 7589, Sorbonne Universités,
UPMC Paris 6, F-75005 Paris, France
Albert Einstein Center, Institute for Theoretical Physics,
Bern University, Sidlerstrasse 5, CH-3012 Bern, Switzerland
ab_antoniadis@itp.unibe.ch

I describe the phenomenology of a model of supersymmetry breaking in the presence of a tiny (tuneable) positive cosmological constant. It utilizes a single chiral multiplet with a gauged shift symmetry, that can be identified with the string dilaton (or an appropriate compactification modulus). The model is coupled to the MSSM, leading to calculable soft supersymmetry breaking masses and a distinct low energy phenomenology that allows to differentiate it from other models of supersymmetry breaking and mediation mechanisms. We also study the question if this model can lead to inflation by identifying the dilaton with the inflaton. We find that this is possible if the Kähler potential is modified by a term that has the form of NS5-brane instantons, leading to an appropriate inflationary plateau around the maximum of the scalar potential, depending on two extra parameters.

1. Introduction

If String Theory is a fundamental theory of Nature and not just a tool for studying systems with strongly coupled dynamics, it should be able to describe at the same time particle physics and cosmology, which are phenomena that involve very different scales from the microscopic four-dimensional (4d) quantum gravity length of 10^{-33} cm to large macroscopic distances of the size of the observable universe $\sim 10^{28}$ cm spanned a region of about 60 orders of magnitude. In particular, besides the 4d Planck mass, there are three very different scales with very different physics corresponding to the electroweak, dark energy and inflation. These scales might be related via the scale of the underlying fundamental theory, such as string theory, or they might be independent in the sense that their origin could be based on different and independent dynamics. An example of the former constraint and more predictive possibility is provided by TeV strings with a fundamental scale at low energies due for instance to large extra dimensions transverse to a four-dimensional braneworld forming our Universe.[1] In this case, the 4d Planck mass is emergent from the fundamental string scale and inflation should also happen around the same scale.[2]

Here, we will adopt the second more conservative approach, assuming that all three scales have an independent dynamical origin. Moreover, we will assume the presence of low energy supersymmetry that allows for an elegant solution of the mass hierarchy problem, a unification of fundamental forces as indicated by low energy data and a natural dark matter candidate due to an unbroken R-parity. The assumption of independent scales implies that supersymmetry breaking should be realized in a metastable de Sitter vacuum with an infinitesimally small (tunable) cosmological constant independent of the supersymmetry breaking scale that should be in the TeV region. In a recent work,[3] we studied a simple $N = 1$ supergravity model having this property and motivated by string theory. Besides the gravity multiplet, the minimal field content consists of a chiral multiplet with a shift symmetry promoted to a gauged R-symmetry using a vector multiplet. In the string theory context, the chiral multiplet can be identified with the string dilaton (or an appropriate compactification modulus) and the shift symmetry associated to the gauge invariance of a two-index antisymmetric tensor that can be dualized to a (pseudo)scalar. The shift symmetry fixes the form of the superpotential and the gauging allows for the presence of a Fayet–Iliopoulos (FI) term, leading to a supergravity action with two independent parameters that can be tuned so that the scalar potential possesses a metastable de Sitter minimum with a tiny vacuum energy (essentially the relative strength between the F- and D-term contributions). A third parameter fixes the Vacuum Expectation Value (VEV) of the string dilaton at the desired (phenomenologically) weak coupling regime. An important consistency constraint of our model is anomaly cancellation which has been studied in Ref. 5 and implies the existence of additional charged fields under the gauged R-symmetry.

In a more recent work,[6] we analyzed a small variation of this model which is manifestly anomaly free without additional charged fields and allows to couple in a straightforward way a visible sector containing the minimal supersymmetric extension of the Standard Model (MSSM) and studied the mediation of supersymmetry breaking and its phenomenological consequences. It turns out that an additional "hidden sector" field z is needed to be added for the matter soft scalar masses to be nontachyonic; although this field participates in the supersymmetry breaking and is similar to the so-called Polonyi field, it does not modify the main properties of the metastable de Sitter (dS) vacuum. All soft scalar masses, as well as trilinear A-terms, are generated at the tree level and are universal under the assumption that matter kinetic terms are independent of the "Polonyi" field, since matter fields are neutral under the shift symmetry and supersymmetry breaking is driven by a combination of the $U(1)$ D-term and the dilaton and z-field F-term. Alternatively, a way to avoid the tachyonic scalar masses without adding the extra field z is to modify the matter kinetic terms by a dilaton dependent factor.

A main difference of the second analysis from the first work is that we use a field representation in which the gauged shift symmetry corresponds to an ordinary $U(1)$ and not an R-symmetry. The two representations differ by a Kähler transformation that leaves the classical supergravity action invariant. However, at the quantum level, there is a Green–Schwarz term generated that amounts an extra dilaton dependent contribution to the gauge kinetic terms needed to cancel the anomalies of the R-symmetry. This creates an apparent puzzle with the gaugino masses that vanish in the first representation but not in the latter. The resolution to the puzzle is based to the so-called anomaly mediation contributions[7,8] that explain precisely the above apparent discrepancy. It turns out that gaugino masses are generated at the quantum level and are thus suppressed compared to the scalar masses (and A-terms).

This model has the necessary ingredients to be obtained as a remnant of moduli stabilization within the framework of internal magnetic fluxes in type I string theory, turned on along the compact directions for several Abelian factors of the gauge group. All geometric moduli can in principle be fixed in a supersymmetric way, while the shift symmetry is associated to the 4d axion and its gauging is a consequence of anomaly cancellation.[9,10]

We then make an attempt to connect the scale of inflation with the electroweak and supersymmetry breaking scales within the same effective field theory, that at the same time allows the existence of an infinitesimally small (tuneable) positive cosmological constant describing the present dark energy of the universe. We thus address the question whether the same scalar potential can provide inflation with the dilaton playing also the role of the inflaton at an earlier stage of the universe evolution.[11] We show that this is possible if one modifies the Kähler potential by a correction that plays no role around the minimum, but creates an appropriate plateau around the maximum. In general, the Kähler potential receives perturbative and nonperturbative corrections that vanish in the weak coupling limit. After analyzing all such corrections, we find that only those that have the form of (Neveu–Schwarz) NS5-brane instantons can lead to an inflationary period compatible with cosmological observations. The scale of inflation turns out then to be of the order of low energy supersymmetry breaking, in the TeV region. On the other hand, the predicted tensor-to-scalar ratio is too small to be observed.

2. Conventions

Throughout this paper we use the conventions of Ref. 12. A supergravity theory is specified (up to Chern–Simons terms) by a Kähler potential $\mathcal{K}$, a superpotential W, and the gauge kinetic functions $f_{AB}(z)$. The chiral multiplets z^α, χ^α are enumerated by the index α and the indices A, B indicate the different gauge groups. Classically, a supergravity theory is invariant under Kähler tranformations, viz.

$$\begin{aligned} \mathcal{K}(z,\bar{z}) &\longrightarrow \mathcal{K}(z,\bar{z}) + J(z) + \bar{J}(\bar{z})\,, \\ W(z) &\longrightarrow e^{-\kappa^2 J(z)} W(z)\,, \end{aligned} \tag{1}$$

where κ is the inverse of the reduced Planck mass, $m_p = \kappa^{-1} = 2.4 \times 10^{15}$ TeV. The gauge transformations of chiral multiplet scalars are given by holomorphic Killing vectors, i.e. $\delta z^\alpha = \theta^A k_A^\alpha(z)$, where θ^A is the gauge parameter of the gauge group A. The Kähler potential and superpotential need not be invariant under this gauge transformation, but can change by a Kähler transformation

$$\delta \mathcal{K} = \theta^A \left[r_A(z) + \bar{r}_A(\bar{z}) \right], \tag{2}$$

provided that the gauge transformation of the superpotential satisfies $\delta W = -\theta^A \kappa^2 r_A(z) W$. One then has from $\delta W = W_\alpha \delta z^\alpha$

$$W_\alpha k_A^\alpha = -\kappa^2 r_A W, \tag{3}$$

where $W_\alpha = \partial_\alpha W$ and α labels the chiral multiplets. The supergravity theory can then be described by a gauge invariant function

$$\mathcal{G} = \kappa^2 \mathcal{K} + \log(\kappa^6 W \bar{W}). \tag{4}$$

The scalar potential is given by

$$\begin{aligned} V &= V_F + V_D, \\ V_F &= e^{\kappa^2 \mathcal{K}} \left(-3\kappa^2 W \bar{W} + \nabla_\alpha W g^{\alpha\bar{\beta}} \bar{\nabla}_{\bar{\beta}} \bar{W} \right), \\ V_D &= \frac{1}{2} (\mathrm{Re}\, f)^{-1\ AB} \mathcal{P}_A \mathcal{P}_B, \end{aligned} \tag{5}$$

where W appears with its Kähler covariant derivative

$$\nabla_\alpha W = \partial_\alpha W(z) + \kappa^2 (\partial_\alpha \mathcal{K}) W(z). \tag{6}$$

The moment maps $\mathcal{P}_A$ are given by

$$\mathcal{P}_A = i \left(k_A^\alpha \partial_\alpha \mathcal{K} - r_A \right). \tag{7}$$

In this paper we will be concerned with theories having a gauged R-symmetry, for which $r_A(z)$ is given by an imaginary constant $r_A(z) = i\kappa^{-2}\xi$. In this case, $\kappa^{-2}\xi$ is a Fayet–Iliopoulos[13] constant parameter.

3. The Model

The starting point is a chiral multiplet S and a vector multiplet associated with a shift symmetry of the scalar component s of the chiral multiplet S

$$\delta s = -ic\theta, \tag{8}$$

and a string-inspired Kähler potential of the form $-p \log(s + \bar{s})$. The most general superpotential is either a constant $W = \kappa^{-3} a$ or an exponential superpotential $W = \kappa^{-3} a e^{bs}$ (where a and b are constants). A constant superpotential is (obviously) invariant under the shift symmetry, while an exponential superpotential transforms as $W \to W e^{-ibc\theta}$, as in Eq. (3). In this case the shift symmetry becomes a gauged R-symmetry and the scalar potential contains a Fayet–Iliopoulos term. Note however that by performing a Kähler transformation (1) with $J = \kappa^{-2} bs$, the model can be recast into a constant superpotential at the cost of introducing a linear term in the Kähler potential $\delta K = b(s + \bar{s})$. Even though in this representation, the shift symmetry is not an R-symmetry, we will still refer to it as $U(1)_R$. The most general gauge kinetic function has a constant term and a term linear in s, $f(s) = \delta + \beta s$.

To summarize,[a]

$$\begin{aligned} \mathcal{K}(s, \bar{s}) &= -p \log(s + \bar{s}) + b(s + \bar{s}) \,, \\ W(s) &= a \,, \\ f(s) &= \delta + \beta s \,, \end{aligned} \tag{9}$$

where we have set the mass units $\kappa = 1$. The constants a and b together with the constant c in Eq. (8) can be tuned to allow for an infinitesimally small cosmological constant and a TeV gravitino mass. For $b > 0$, there always exists a supersymmetric AdS (anti-de Sitter) vacuum at $\langle s + \bar{s} \rangle = b/p$, while for $b = 0$ (and $p < 3$) there is an AdS vacuum with broken supersymmetry. We therefore focus on $b < 0$. In the context of string theory, S can be identified with a compactification modulus or the universal dilaton and (for negative b) the exponential superpotential may be generated by nonperturbative effects.

The scalar potential is given by

$$\begin{aligned} V &= V_F + V_D \,, \\ V_F &= a^2 e^{\frac{b}{l}} l^{p-2} \left\{ \frac{1}{p} (pl - b)^2 - 3l^2 \right\}, \quad l = 1/(s + \bar{s}) \,, \\ V_D &= c^2 \frac{l}{\beta + 2\delta l} (pl - b)^2 \,. \end{aligned} \tag{10}$$

In the case where S is the string dilaton, V_D can be identified as the contribution of a magnetized D-brane, while V_F for $b = 0$ and $p = 2$ coincides with the tree-level dilaton potential obtained by considering string theory away its critical dimension.[14] For $p \geq 3$ the scalar potential V is positive and monotonically decreasing, while for $p < 3$, its F-term part V_F is unbounded from below when $s + \bar{s} \to 0$. On the other hand, the D-term part of the scalar potential V_D is positive and diverges when $s + \bar{s} \to 0$ and for various values for the parameters an (infinitesimally small) positive (local) minimum of the potential can be found.

If we restrict ourselves to integer p, tunability of the vacuum energy restricts $p = 2$ or $p = 1$ when $f(s) = s$, or $p = 1$ when the gauge kinetic function is constant.

[a] In superfields the shift symmetry (8) is given by $\delta S = -ic\Lambda$, where Λ is the superfield generalization of the gauge parameter. The gauge invariant Kähler potential is then given by $\mathcal{K}(S, \bar{S}) = -p\kappa^{-2} \log(S + \bar{S} + cV_R) + \kappa^{-2} b(S + \bar{S} + cV_R)$, where V_R is the gauge superfield of the shift symmetry.

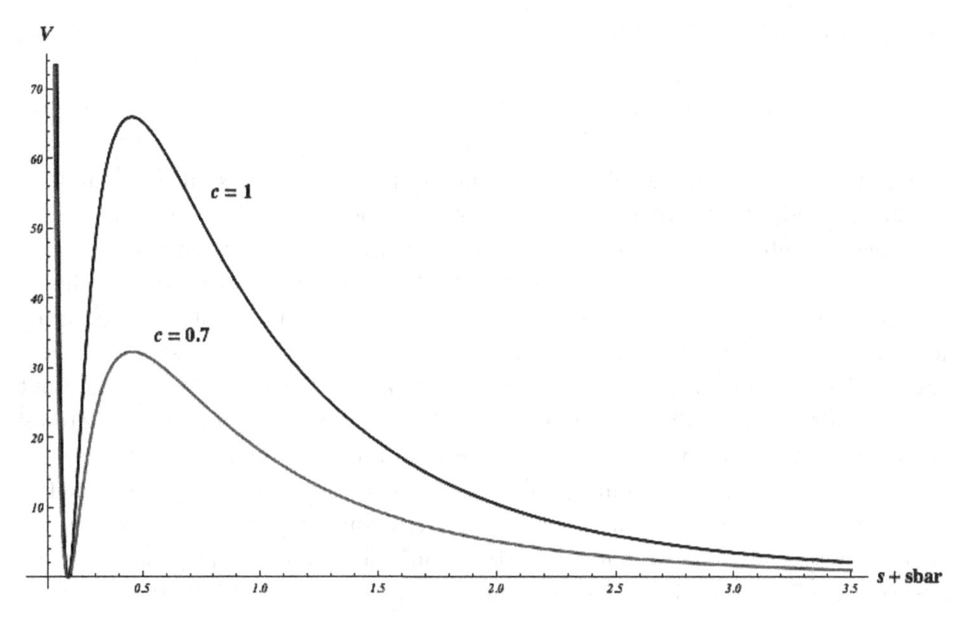

Fig. 1. (Color online) A plot of the scalar potential for $p = 2$, $b = -1$, $\delta = 0$, $\beta = 1$ and a given by Eq. (12) for $c = 1$ (black curve) and $c = 0.7$ (red curve).

For $p = 2$ and $f(s) = s$, the minimization of V yields

$$b/l = -\rho_0 \approx -0.183268\,, \quad p = 2\,, \tag{11}$$

$$\frac{a^2}{bc^2} = A_2(-\rho_0) + B_2(-\rho_0)\frac{\Lambda}{b^3c^2} \approx -50.6602 + \mathcal{O}(\Lambda)\,, \tag{12}$$

where Λ is the value of V at the minimum (i.e. the cosmological constant), $-\rho_0$ is the negative root of the polynomial $-x^5 + 7x^4 - 10x^3 - 22x^2 + 40x + 8$ compatible with (12) for $\Lambda = 0$ and $A_2(\alpha)$, $B_2(\alpha)$ are given by

$$A_2(\alpha) = 2e^{-\alpha}\frac{-4 + 4\alpha - \alpha^2}{\alpha^3 - 4\alpha^2 - 2\alpha}\,; \quad B_2(\alpha) = 2\frac{\alpha^2 e^{-\alpha}}{\alpha^2 - 4\alpha - 2}\,. \tag{13}$$

It follows that by carefully tuning a and c, Λ can be made positive and arbitrarily small independently of the supersymmetry breaking scale. A plot of the scalar potential for certain values of the parameters is shown in Fig. 1.

At the minimum of the scalar potential, for nonzero a and $b < 0$, supersymmetry is broken by expectation values of both an F- and D-term. Indeed the F-term and D-term contributions to the scalar potential are

$$\begin{aligned} V_F|_{s+\bar{s}=\frac{-\rho_0}{b}} &= \frac{1}{2}a^2b^2e^{-}\rho_0\left(1 + \frac{2}{\rho_0}\right)^2 > 0\,, \\ V_D|_{s+\bar{s}=\frac{-\rho_0}{b}} &= -\frac{b^3c^2}{\rho_0}\left(1 + \frac{2}{\rho_0}\right)^2 > 0\,. \end{aligned} \tag{14}$$

The gravitino mass term is given by

$$(m_{3/2})^2 = e^{\mathcal{G}} = \frac{a^2 b^2}{\rho_0^2} e^{-\rho_0} . \tag{15}$$

Due to the Stueckelberg coupling, the imaginary part of s (the axion) gets eaten by the gauge field, which acquires a mass. On the other hand, the Goldstino, which is a linear combination of the fermion of the chiral multiplet χ and the gaugino λ gets eaten by the gravitino. As a result, the physical spectrum of the theory consists (besides the graviton) of a massive scalar, namely the dilaton, a Majorana fermion, a massive gauge field and a massive gravitino. All the masses are of the same order of magnitude as the gravitino mass, proportional to the same constant a (or c related by Eq. (12) where b is fixed by Eq. (11)), which is a free parameter of the model. Thus, they vanish in the same way in the supersymmetric limit $a \to 0$.

The local dS minimum is metastable since it can tunnel to the supersymmetric ground state at infinity in the s-field space (zero coupling). It turns out however that it is extremely long lived for realistic perturbative values of the gauge coupling $l \simeq 0.02$ and TeV gravitino mass and, thus, practically stable; its decay rate is:[5]

$$\Gamma \sim e^{-B} \quad \text{with} \quad B \approx 10^{300} . \tag{16}$$

4. Coupling a Visible Sector

The guideline to construct a realistic model keeping the properties of the toy model described above is to assume that matter fields are invariant under the shift symmetry (8) and do not participate in the supersymmetry breaking. In the simplest case of a canonical Kähler potential, MSSM-like fields ϕ can then be added as:

$$\begin{aligned} \mathcal{K} &= -\kappa^{-2} \log(s+\bar{s}) + \kappa^{-2} b(s+\bar{s}) + \sum \varphi\bar{\varphi} , \\ W &= \kappa^{-3} a + W_{\text{MSSM}} , \end{aligned} \tag{17}$$

where $W_{\text{MSSM}}(\phi)$ is the usual MSSM superpotential. The squared soft scalar masses of such a model can be shown to be positive and close to the square of the gravitino mass (TeV^2). On the other hand, for a gauge kinetic function with a linear term in s, $\beta \neq 0$ in Eq. (9), the Lagrangian is not invariant under the shift symmetry

$$\delta\mathcal{L} = -\theta \frac{\beta c}{8} \epsilon^{\mu\nu\rho\sigma} F_{\mu\nu} F_{\rho\sigma} \tag{18}$$

and its variation should be canceled. As explained in Ref. 5, in the "frame" with an exponential superpotential the R-charges of the fermions in the model can give an anomalous contribution to the Lagrangian. In this case the "Green–Schwarz" term $\text{Im}\, sF\tilde{F}$ can cancel quantum anomalies. However as shown in Ref. 5, with the minimal MSSM spectrum, the presence of this term requires the existence of additional fields in the theory charged under the shift symmetry.

Instead, to avoid the discussion of anomalies, we focus on models with a constant gauge kinetic function. In this case the only (integer) possibility[b] is $p = 1$. The scalar potential is given by (10) with $\beta = 0$, $\delta = p = 1$. The minimization yields to equations similar to (11), (12) and (13) with a different value of ρ_0 and functions A_1 and B_1 given by

$$\begin{aligned} b\langle s+\bar{s}\rangle &= -\rho_0 \approx -0.233153\,, \\ \frac{bc^2}{a^2} &= A_1(-\rho_0) + B_1(-\rho_0)\frac{\Lambda}{a^2 b} \approx -0.359291 + \mathcal{O}(\Lambda)\,, \\ A_1(\alpha) &= 2e^{\alpha}\alpha\frac{3-(\alpha-1)^2}{(\alpha-1)^2}\,, \quad B_1(\alpha) = \frac{2\alpha^2}{(\alpha-1)^2}\,, \end{aligned} \tag{19}$$

where $-\rho_0$ is the negative root of $-3 + (\rho-1)^2(2-\rho^2/2) = 0$ close to -0.23, compatible with the second constraint for $\Lambda = 0$. However, this model suffers from tachyonic soft masses when it is coupled to the MSSM, as in (17). To circumvent this problem, one can add an extra hidden sector field which contributes to (F-term) supersymmetry breaking. Alternatively, the problem of tachyonic soft masses can also be solved if one allows for a noncanonical Kähler potential in the visible sector, which gives an additional contribution to the masses through the D-term.

Let us discuss first the addition of an extra hidden sector field z (similar to the so-called Polonyi field[15]). The Kähler potential, superpotential and gauge kinetic function are given by

$$\begin{aligned} \mathcal{K} &= -\kappa^{-2}\log(s+\bar{s}) + \kappa^{-2}b(s+\bar{s}) + z\bar{z} + \sum\varphi\bar{\varphi}\,, \\ W &= \kappa^{-3}a(1+\gamma\kappa z) + W_{\text{MSSM}}(\varphi)\,, \\ f(s) &= 1\,, \quad f_A = 1/g_A^2\,, \end{aligned} \tag{20}$$

where A labels the Standard Model gauge group factors and γ is an additional constant parameter. The existence of a tuneable dS vacuum with supersymmetry breaking and nontachyonic scalar masses implies that γ must be in a narrow region:

$$0.5 \lesssim \gamma \lesssim 1.7\,. \tag{21}$$

In the above range of γ the main properties of the toy model described in the previous section remain, while $\operatorname{Re} z$ and its F-auxiliary component acquire non-vanishing VEVs. All MSSM soft scalar masses are then equal to a universal value m_0 of the order of the gravitino mass, while the B_0 Higgs mixing parameter

[b] If $f(s)$ is constant, the leading contribution to V_D when $s+\bar{s}\to 0$ is proportional to $1/(s+\bar{s})^2$, while the leading contribution to V_F is proportional to $1/(s+\bar{s})^p$. It follows that $p < 2$; if $p > 2$, the potential is unbounded from below, while if $p = 2$, the potential is either positive and monotonically decreasing or unbounded from below when $s+\bar{s}\to 0$ depending on the values of the parameters.

is also of the same order:

$$
\begin{aligned}
m_0^2 &= m_{3/2}^2\left[(\sigma_s+1)+\frac{(\gamma+t+\gamma t)^2}{(1+\gamma t)^2}\right],\\
A_0 &= m_{3/2}\left[(\sigma_s+3)+t\frac{(\gamma+t+\gamma t^2)}{1+\gamma t}\right],\\
B_0 &= m_{3/2}\left[(\sigma_s+2)+t\frac{(\gamma+t+\gamma t^2)}{(1+\gamma t)}\right],
\end{aligned}
\tag{22}
$$

where $\sigma_s = -3 + (\rho+1)^2$ with $\rho = -b(s+\bar{s})$ and $t \equiv \langle \mathrm{Re}\, z\rangle$ determined by the minimization conditions as functions of γ. Also, A_0 is the soft trilinear scalar coupling in the standard notation, satisfying the relation[16]

$$
A_0 = B_0 + m_{3/2}\,. \tag{23}
$$

On the other hand, the gaugino masses appear to vanish at tree-level since the gauge kinetic functions are constants (see (20)). However, as mentioned in Sec. 3, this model is classically equivalent to the theory[c]

$$
\begin{aligned}
\mathcal{K} &= -\kappa^{-2}\log(s+\bar{s}) + z\bar{z} + \sum_{\varphi}\varphi\bar{\varphi}\,,\\
W &= \left(\kappa^{-3}a(1+z) + W_{\mathrm{MSSM}}(\varphi)\right)e^{bs}\,,
\end{aligned}
\tag{24}
$$

obtained by applying a Kähler transformation (1) with $J = -\kappa^{-2}bs$. All classical results remain the same, such as the expressions for the scalar potential and the soft scalar masses (22), but now the shift symmetry (8) of s became a gauged R-symmetry since the superpotential transforms as $W \longrightarrow We^{-ibc\theta}$. Therefore, all fermions (including the gauginos and the gravitino) transform[d] as well under this $U(1)_R$, leading to cubic $U(1)_R^3$ and mixed $U(1)\times G_{\mathrm{MSSM}}$ anomalies. These anomalies are canceled by a Green–Schwarz (GS) counterterm that arises from a quantum correction to the gauge kinetic functions:

$$
f_A(s) = 1/g_A^2 + \beta_A s \quad \text{with} \quad \beta_A = \frac{b}{8\pi^2}\left(T_{R_A} - T_{G_A}\right), \tag{25}
$$

where T_G is the Dynkin index of the adjoint representation, normalized to N for $SU(N)$, and T_R is the Dynkin index associated with the representation R of dimension d_R, equal to $1/2$ for the $SU(N)$ fundamental. An implicit sum over all matter representations is understood. It follows that gaugino masses are nonvanishing in this representation, creating a puzzle on the quantum equivalence of the two classically equivalent representations. The answer to this puzzle is based on the fact that gaugino masses are present in both representations and are generated at

[c] This statement is only true for supergravity theories with a nonvanishing superpotential where everything can be defined in terms of a gauge invariant function $G = \kappa^2\mathcal{K} + \log(\kappa^6 W\bar{W})$.[17]

[d] The chiral fermions, the gauginos and the gravitino carry a charge $bc/2$, $-bc/2$ and $-bc/2$, respectively.

one-loop level by an effect called Anomaly Mediation.[7,8] Indeed, it has been argued that gaugino masses receive a one-loop contribution due to the super-Weyl–Kähler and sigma-model anomalies, given by[8]

$$M_{1/2} = -\frac{g^2}{16\pi^2}\left[(3T_G - T_R)m_{3/2} + (T_G - T_R)\mathcal{K}_\alpha F^\alpha + 2\frac{T_R}{d_R}(\log\det\mathcal{K}|_R{''})_{,\alpha}F^\alpha\right]. \tag{26}$$

The expectation value of the auxiliary field F^α, evaluated in the Einstein frame is given by

$$F^\alpha = -e^{\kappa^2\mathcal{K}/2}g^{\alpha\bar{\beta}}\bar{\nabla}_{\bar{\beta}}\bar{W}. \tag{27}$$

Clearly, for the Kähler potential (20) or (24) the last term in Eq. (26) vanishes. However, the second term survives due to the presence of Planck scale VEVs for the hidden sector fields s and z. Since the Kähler potential between the two representations differs by a linear term $b(s+\bar{s})$, the contribution of the second term in Eq. (26) differs by a factor

$$\delta m_A = \frac{g_A^2}{16\pi^2}(T_G - T_R)be^{\kappa^2\mathcal{K}/2}g^{\alpha\bar{\beta}}\bar{\nabla}_{\bar{\beta}}\bar{W}, \tag{28}$$

which exactly coincides with the "direct" contribution to the gaugino masses due to the field dependent gauge kinetic function (25) (taking into account a rescaling proportional to g_A^2 due to the noncanonical kinetic terms).

We conclude that even though the models (20) and (24) differ by a (classical) Kähler transformation, they generate the same gaugino masses at one-loop. While the one-loop gaugino masses for the model (20) are generated entirely by Eq. (26), the gaugino masses for the model (24) after a Kähler transformation have a contribution from Eq. (26) as well as from a field dependent gauge kinetic term whose presence is necessary to cancel the mixed $U(1)_R \times G$ anomalies due to the fact that the extra $U(1)$ has become an R-symmetry giving an R-charge to all fermions in the theory. Using (26), one finds:

$$M_{1/2} = -\frac{g^2}{16\pi^2}m_{3/2}\left[(3T_G - T_R) - (T_G - T_R)\left((\rho+1)^2 + t\frac{\gamma + t + \gamma t^2}{1+\gamma t}\right)\right]. \tag{29}$$

For $U(1)_Y$ we have $T_G = 0$ and $T_R = 11$, for $SU(2)$ we have $T_G = 2$ and $T_R = 7$, and for $SU(3)$ we have $T_G = 3$ and $T_R = 6$, such that for the different gaugino masses this gives (in a self-explanatory notation):

$$\begin{aligned}
M_1 &= 11\frac{g_Y^2}{16\pi^2}m_{3/2}\left[1 - (\rho+1)^2 - \frac{t(\gamma + t + \gamma t)}{1+\gamma t}\right], \\
M_2 &= \frac{g_2^2}{16\pi^2}m_{3/2}\left[1 - 5(\rho+1)^2 - 5\frac{t(\gamma + t + \gamma t^2)}{1+\gamma t}\right], \\
M_3 &= -3\frac{g_3^2}{16\pi^2}m_{3/2}\left[1 + (\rho+1)^2 + \frac{t(\gamma + t + \gamma t^2)}{1+\gamma t}\right].
\end{aligned} \tag{30}$$

Table 1. The soft terms (in terms of $m_{3/2}$) for various values of γ. If a solution to the RGE exists, the value of $\tan\beta$ is shown in the last columns for $\mu > 0$ and $\mu < 0$.

γ	t	ρ	m_0	A_0	M_1	M_2	M_3	$\tan\beta$ ($\mu > 0$)	$\tan\beta$ ($\mu < 0$)
0.6	0.446	0.175	0.475	1.791	0.017	0.026	0.027		
1	0.409	0.134	0.719	1.719	0.015	0.025	0.026		
1.1	0.386	0.120	0.772	1.701	0.015	0.024	0.026	46	29
1.4	0.390	0.068	0.905	1.646	0.014	0.023	0.026	40	23
1.7	0.414	0.002	0.998	1.588	0.013	0.022	0.025	36	19

5. Phenomenology

The results for the soft terms calculated in the previous section, evaluated for different values of the parameter γ are summarized in Table 1. For every γ, the corresponding t and ρ are calculated by imposing a vanishing cosmological constant at the minimum of the potential. The scalar soft masses and trilinear terms are then evaluated by Eqs. (22) and the gaugino masses by Eqs. (30). Note that the relation (23) is valid for all γ. We therefore do not list the parameter B_0.

In most phenomenological studies, B_0 is substituted for $\tan\beta$, the ratio between the two Higgs VEVs, as an input parameter for the renormalization group equations (RGE) that determine the low energy spectrum of the theory. Since B_0 is not a free parameter in our theory, but is fixed by Eq. (23), this corresponds to a definite value of $\tan\beta$. For more details see Ref. 18 and references therein. The corresponding $\tan\beta$ for a few particular choices for γ are listed in the last two columns of Table 1 for $\mu > 0$ and $\mu < 0$ respectively. No solutions were found for $\gamma \lesssim 1.1$, for both signs of μ. The lightest supersymmetric particle (LSP) is given by the lightest neutralino and since $M_1 < M_2$ (see Table 1) the lightest neutralino is mostly Bino-like, in contrast with a typical mAMSB (minimal anomaly mediation supersymmetry breaking) scenario, where the lightest neutralino is mostly Wino-like.[19]

To get a lower bound on the stop mass, the sparticle spectrum is plotted in Fig. 2 as a function of the gravitino mass for $\gamma = 1.1$ and $\mu > 0$ (for $\mu < 0$ the bound is higher). The experimental limit on the gluino mass forces $m_{3/2} \gtrsim 15$ TeV. In this limit the stop mass can be as low as 2 TeV. To conclude, the lower end mass spectrum consists of (very) light charginos (with a lightest chargino between 250 and 800 GeV) and neutralinos, with a mostly Bino-like neutralino as LSP (80–230 GeV), which would distinguish this model from the mAMSB where the LSP is mostly Wino-like. These upper limits on the LSP and the lightest chargino imply that this model could in principle be excluded in the next LHC run. In order for the gluino to escape experimental bounds, the lower limit on the gravitino mass is about 15 TeV. The gluino mass is then between 1–3 TeV. This however forces the squark masses to be very high (10–35 TeV), with the exception of the stop mass which can be relatively light (2–15 TeV).

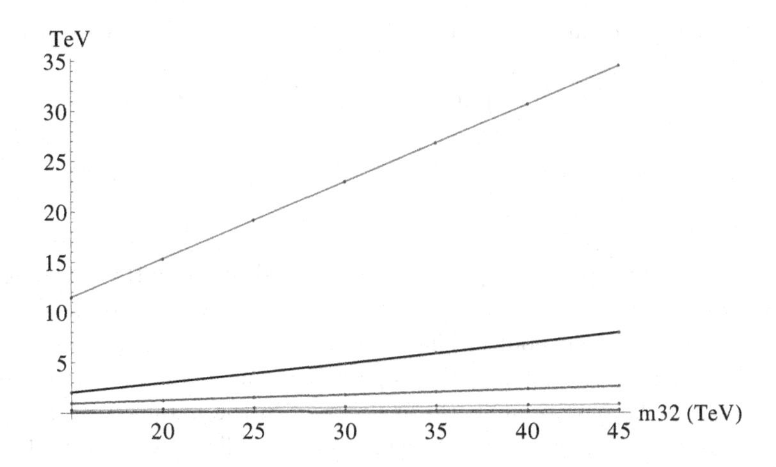

Fig. 2. (Color online) The masses of the sbottom (yellow), stop (black), gluino (red), lightest chargino (green) and lightest neutralino (blue) as a function of $m_{3/2}$ for $\gamma = 1.1$ and for $\mu > 0$. No solutions to the RGE were found when $m_{3/2} \gtrsim 45$ TeV. The lower bound corresponds to a gluino mass of 1 TeV.

6. Noncanonical Kähler Potential for the Visible Sector

As mentioned already in Sec. 4, an alternative way to avoid tachyonic soft scalar masses for the MSSM fields in the model (17), instead of adding the extra Polonyi-type field z in the hidden sector, is by introducing noncanonical kinetic terms for the MSSM fields, such as

$$\begin{aligned} \mathcal{K} &= -\kappa^{-2}\log(s+\bar{s}) + \kappa^{-2}b(s+\bar{s}) + (s+\bar{s})^{-\nu}\sum\varphi\bar{\varphi}\,, \\ W &= \kappa^{-3}a + W_{\text{MSSM}}\,, \\ f(s) &= 1\,, \quad f_A(s) = 1/g_A^2\,, \end{aligned} \tag{31}$$

where ν is an additional parameter of the theory, with $\nu = 1$ corresponding to the leading term in the Taylor expansion of $-\log(s+\bar{s}-\varphi\bar{\varphi})$. Since the visible sector fields appear only in the combination $\varphi\bar{\varphi}$, their VEVs vanish provided that the scalar soft masses squared are positive. Moreover, for vanishing visible sector VEVs, the scalar potential and is minimization remains the same as in Eqs. (19). Therefore, the noncanonical Kähler potential does not change the fact that the F-term contribution to the soft scalar masses squared is negative. On the other hand, the visible fields enter in the D-term scalar potential through the derivative of the Kähler potential with respect to s. Even though this has no effect on the ground state of the potential, the φ-dependence of the D-term scalar potential does result in an extra contribution to the scalar masses squared which become positive

$$\nu > -\frac{e^{\alpha}(\sigma_s+1)\alpha}{A(\alpha)(1-\alpha)} \approx 2.6\,. \tag{32}$$

The soft MSSM scalar masses and trilinear couplings in this model are:

$$\begin{aligned} m_0^2 &= \kappa^2 a^2 \left(\frac{b}{\alpha}\right)\left(e^{\alpha}(\sigma_s+1)+\nu\frac{A(\alpha)}{\alpha}(1-\alpha)\right), \\ A_0 &= m_{3/2}(s+\bar{s})^{\nu/2}(\sigma_s+3), \\ B_0 &= m_{3/2}(s+\bar{s})^{\nu/2}(\sigma_s+2), \end{aligned} \tag{33}$$

where σ_s is defined as in (22), Eq. (19) has been used to relate the constants a and c, and corrections due to a small cosmological constant have been neglected. A field redefinition due to a noncanonical kinetic term $g_{\varphi\bar{\varphi}} = (s+\bar{s})^{-\nu}$ is also taken into account. The main phenomenological properties of this model are not expected to be different from the one we analyzed in Sec. 5 with the parameter ν replacing γ. Gaugino masses are still generated at one-loop level while mSUGRA applies to the soft scalar sector. We therefore do not repeat the phenomenological analysis for this model.

7. Identifying the Dilaton with the Inflaton

In the following, we study the possibility to identify the dilaton with the inflaton. We will show first that the above model does not allow slow-roll inflation.

Indeed, the kinetic terms in the model (9)–(10) for the scalar $\phi \equiv s+\bar{s} = 1/l$ are given by

$$\mathcal{L}_s/e = -g_{s\bar{s}}\partial_\mu s \partial^\mu \bar{s} = -\frac{p\kappa^{-2}}{4}\frac{1}{\phi^2}\partial_\mu\phi\partial^\mu\phi\,. \tag{34}$$

The canonically normalized field χ therefore satisfies $\chi = \kappa^{-1}\sqrt{\frac{p}{2}}\log\phi$, where we reintroduce the gravitational coupling κ.

The slow-roll parameters are given by

$$\begin{aligned} \epsilon &= \frac{1}{2\kappa^2}\left(\frac{dV/d\chi}{V}\right)^2 = \frac{1}{2\kappa^2}\left[\frac{1}{V}\frac{dV}{d\phi}\left(\frac{d\chi}{d\phi}\right)^{-1}\right]^2, \\ \eta &= \frac{1}{\kappa^2}\frac{V''(\chi)}{V} = \frac{1}{\kappa^2}\frac{1}{V}\left[\frac{d^2V}{d\phi^2}\left(\frac{d\chi}{d\phi}\right)^{-2} - \frac{dV}{d\phi}\frac{d^2\chi}{d\phi^2}\left(\frac{d\chi}{d\phi}\right)^{-3}\right]. \end{aligned} \tag{35}$$

It can be shown that, when the conditions (11) and (12) are satisfied, the slow-roll parameters and the potential depend only on $\rho = -b\phi$; indeed

$$\frac{\kappa^4 V(\rho)}{b^3c^2} = \frac{e^{-\rho}\left(A_2(\alpha)\rho(\rho^2+4\rho-2) - 2e^{\rho}(\rho+2)^2\right)}{2\rho^3}, \tag{36}$$

where $A_2(\alpha) \approx -50.66$ as in Eq. (12). In Fig. 3, a plot is shown of $\frac{\kappa^4 V(\rho)}{|b|^3c^2}$ as a function of ρ. The minimum of the potential is at $\rho_{\min} \approx 0.1832$ (see Eq. (11)), while the potential has a local maximum at $\rho_{\max} \approx 0.4551$. A plot of the slow-roll parameter η (also in Fig. 3) shows that $|\eta| \ll 1$ is not satisfied. This result holds

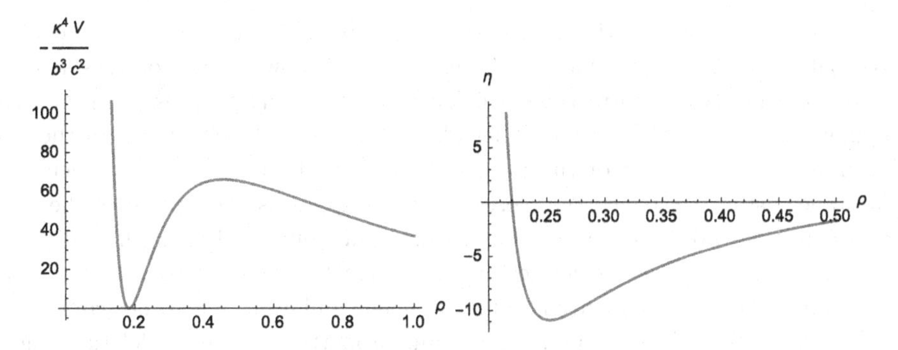

Fig. 3. A plot of $-\frac{\kappa^4 V(\rho)}{b^3 c^2}$ as a function of $\rho = -b\phi$ (left), and a plot of the slow-roll parameter η as a function of ρ (right). The slow-roll condition $|\eta| \ll 1$ is not satisfied for any value of the parameters a, b, c.

for any parameters a, b, c satisfying Eqs. (11) and (12). A similar analysis to the one above can be performed for $p = 1$, showing that the slow-roll condition $\eta \ll 1$ cannot be satisfied.

8. Extensions of the Model that Satisfy the Slow-roll Conditions

In the previous section we showed that the slow-roll conditions cannot be satisfied in the minimal versions of the model. In this section we modify the above model by modifying the Kähler potential. While the superpotential is uniquely fixed (up to a Kähler transformation), the Kähler potential admits corrections that can always be put in the form

$$\mathcal{K} = -p\kappa^{-2} \log\left(s + \bar{s} + \frac{\xi}{b} F(s + \bar{s}) \right) + \kappa^{-2} b(s + \bar{s}) , \tag{37}$$

while the superpotential, the gauge kinetic function and moment map are given by

$$\begin{gathered} W = \kappa^{-3} a , \quad f(s) = \delta + \beta s , \\ \mathcal{P} = \kappa^{-2} c \left(b - p \frac{1 + \frac{\xi}{b} F_s}{s + \bar{s} + \frac{\xi}{b} F} \right) , \end{gathered} \tag{38}$$

where $\mathcal{P}$ is the $U(1)$ moment map (7) and $F_s = \partial_s F(s + \bar{s})$. The scalar potential is given by ($\phi = s + \bar{s}$)

$$\begin{aligned} V &= V_F + V_D , \\ V_F &= \kappa^{-4} \frac{|a|^2 e^{b\phi}}{(\phi + \frac{\xi}{b} F)^p} \left[-3 - \frac{1}{p} \frac{(b(b\phi + \xi F) - p(b + \xi F_\phi))^2}{\xi F_{\phi\phi}(b\phi + \xi F) - (b + \xi F_\phi)^2} \right] , \\ V_D &= \kappa^{-4} \frac{b^2 c^2}{2\delta + \beta\phi} \left[1 - p \frac{1 + \frac{\xi}{b} F_\phi}{b\phi + \xi F} \right]^2 . \end{aligned} \tag{39}$$

As was discussed above, we take $\delta = 1$, $\beta = 0$ for $p = 1$ and $\delta = 0$, $\beta = 1$ for $p = 2$.

Identifying Re(s) with the inverse string coupling, the function F may contain perturbative contributions that can be expressed as power series of $1/(s+\bar{s})$, as well as nonperturbative corrections which are exponentially suppressed in the weak coupling limit. The later can be either of the form $e^{-\lambda(s+\bar{s})}$ for $\lambda > 0$ in the case of D-brane instantons, or of the form $e^{-\lambda(s+\bar{s})^2}$ in the case of (Neveu–Schwarz) NS5-brane instantons (since the closed string coupling is the square of the open string coupling). We have considered a generic contribution of these three different types of corrections and we found that only the last type of contributions can lead to an inflationary plateau providing sufficient inflation. The other corrections can be present but do not modify the main properties of the model (as long as weak coupling description holds). In the following section, we analyze in detailed a function F describing a generic NS5-brane instanton correction to the Kähler potential.

9. Slow-roll Inflation

9.1. $p = 2$ *case*

We now consider the case with

$$F(\phi) = \exp(\alpha b^2\phi^2)\,, \tag{40}$$

where $b < 0$ and $\alpha < 0$. $F(\phi)$ vanishes asymptotically at large ϕ. In this case, we obtain

$$V_D = \frac{\kappa^{-4}b^3c^2}{b\phi}\left[\frac{b\phi - 2 + \xi e^{\alpha b^2\phi^2}(1-4\alpha b\phi)}{b\phi + \xi e^{\alpha b^2\phi^2}}\right]^2, \tag{41}$$

and

$$V_F = -\frac{\kappa^{-4}|a|^2b^2e^{b\phi}}{2\left(\xi e^{\alpha b^2\phi^2}+b\phi\right)^2}\left[\frac{\left(b\phi+\xi e^{\alpha b^2\phi^2}(1-4\alpha b\phi)-2\right)^2}{2\alpha\xi e^{\alpha b^2\phi^2}\left(2\alpha b^3\phi^3+\xi e^{\alpha b^2\phi^2}-b\phi\right)-1}+6\right]. \tag{42}$$

There are four parameters in this model namely α, ξ, b and c. The first two parameters α and ξ control the shape of the potential. There are some regions in the parameter space of α and ξ that the potential satisfies the slow-roll conditions i.e. $\epsilon \ll 1$ and $|\eta| \ll 1$. In order to obtain the potential with flat plateau shape which is suitable for inflation and in agreement with Planck '15 data, we choose

$$\alpha \simeq -4.84 \quad \text{and} \quad \xi \simeq 0.025\,. \tag{43}$$

Note that in the case of $\xi = 0$ and $b < 0$, we can find the Minkowski minimum by solving the equations $V(\phi_{\text{min}}) = 0$ and $dV(\phi_{\text{min}})/d\phi = 0$, where $\phi_{\text{min}} = s_{\text{min}} + \bar{s}_{\text{min}}$ is the value of ϕ at the minimum of the potential. In the case of $\xi \neq 0$, we cannot solve the equations analytically and the relations (11), (12) are not valid. We can always assume that they are modified into

$$b\phi_{\text{min}} = -\rho(\xi,\alpha) \quad \text{and} \quad \frac{a^2}{bc^2} = -50.66\times\lambda(\xi,\alpha,\Lambda)^2\,, \tag{44}$$

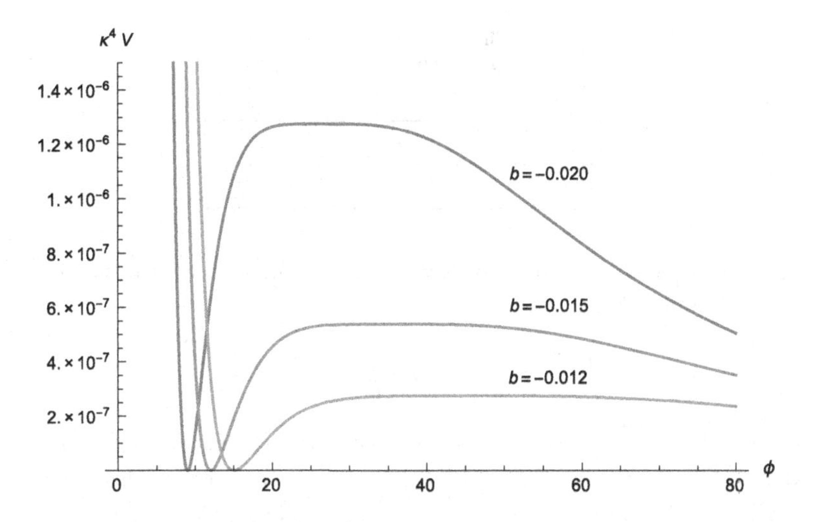

Fig. 4. A plot of the scalar potential for $p = 2$, with $b = -0.020$, $b = -0.015$ and $b = -0.012$. Note that we choose the parameters α and ξ as in Eq. (43) with $c = 0.06$.

where λ takes positive values and satisfies $|\lambda - 1| \ll 1$. For any given value of parameters ξ, α and the cosmological constant Λ, one can numerically fix the value of ρ and λ. By fine-tuning the cosmological constant Λ to be very close to zero, we can numerically solve the equations $V = 0$ and $dV/d\phi = 0$ for the value of ρ and λ in (44) as

$$\rho \approx 0.18\,, \tag{45}$$

$$\lambda \approx 1.017\,. \tag{46}$$

From Eq. (44), we can see that the third parameter, b, controls the vacuum expectation value $\phi_{\rm min}$. This can be shown in Fig. 4 where we compare the scalar potential for different values of b. Motivated by string theory, we have the identification $\phi \sim 1/g_s$. We can choose the value of the parameter b such that $\phi_{\rm min}$ is of the order of 10 to make sure that we are in the perturbative regime in g_s. The last parameter, c, controls the overall scale of the potential but does not change its minimum and its shape. In the following, we will fix b and c by using the cosmological data.

In order to compare the predictions of our models with Planck '15 data, we choose the following boundary conditions:

$$\phi_{\rm int} = 27.32\,, \quad \phi_{\rm end} = 22.68\,. \tag{47}$$

The initial conditions are chosen very near the maximum on the (left) side, so that the field rolls down towards the electroweak minimum. Any initial condition on the right of the maximum may produce also inflation, but the field will roll towards the SUSY vacuum at infinity. The results are therefore very sensitive to the initial conditions (47) of the inflaton field.

Table 2. The theoretical predictions for $p = 2$, with $b = -0.0182$, $c = 0.61 \times 10^{-13}$, and α, ξ given in Eq. (43).

n_s	r	A_s
0.965	2.969×10^{-23}	2.259×10^{-9}

The slow-roll parameters are given as in Eq. (35). The total number of e-folds N can be determined by

$$N = \kappa^2 \int_{\chi_{\text{end}}}^{\chi_{\text{int}}} \frac{V}{\partial_\chi V} d\chi = \kappa^2 \int_{\phi_{\text{end}}}^{\phi_{\text{int}}} \frac{V}{\partial_\phi V} \left(\frac{d\chi}{d\phi}\right)^2 d\phi \,. \tag{48}$$

Note that we choose $|\eta(\chi_{\text{end}})| = 1$. We can compare the theoretical predictions of our model to the experimental results via the power spectrum of scalar perturbations of the CMB, namely the amplitude A_s and tilt n_s, and the relative strength of tensor perturbations, i.e. the tensor-to-scalar ratio r. In terms of slow-roll parameters, these are given by

$$A_s = \frac{\kappa^4 V_*}{24\pi^2 \epsilon_*} \,, \tag{49}$$

$$n_s = 1 + 2\eta_* - 6\epsilon_* \,, \tag{50}$$

$$r = 16\epsilon_* \,, \tag{51}$$

where all parameters are evaluated at the field value χ_{int}.

In order to satisfy Planck '15 data, we choose the parameters $b = -0.0182$, $c = 0.61 \times 10^{-13}$. The value of the slow-roll parameters at the beginning of inflation are

$$\epsilon(\phi_{\text{int}}) \simeq 1.86 \times 10^{-24} \quad \text{and} \quad \eta(\phi_{\text{int}}) \simeq -1.74 \times 10^{-2} \,. \tag{52}$$

The total number of e-folds N, the scalar power spectrum amplitude A_s, the spectral index of curvature perturbation n_s and the tensor-to-scalar ratio r are calculated and summarized in Table 2, in agreement with Planck '15 data.[20] Figure 5 shows that our predictions for n_s and r are within 1σ C.L. of Planck '15 contours with the total number of e-folds $N \approx 1075$. Note that N is the total number of e-folds from ϕ_{int} to ϕ_{end}. However the number of e-folds associated with the CMB observation corresponds to a period between the time of horizon crossing and the end of inflation, which is much smaller than 1075. According to general formula in Ref. 20, the number of e-folds between the horizon crossing and the end of inflation is roughly estimated to be around 50–60.

We would like to remark that the parameter c also controls the gravitino mass at the minimum of the potential around $O(10)$ TeV. Indeed, the gravitino mass is written as

$$m_{3/2} = \kappa^2 e^{\kappa^2 K/2} W = \frac{1}{\kappa} \left(\frac{a b e^{b\phi/2}}{b\phi + \xi F(\phi)} \right) . \tag{53}$$

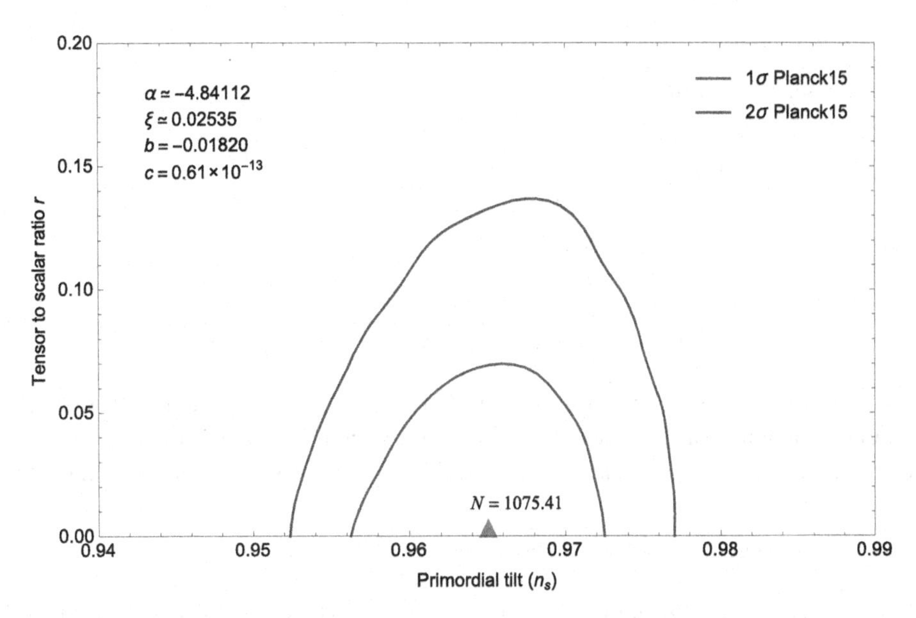

Fig. 5. We plot the theoretical predictions for the case $p = 2$, shown in Table 2, in the $n_s - r$ plane together with the Planck '15 results for TT, TE, EE, + lowP and assuming $\Lambda\text{CDM} + r$.

For $b = -0.0182$, we get $\phi_{\text{min}} \approx 9.91134$ and the gravitino mass at the minimum of the potential

$$\langle m_{3/2} \rangle \approx 14.98 \text{ TeV}\,. \tag{54}$$

The Hubble parameter during inflation (evaluated at $\phi_* = \phi_{\text{int}}$) is

$$H_* = \kappa\sqrt{V_*/3} = 1.38 \text{ TeV}\,. \tag{55}$$

This shows that our predicted scale for inflation is of the order of TeV. The mass of gravitino during the inflation $m^*_{3/2} = 4.15$ TeV is higher than the inflation scale, and the gauge boson mass is $M^*_{A_\mu} = 3.12$ TeV.[e] In fact, the gauge boson acquires a mass due to a Stueckelberg mechanism by eating the imaginary component of s, where its mass at the minimum of the potential is given by

$$\langle M_{A_\mu} \rangle = 15.48 \text{ TeV}\,. \tag{56}$$

As a result, the model essentially contains only one scalar field $\text{Re}(s)$, which is the inflaton. This is in contrast with other supersymmetric models of inflation, which usually contain at least two real scalars.[21,f]

[e]The gauge boson mass is given by $m_{A_\mu} = \sqrt{2g_{s\bar{s}}c^2/\text{Re}(s)}$.
[f]This is because a chiral multiplet contains a complex scalar.

9.2. $p = 1$ *case*

In this case, we obtain

$$V_D = \frac{\kappa^{-4}b^2c^2}{2}\left[\frac{b\phi - 1 + \xi e^{\alpha b^2\phi^2}(1 - 2\alpha b\phi)}{b\phi + \xi e^{\alpha b^2\phi^2}}\right]^2, \tag{57}$$

and

$$V_F = -\frac{\kappa^{-4}|a|^2 b e^{b\phi}}{\xi e^{\alpha b^2\phi^2} + b\phi}\left[\frac{\left(b\phi + \xi e^{\alpha b^2\phi^2}(1 - 2\alpha b\phi) - 1\right)^2}{2\alpha\xi e^{\alpha b^2\phi^2}\left(2\alpha b^3\phi^3 + \xi e^{\alpha b^2\phi^2} - b\phi\right) - 1} + 3\right]. \tag{58}$$

The potential has similar properties with the $p = 2$ case although it may give different phenomenological results at low energy. Similar to the previous case, the relations (19) are not valid when $\xi \neq 0$ and we assume that they are modified into

$$b\phi_{\min} = -\rho(\xi,\alpha) \quad \text{and} \quad \frac{bc^2}{a^2} \simeq -0.359 \times \lambda(\xi,\alpha,\Lambda)^{-2}. \tag{59}$$

By choosing $\alpha = -0.781$ and $\xi = 0.3023$ and tuning the cosmological constant Λ to be very close to zero, we can numerically fix $\rho \approx 0.56$ and $\lambda \approx 1.29$ for this case. The gravitino mass for $p = 1$ case can be written as

$$m_{3/2} = \kappa^2 e^{\kappa^2 K/2} W = \frac{1}{\kappa}\left(\frac{a\sqrt{b}e^{b\phi/2}}{\sqrt{b\phi + \xi F(\phi)}}\right). \tag{60}$$

By choosing the parameters $b = -0.0234$, $c = 1 \times 10^{-13}$, the gravitino mass at the minimum of the potential is

$$\langle m_{3/2}\rangle = 18.36 \text{ TeV} \tag{61}$$

with $\phi_{\min} \approx 21.53$, and

$$\langle M_{A_\mu}\rangle = 36.18 \text{ TeV}. \tag{62}$$

By choosing appropriate boundary conditions, we find

$$\phi_{\text{int}} = 64.53 \quad \text{and} \quad \phi_{\text{end}} = 50.99. \tag{63}$$

As summarized in Table 3, the predictions for the $p = 1$ case are similar to those of $p = 2$, in agreement with Planck '15 data with the total number of e-folds $N \approx 888$. In this case, the Hubble parameter during inflation is

$$H_* = \kappa\sqrt{V_*/3} = 5.09 \text{ TeV}. \tag{64}$$

Note that for the $p = 1$ case, the mass of the gauge boson is $M^*_{A_\mu} = 6.78$ TeV, and the mass of the gravitino during inflation is $m^*_{3/2} = 4.72$ TeV.

Table 3. The theoretical predictions for $p = 1$ with $b = -0.0234$, $c = 1 \times 10^{-13}$, $\alpha = -0.781$ and $\xi = 0.3023$.

n_s	r	A_s
0.959	4.143×10^{-22}	2.205×10^{-9}

9.3. *SUGRA spectrum*

The above model can be coupled to MSSM, as described in Sec. 4:

$$\begin{aligned} \mathcal{K} &= \mathcal{K}(s+\bar{s}) + \sum \varphi\bar{\varphi}\,, \\ W &= W_h(s) + W_{\text{MSSM}}\,. \end{aligned} \tag{65}$$

The soft supersymmetry breaking terms can then be calculated as follows:

$$\begin{aligned} m_0^2 &= e^{\kappa^2\mathcal{K}}\left(-2\kappa^4 W_h(s)\bar{W}_h(s) + \kappa^2 g^{s\bar{s}}|\nabla_s W_h|^2\right), \\ A_0 &= \kappa^2 e^{\kappa^2\mathcal{K}/2} g^{s\bar{s}} K_s\left(\bar{W}_{\bar{s}} + \kappa^2 K_s \bar{W}\right), \\ B_0 &= \kappa^2 e^{\kappa^2\mathcal{K}/2}\left(g^{s\bar{s}} K_s(\bar{W}_{\bar{s}} + \kappa^2 K_s\bar{W}) - \bar{W}\right). \end{aligned} \tag{66}$$

For $p = 2$ the Lagrangian contains a Green–Schwarz term Eq. (18), and the theory is not gauge invariant (without the inclusion of extra fields that are charged under the $U(1)$). We therefore focus on $p = 1$. The soft terms can be written in terms of the gravitino mass (see Eq. (53))

$$\begin{aligned} m_0^2 &= m_{3/2}^2[-2 + \mathcal{C}]\,, \\ A_0 &= m_{3/2}\mathcal{C}\,, \\ B_0 &= A_0 - m_{3/2}\,, \end{aligned} \tag{67}$$

where

$$\mathcal{C} = -\frac{\left(-\xi e^{\alpha b^2\phi^2} + b\phi\left(4\alpha\xi e^{\alpha b^2\phi^2} - 1\right) + 2\right)^2}{4\alpha\xi^2 e^{2\alpha b^2\phi^2} - 4\alpha b\xi\phi e^{\alpha b^2\phi^2} + 8\alpha^2 b^3\xi\phi^3 e^{\alpha b^2\phi^2} - 2}\Bigg|_{\phi=\phi_{\min}}\,. \tag{68}$$

Using the parameters presented in Subsec. 9.2, we find $m_{3/2} = 18.36$ TeV and $\mathcal{C} = 1.53$. For $\xi = 0$ the model reduces to the one analyzed in Sec. 4, where one has $\mathcal{C} = 1.52$ and $m_{3/2} = 17.27$ TeV (with $\phi_{\min} = 9.96$). Moreover, the scalar soft mass is tachyonic. This can be solved either by introducing an extra Polonyi-like field, or by allowing a noncanonical Kähler potential for the MSSM-like fields φ. The resulting low energy spectrum is expected to be similar to the one described in Secs. 4 and 5. We do not perform this analysis, but only summarize the results.

Since the tree-level contribution to the gaugino masses vanishes, their mass is generated at one-loop by the so-called "Anomaly Mediation" contribution (26). As a result, the spectrum consists of very light neutralinos ($O(10^2)$ GeV), of which the lightest (a mostly Bino-like neutralino) is the LSP dark matter candidate, slightly heavier charginos and a gluino in the 1–3 TeV range. The squarks are of the order of the gravitino mass ($\sim$10 TeV), with the exception of the stop squark which can be as light as 2 TeV.

References

1. I. Antoniadis, N. Arkani-Hamed, S. Dimopoulos and G. R. Dvali, *Phys. Lett. B* **436**, 257 (1998), arXiv:hep-ph/9804398.
2. I. Antoniadis and S. P. Patil, *Eur. Phys. J. C* **75**, 182 (2015), arXiv:1410.8845 [hep-th].
3. I. Antoniadis and R. Knoops, *Nucl. Phys. B* **886**, 43 (2014), arXiv:1403.1534 [hep-th].
4. F. Catino, G. Villadoro and F. Zwirner, *J. High Energy Phys.* **1201**, 002 (2012), arXiv:1110.2174 [hep-th]; G. Villadoro and F. Zwirner, *Phys. Rev. Lett.* **95**, 231602 (2005), arXiv:hep-th/0508167.
5. I. Antoniadis, D. M. Ghilencea and R. Knoops, *J. High Energy Phys.* **1502**, 166 (2015), arXiv:1412.4807 [hep-th].
6. I. Antoniadis and R. Knoops, *Nucl. Phys. B* **902**, 69 (2016), arXiv:1507.06924 [hep-ph].
7. L. Randall and R. Sundrum, *Nucl. Phys. B* **557**, 79 (1999), arXiv:hep-th/9810155; G. F. Giudice, M. A. Luty, H. Murayama and R. Rattazzi, *J. High Energy Phys.* **9812**, 027 (1998), arXiv:hep-ph/9810442.
8. J. A. Bagger, T. Moroi and E. Poppitz, *J. High Energy Phys.* **0004**, 009 (2000), arXiv:hep-th/9911029.
9. I. Antoniadis and T. Maillard, *Nucl. Phys. B* **716**, 3 (2005), arXiv:hep-th/0412008; I. Antoniadis, A. Kumar and T. Maillard, *Nucl. Phys. B* **767**, 139 (2007), arXiv: hep-th/0610246.
10. I. Antoniadis, J.-P. Derendinger and T. Maillard, *Nucl. Phys. B* **808**, 53 (2009), arXiv:0804.1738 [hep-th].
11. I. Antoniadis, A. Chatrabhuti, H. Isono and R. Knoops, *Eur. Phys. J. C* **76**, 680 (2016), arXiv:1608.02121 [hep-ph].
12. D. Z. Freedman and A. Van Proeyen, *Supergravity* (Cambridge University Press, 2012).
13. P. Fayet and J. Iliopoulos, *Phys. Lett. B* **51**, 461 (1974); P. Fayet, *Phys. Lett. B* **69**, 489 (1977).
14. I. Antoniadis, J.-P. Derendinger and T. Maillard, *Nucl. Phys. B* **808**, 53 (2009), arXiv:0804.1738 [hep-th].
15. J. Polonyi, Hungary Central Inst Res — KFKI-77-93 (77, REC. JUL 78).
16. H. P. Nilles, *Phys. Rep.* **110**, 1 (1984).
17. S. Ferrara, L. Girardello, T. Kugo and A. Van Proeyen, *Nucl. Phys. B* **223**, 191 (1983).
18. J. R. Ellis, K. A. Olive, Y. Santoso and V. C. Spanos, *Phys. Lett. B* **573**, 162 (2003), arXiv:hep-ph/0305212.
19. T. Gherghetta, G. F. Giudice and J. D. Wells, *Nucl. Phys. B* **559**, 27 (1999), arXiv:hep-ph/9904378.
20. Planck Collab. (P. A. R. Ade *et al.*), Planck 2015 results. XX. Constraints on inflation, arXiv:1502.02114 [astro-ph.CO].
21. D. Baumann and D. Green, *Phys. Rev. D* **85**, 103520 (2012), arXiv:1109.0292 [hep-th].

Implications of Higgs' Universality for Physics Beyond the Standard Model*

T. Goldman† and G. J. Stephenson Jr.‡

Department of Physics and Astronomy, University of New Mexico,
Albuquerque, NM 87501, USA
Theoretical Division, MS-B283, Los Alamos National Laboratory,
Los Alamos, NM 87545, USA
† *tjgoldman@post.harvard.edu*
‡ *gjs@phys.unm.edu*

We emulate Cabibbo by assuming a kind of universality for fermion mass terms in the Standard Model. We show that this is consistent with all current data and with the concept that deviations from what we term Higgs' universality are due to corrections from currently unknown physics of nonetheless conventional form. The application to quarks is straightforward, while the application to leptons makes use of the recognition that Dark Matter can provide the "sterile" neutrinos needed for the seesaw mechanism. Requiring agreement with neutrino oscillation results leads to the prediction that the mass eigenstates of the sterile neutrinos are separated by quadratically larger ratios than for the charged fermions. Using consistency with the global fit to LSND-like, short-baseline oscillations to determine the scale of the lowest mass sterile neutrino strongly suggests that the recently observed astrophysical 3.55 keV γ-ray line is also consistent with the mass expected for the second most massive sterile neutrino in our analysis.

1. Introduction

We first recall that Cabibbo[1] began the resolution of problems with the weak interaction posed by the difference between kaon decay rates and the otherwise universal strength found in nuclear beta decay, muon decay and pion decay. By introducing an angle between the weak current for weak interactions involving kaons and those that involved all other hadrons known at the time, which negligibly changed the other rates, he successfully restored universality of the weak current which was crucial to the development of the Standard Model (SM).

*LA-UR-17-23677

Mass in the SM poses a similar but even more extreme problem with the wide range of values for the Yukawa coupling of the Higgs to the various fermions needed to account for the wide range of mass values. We seek a basis in which this coupling will be universal for all fermions of a given charge, but that will nonetheless reproduce the observed mass spectrum in each case. To do so, we first revisit how mass is constructed in the SM.

1.1. *Mass in the SM*

The structure of SM mass terms is conventionally defined by (arbitrary) Yukawa couplings of the Higgs boson to the weak interaction active left-chiral doublets and weak interaction "sterile" (except for their $U(1)_B$ quantum numbers) singlets. The form for the quarks is

$$\mathcal{L}_m = \Sigma_i Y_{Ui} \overline{\left(\frac{1+\gamma_5}{2}\right)\Psi_{Ui}} \langle \phi^0, \phi^+ \rangle \left(\frac{1-\gamma_5}{2}\right) \begin{bmatrix} \Psi_{Ui} \\ \Psi_{Di} \end{bmatrix} + \text{h.c.}$$

$$+ \Sigma_i Y_{Di} \overline{\left(\frac{1+\gamma_5}{2}\right)\Psi_{Di}} \langle \phi^-, \phi^{0*} \rangle \left(\frac{1-\gamma_5}{2}\right) \begin{bmatrix} \Psi_{Ui} \\ \Psi_{Di} \end{bmatrix} + \text{h.c.} \tag{1}$$

Unlike the comparable terms for the interaction with the weak gauge bosons, which is naturally in the current basis, this Dirac notation is in the mass eigenstate basis and suppresses the information that pairs of independent Weyl spinors are involved. In the SM, the Y_i are taken to equal the mass of the appropriate fermion divided by the vacuum expectation value of the Higgs boson.

We recall, however, that a Dirac bispinor is composed of two Weyl spinors[2]

$$\Psi = \begin{bmatrix} \xi \\ \chi \end{bmatrix}, \tag{2}$$

where these two chiral spinors are constructed from another pair (ζ_a and ζ_b) with a fixed phase relation, as

$$\xi = (\zeta_a + \zeta_b), \tag{3}$$

$$\chi = -\sigma_2(\zeta_a - \zeta_b)^* \tag{4}$$

so that the Dirac mass term appears as

$$\bar{\Psi}\Psi = -(\chi^\dagger \xi + \xi^\dagger \chi). \tag{5}$$

(This notation has been essential in Grand Unification efforts.) This makes it clear that the Dirac mass term couples two independent Weyl spinors (left- and right-chiral) one of which is weak interaction active and the other sterile in the sense of being a singlet under the weak interaction. Except in the case of neutrinos as seen below, the right-chiral Weyl spinors do carry a weak hypercharge quantum number, but, as for the weak isospin quantum numbers, it is identical for all of the members of these triplets.

In this form, rather than 3×3 diagonal matrices for fermion masses, we obtain 6×6 matrices with off-diagonal blocks of mass values and 3×3 on-diagonal blocks of zeroes. If the off-diagonal blocks were still themselves diagonal, we would still have the nonuniversality problem. For example, for the up-type quarks, we would have

$$m_u = \begin{bmatrix} 2.3 & 0 & 0 \\ 0 & 1275 & 0 \\ 0 & 0 & 173500 \end{bmatrix}, \tag{6}$$

where all values are expressed in MeV$/c^2$. We will ignore the significant uncertainties and variation with scale of these masses[3] as the ratios vary less dramatically, although the values of even the ratios are not known to very high accuracy. We use here typical values recently reported by the Particle Data Group.[4] Similar results hold for the down-type quarks and for the charged leptons.

For several decades, the quantum numbers and corresponding gauge interactions that distinguish the different "generations" of fermions have been sought by starting from various symmetry assumptions without overt success. The various efforts to understand the fermion masses have ranged from substructure to Grand Unification. The former approaches include rishons[5] and technicolor.[6] Some interesting relations between quarks and leptons[7,8] have been found and even some mixing angles[9] but no convincing overall solution has been obtained. A few others of which we are aware are referenced here.[10] Efforts along these lines have continued.[a] They are all "top-down" approaches, making initial symmetry assumptions.

2. Higgs Universality

Seeking universality, we instead recall the pairing (also called "democratic"[12]) matrix of nuclear physics. This matrix has the same elements throughout. It is diagonalized, in the 3×3 case, by the matrix known as tri-bimaximal (TBM)

$$\text{TBM} = \begin{bmatrix} \frac{1}{\sqrt{6}} & -\frac{1}{\sqrt{2}} & \frac{1}{\sqrt{3}} \\ \frac{1}{\sqrt{6}} & \frac{1}{\sqrt{2}} & \frac{1}{\sqrt{3}} \\ \frac{-2}{\sqrt{6}} & 0 & \frac{1}{\sqrt{3}} \end{bmatrix} \tag{7}$$

to the eigenvalues 0, 0 and 1 when the elements of the mass matrix are normalized to $1/3$. (The eigenvectors for the two degenerate zero mass eigenvalues can, of course, be transformed to any mixture of the conventional two chosen here.)

Recalling the relative smallness of the first two of the three masses for each triple of charged fermions, we apply the inverse transformation to the m_u mass matrix

[a]There is no recent review and too many papers to refer, so we list only a few recent publications as examples.[11]

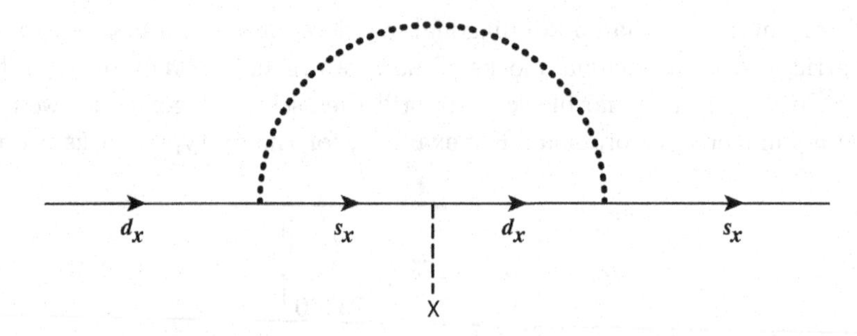

Fig. 1. BSM scalar loop correction to the Higgs vertex coupling a member of a weak doublet (d_x) to a weak singlet (s_x) fermion. A vector loop correction appears similarly with s_x and d_x interchanged at the Higgs vertex.

above and find

$$m_{u\text{-TBM}} = \text{TBM} \times m_u \times \text{TBM}^\dagger$$
$$= (173500) \times \begin{bmatrix} 0.33701 & 0.32966 & 0.33333 \\ 0.32966 & 0.33701 & 0.33333 \\ 0.33333 & 0.33333 & 0.33334 \end{bmatrix} \tag{8}$$

on scaling out the mass of the top quark as the overall scale. Similar results obtain for the down-type quarks and charged leptons. The deviations from universality are on the scale of a few percent in all cases.

The current basis for the (Weyl) fermions is defined by the alignment of the weak current and is also the basis for the Higgs boson doublet. So we recognize that in this Weyl spinor basis, that is, in the current basis, the mass matrices may be assumed to be "democratic" in the absence of any other quantum numbers that distinguish what have been called the "generations" of the "families" of fermions.

2.1. *BSM corrections*

Of course, the two smaller eigenvalues in each set of 3 are not zero, which means that deviations from universality due to something else must exist. In fact, loop corrections from physics beyond the SM (BSM) can provide the needed corrections. Whether vector or scalar, loop corrections (see Fig. 1) to the Higgs coupling to the Weyl spinors can have no effect on the fermion mass terms until the Higgs field acquires its vacuum expectation value. As shown long ago in Ref. 13, all the corrections corresponding to the "SU(3)" structure of the matrices will even be finite.

Thus, we take a "bottom-up" starting point focusing on the *absence* of a known symmetry or quantum numbers. This may be viewed as an accidental S_3 symmetry but that is not the fundamental nature of our basic assumption. Rather, we assume that the final form of the fermion mass matrix is determined by all possible (loop) corrections to the SM from BSM physics and invert the relation to extract information on some matrix elements of BSM physics. So we revise the SM, in the fermion mass sector only, by assuming Higgs universality: all of the fermions with a given electric charge should be viewed as having *nothing* that makes their right-chiral, weak interaction singlet components distinguishable to the Higgs boson. The Higgs coupling is completely insensitive to "generation" and only requires one Yukawa coupling constant for each set of fermions of a given charge.

In Ref. 14, we applied the concept of Higgs Universality to the problem of quark masses and mixings in the weak interaction currents, the Cabibbo–Kobayashi–Maskawa (CKM) matrix.[4] There we showed that the features of the mass spectra can be reproduced in a manner that naturally makes sensible predictions for the form of the CKM matrix and some of its parameters. While the charged leptons follow the discussion of quarks faithfully, the Majorana nature of the neutrinos requires a more complex treatment. For the extension to lepton masses and mixings, we find that the known form of the Pontecorvo–Maki–Nakagawa–Sakata (PMNS) matrix[4] and the allowed hierarchies for the active neutrino masses, used in conjunction with Higgs universality, provide constraints on the masses of sterile neutrinos. The results are consistent with some recent analyses of experiments reporting possible observations of the effects of sterile neutrinos.[15,16]

2.2. *Quark implementation*

For each type of quark, there are two known small quantities: ϵ, the ratio of the middle mass to the largest mass, and δ, the ratio of the smallest mass to the middle mass. We parametrized the BSM deviations in terms of ϵ times 10 parameters characterizing the BSM corrections. These are displayed in the following full Higgs plus BSM-loop-corrected 3×3 mass matrix:

$$\mathcal{M}_{\rm mrSM} = m \times [\mathcal{M}_{\rm dem} + \epsilon \mathcal{M}_{\rm BSM}] \tag{9}$$

but now in the *current* quark basis consistently defined by the Higgs and weak vector boson couplings. That is, we define the mass matrix for each set of 3 quarks of a given electric charge as m (an overall scale which is approximately one-third of the mass of the most massive of each triple of the fermions of a given nonzero electric charge) times the matrix $\mathcal{M}$, where

$$\begin{aligned}\mathcal{M} = \frac{(1+\epsilon\xi)}{3} &\times \begin{bmatrix} 1 & 1 & 1 \\ 1 & 1 & 1 \\ 1 & 1 & 1 \end{bmatrix} \\ &+ \epsilon \times \begin{bmatrix} \sqrt{\frac{2}{3}}y_0 + y_3 + \frac{1}{\sqrt{3}}y_8 & y_1 - Iy_2 & y_4 - Iy_5 \\ y_1 + Iy_2 & \sqrt{\frac{2}{3}}y_0 - y_3 + \frac{1}{\sqrt{3}}y_8 & y_6 - Iy_7 \\ y_4 + Iy_5 & y_6 + Iy_7 & \sqrt{\frac{2}{3}}y_0 - \frac{2}{\sqrt{3}}y_8 \end{bmatrix}. \end{aligned} \tag{10}$$

We diagonalized the matrices by first applying TBM, then block diagonalized (with $X_{3\to 2}$) to a 2×2 submatrix for the lighter quarks making use of the fact that the large eigenvalue and eigenvector were already determined to a good approximation (i.e. to $\mathcal{O}(\epsilon)$). Avoiding large CP violation in the 2×2 submatrix requires setting $y_2 = y_5 - y_7$. A final simple rotation (X_ω) about the axis defined by the most massive state diagonalizes the 2×2 submatrix. Thus, in addition to matching the eigenvalues, we found the matrices, $X_{tot} = \mathrm{TBM} \times X_{3\to 2} \times X_\omega$ that separately diagonalize the up-type and down-type quarks.

The product of the adjoint of $X_{\text{tot-}u}$ with $X_{\text{tot-}d}$ produces the CKM matrix[4] for the charge-raising weak current in terms of the parameters. This appears contrary to the PDG definition ($UV^\dagger$) but is due to the fact that we have started from the current eigenstates, whereas the PDG defines U and V as transforming from the mass to the current eigenstates. Note also that the TBM factors exactly cancel out here. We found that this form allowed us to relate PDG values of the moduli of the CKM matrix elements to the observed mass ratios and to characterize the remaining ambiguities in a simple form.

In particular, we found that, to leading approximation, the Cabibbo angle is equal to the difference between the 2×2 rotation angles (ω) for the up-type quarks and the down-type quarks, and so is independent of the initial basis for the 2×2 subsector as defined by the particular choice for the TBM matrix. However, we found that we could not consistently match all of the moduli of the entries in the CKM matrix without a nonzero value for the parameters related to CP-violation (y_5 and y_7). Note also that the small size of the mixing from the so-called "third generation" to the first two is immediate due to the large mass ratio characterized by ϵ.

The full details may be found in Ref. 14. We show here only a plot (Fig. 2) of three combinations of ϵ-weighted y_j parameters that remain dependent upon the

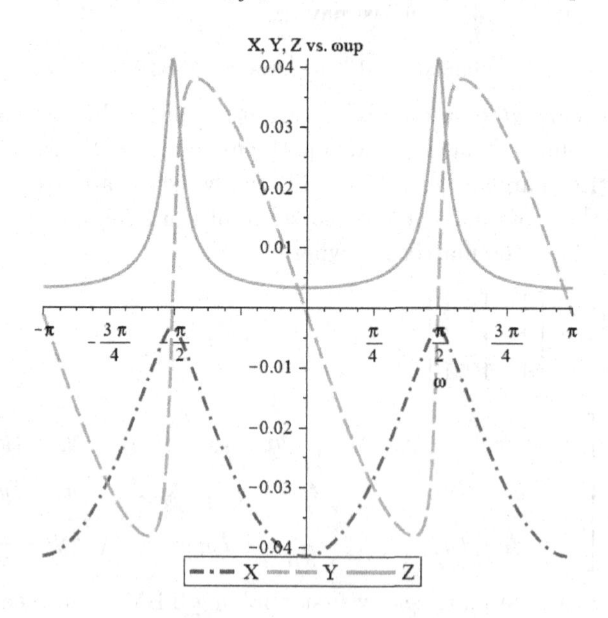

Fig. 2. Combinations of quark BSM parameters as functions of ω_u.

undetermined 2×2 rotation angle, ω_u, after applying all the constraints available from matching the moduli of the CKM entries and the value of the CP-violating Jarlskog invariant.[17] We view it as a significant result that these combinations describing the BSM physics remain of order ϵ, the so-called "natural" result.

3. Leptons

3.1. *Charged leptons*

Consider the mass matrix of the charged leptons in light of the discussion above. The description of a Dirac bispinor (required by charge conservation) is again expanded in terms of basic Weyl spinors so that in the Weyl basis, the form of the mass matrix is again

$$M_\ell = \left(\imath(\eta_a^*)^T, \imath(\eta_s^*)^T \right) \begin{pmatrix} 0 & m_D \\ m_D^\dagger & 0 \end{pmatrix} \begin{pmatrix} \eta_a \\ \eta_s \end{pmatrix}, \tag{11}$$

where m_D is a 3×3 Hermitian matrix, η_a is a three component vector of the left-chiral representation of (active) Weyl spinors carrying negative electric charge and weak charge $-\frac{1}{2}$ and η_s is a three component vector of the left-chiral representation of (sterile) Weyl spinors carrying positive electric charge and weak charge 0. In the mass eigenbasis,

$$\eta_a = \begin{pmatrix} \eta_{e^-} \\ \eta_{\tau^-} \\ \eta_{\mu^-} \end{pmatrix} \tag{12}$$

and

$$\eta_s = \begin{pmatrix} \eta_{e^+} \\ \eta_{\tau^+} \\ \eta_{\mu^+} \end{pmatrix}. \tag{13}$$

The ordering choice in Eqs. (12) and (13) is different from above (and that in Ref. 14) and is influenced by the PMNS matrix as described by the Particle Data Group.[4] Defining the parameters $(m_\ell, \epsilon, \delta)$ by $m_\ell = m_\tau$, $\epsilon = \frac{m_\mu}{m_\tau}$ and $\delta = \frac{m_e}{m_\mu}$ leads to the values,[4] $\epsilon = 0.059464$ and $\delta = 0.004836$, using $m_\ell = 1776.86\ \mathrm{MeV}/c^2$. These are very similar to what obtains in the quark sector.

In the mass basis, the 6×6 mass matrix becomes

$$M_\ell = m_\ell \times \begin{pmatrix} 0 & 0 & 0 & \epsilon\delta & 0 & 0 \\ 0 & 0 & 0 & 0 & 1 & 0 \\ 0 & 0 & 0 & 0 & 0 & \epsilon \\ \epsilon\delta & 0 & 0 & 0 & 0 & 0 \\ 0 & 1 & 0 & 0 & 0 & 0 \\ 0 & 0 & \epsilon & 0 & 0 & 0 \end{pmatrix}. \tag{14}$$

To bring this back into the current basis, as defined in the last section, we carry out the following operation. We define a slightly different TBM matrix, TBM_ℓ, as

$$\mathrm{TBM}_\ell = \begin{pmatrix} \frac{2}{\sqrt{6}} & \frac{1}{\sqrt{3}} & 0 \\ \frac{-1}{\sqrt{6}} & \frac{1}{\sqrt{3}} & \frac{-1}{\sqrt{2}} \\ \frac{-1}{\sqrt{6}} & \frac{1}{\sqrt{3}} & \frac{1}{\sqrt{2}} \end{pmatrix} \tag{15}$$

and transform M_ℓ to the current basis with the 6×6 unitary matrix

$$U_{6\times 6} = \begin{pmatrix} 0_{3\times 3} & \mathrm{TBM}_\ell^\dagger \\ \mathrm{TBM}_\ell^\dagger & 0_{3\times 3} \end{pmatrix} \tag{16}$$

giving the form of Eq. (11) with

$$m_D = m_l \times \begin{pmatrix} \left(\frac{1}{3} + \frac{2}{3}\epsilon\delta\right) & \left(\frac{1}{3} - \frac{1}{3}\epsilon\delta\right) & \left(\frac{1}{3} - \frac{1}{3}\epsilon\delta\right) \\ \left(\frac{1}{3} - \frac{1}{3}\epsilon\delta\right) & \left(\frac{1}{3} + \frac{1}{2}\epsilon + \frac{1}{6}\epsilon\delta\right) & \left(\frac{1}{3} - \frac{1}{2}\epsilon + \frac{1}{6}\epsilon\delta\right) \\ \left(\frac{1}{3} - \frac{1}{3}\epsilon\delta\right) & \left(\frac{1}{3} + \frac{1}{2}\epsilon + \frac{1}{6}\epsilon\delta\right) & \left(\frac{1}{3} - \frac{1}{2}\epsilon + \frac{1}{6}\epsilon\delta\right) \end{pmatrix}. \tag{17}$$

Converting from the Weyl representation to the Dirac representations allows us to carry out the rediagonalization as a 3×3 procedure as we did for the quarks in Ref. 14. This is, of course, an approximation, as the TBM_ℓ-matrix may have perturbatively small corrections as we found necessary in the case of the quarks.

3.2. *Active neutrinos*

The mostly active Majorana neutrino mass eigenstates, as constrained by solar and atmospheric oscillation data, may be described by two mass hierarchies.[4] The eigenstates are labeled 1, 2, and 3, where 1 is predominantly composed of the electron neutrino current eigenstate, 2 is an approximately even mixture of the current eigenstates and 3 has little electron current eigenstate admixture. In the limit of the PNMS matrix exactly matching the TBM matrix, these relations would be

$$\begin{aligned} |1\rangle &= \frac{1}{\sqrt{6}}(2|\nu_e\rangle - |\nu_\mu\rangle - |\nu_\tau\rangle)\,, \\ |2\rangle &= \frac{1}{\sqrt{3}}(|\nu_e\rangle + |\nu_\mu\rangle + |\nu_\tau\rangle)\,, \\ |3\rangle &= \frac{1}{2}(-|\nu_\mu\rangle + |\nu_\tau\rangle)\,. \end{aligned} \tag{18}$$

We choose to represent this description in the mass eigenstate representation, using an overall scale, μ, as

$$\mu_a = \mu \times \begin{pmatrix} -f & 0 & 0 \\ 0 & 1 & 0 \\ 0 & 0 & g \end{pmatrix}_{123} \qquad \text{(NO)} \tag{19}$$

for the so-called "Normal Order" (NO) hierarchy. For the "Inverted Order" hierarchy (IO), the diagonal mass matrix would more naturally be written as

$$\mu_a = \mu \times \begin{pmatrix} g & 0 & 0 \\ 0 & -f & 0 \\ 0 & 0 & 1 \end{pmatrix}_{312} \quad \text{(IO)}. \tag{20}$$

(The minus sign is simply for convenience in the calculations; the sign of a Majorana mass has no physical meaning as it can be reversed by an overall phase change of the field.[2])

However, we use only the first formulation for later convenience and account for the mass relations by choosing $-f$ and g in the NO hierarchy as $(|f| < 1 < g)$ whereas in the IO hierarchy, we take $(g < |f| < 1)$. We do this to avoid the need to re-order the eigenvectors related to each mass eigenstate in the matrix that produces the diagonalization and so maintain the PDG convention for the PMNS matrix.

There are two relations among f, g and μ using the experimental results,[4] $(\Delta m^2_{\text{solar}} = 75\ (\text{meV}/c^2)^2)$ and $(\Delta m^2_{\text{atmos}} = 2500\ (\text{meV}/c^2)^2)$. For NO,

$$\begin{aligned} \mu^2(1 - f^2) &= 75\ (\text{meV}/c^2)^2\,, \\ \mu^2(g^2 - 1) &= 2500\ (\text{meV}/c^2)^2 \end{aligned} \tag{21}$$

while, for IO,

$$\begin{aligned} \mu^2(1 - f^2) &= 75\ (\text{meV}/c^2)^2\,, \\ \mu^2(1 - g^2) &= 2575\ (\text{meV}/c^2)^2\,, \end{aligned} \tag{22}$$

where we have arbitrarily chosen which pair to difference for the larger (atmospheric) mass-squared difference. Demanding that the smallest mass^2 be greater than zero requires that, for NO, $\mu > 8.66\ \text{meV}/c^2$, while, for IO, $\mu > 50.74\ \text{meV}/c^2$. Cosmological constraints[4] suggest that $\mu \lesssim 100\ \text{meV}/c^2$. Plotting the values of f and g versus μ shows only slow variation over most of these ranges.

3.3. *Dark matter*

The SM as originally formulated does not include right-chiral projections of neutrino Dirac bispinors, but only the left-chiral Weyl spinors that appear in weak doublets with the left-chiral projections of charged lepton Dirac bispinors. In that case, no renormalizable mass terms can be constructed for the neutrino fields. However, we now know that there is a substantial amount of Dark Matter in the Universe. Although its character is unknown, it would be remarkable if it did not include some number of Weyl spinor fields. These would have no SM quantum numbers and so would have the properties requisite for identification as sterile (Weyl) neutrinos. As such, they provide the secondary set of Weyl spinors needed to construct the Dirac

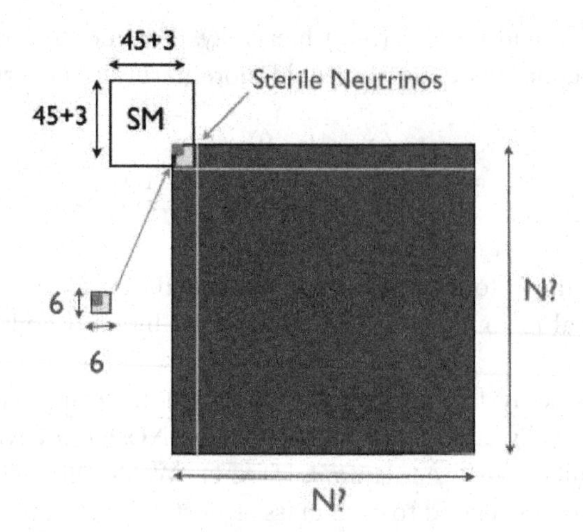

Fig. 3. Block diagonal mass matrices for SM and Dark Matter Weyl spinors.

mass terms involving the active Weyl spinor neutrino fields. Although there may be more than 3 such fields, we can consider the case of just 3 and/or block diagonalize the Dark Matter Weyl spinor mass matrix so that only 3 effective degrees-of-freedom connect with the active neutrino fields. That the latter may be accomplished is an assumption we make for simplicity; experimental results may ultimately require consideration of the more encompassing case.

Figure 3 displays the 45 known Weyl spinor degrees-of-freedom in the SM and N Weyl spinor degrees-of-freedom that we may expect to account for at least some of the Dark Matter that must exist. (This view is, of course, inspired by the efforts at SU(5) but not SO(10) Grand Unification.) The inset 6×6 shows how 3 of these N contribute to neutrino mass terms that are the analog of the Dirac mass terms for the charged fermions. In the next section, we analyze the case of only 3 sterile neutrinos.

Just as for the active neutrino Weyl spinors, there are no renormalizable mass terms that can be constructed for these sterile neutrinos using the (doublet) Higgs mechanism; those mass terms must arise from Dark Matter degrees-of-freedom, such as through a scalar Dark Matter field that acquires a vacuum expectation value similarly to the manner in the known case. This viewpoint thus affords two new possibilities relative to the classic seesaw mechanism:[8] one is that the scale of the Dirac-like mass terms that include the active neutrinos may very be different from that for the quarks and charged leptons (which already span a large range of scales). The other is that the Majorana mass terms for the sterile neutrinos that induce the seesaw mechanism, thus further suppressing the active neutrino Majorana mass eigenvalues, no longer need be terribly large, as on the expected scale of Grand Unification. Indeed, the mass scale for Dark Matter degrees-of-freedom may well be relatively modest.

4. Sterile Masses

Following the discussion in the last section, we assume that the full 6×6 neutrino mass matrix resembles the mass matrix for the charged leptons in the current representation with one important distinction,

$$M_\nu = \left(\imath(\phi_a^*)^T, \imath(\phi_s^*)^T\right) \begin{pmatrix} 0 & m_D \\ m_D^\dagger & M_S \end{pmatrix} \begin{pmatrix} \phi_a \\ \phi_s \end{pmatrix} \tag{23}$$

namely that the 3×3 matrix, M_S, is not zero. This structure is the basis for the seesaw model[8] which, assuming no singularities, led to a prescription for mostly active neutrino masses through the matrix

$$\mu_a = m_D^\dagger M_S^{-1} m_D \,. \tag{24}$$

However, that relation is reflexive so information about the mostly active Majorana masses, coupled with our assumptions about the form of m_D, can be inverted provide to restrictions on the form of M_S through

$$M_S = m_D \mu_a^{-1} m_D^\dagger \,. \tag{25}$$

What may we expect these restrictions to look like in general? Using the facts that the determinant of any mass matrix is the product of its eigenvalues, that determinants are invariant under unitary transformations and that the determinant of a product of matrices is the product of the individual determinants, we may evaluate each factor in its mass eigenstate basis and arrive at (where we extract a factor of m_d as the overall scale of the matrix, m_D, for the neutrino Dirac masses, analogous to the m_l factor for the charged lepton masses)

$$\det(M_S) = \left(\epsilon^2 \delta m_d^3\right) \left(\frac{-1}{fg\mu^3}\right) \left(\epsilon^2 \delta m_d^3\right) \tag{26}$$

or, defining the overall scale for the sterile neutrino mass matrix as $M_s = \frac{m_d^2}{\mu}$, the product of the three eigenvalues becomes

$$\Pi_i \lambda_i = \frac{-(M_s)^3}{fg} \epsilon^4 \delta^2 \,. \tag{27}$$

Whatever M_s, f, g are, this implies that the ratios of sterile neutrino masses should vary as the squares of the ratios of the charged fermion masses, with details depending on the actual hierarchy of active neutrinos and the scale of mostly active masses. The details of the matrix structure can only affect the coefficients of the eigenmass values proportional to M_s, $\epsilon^2 M_s$ and $\epsilon^2 \delta^2 M_s$.

4.1. *PMNS mixing and sterile neutrino eigenfunctions*

We now turn to determining in more detail what we can about the mass matrix and spectrum of the sterile neutrinos using the information available regarding the active neutrino mass spectrum and the PMNS matrix that describes the current mixing in weak leptonic currents, i.e. the analog of the CKM matrix for the weak hadronic currents.

We first recall that the CKM matrix describes the difference between the diagonal weak interaction currents in the (unknown) quark current basis from the (known) nondiagonal currents in the mass basis and that it is defined in terms of the charge-raising weak currents from quarks of charge $-1/3$ to quarks of charge $+2/3$, all of definite mass. The PMNS matrix, however, is defined in terms of the charge-lowering weak currents from charge neutral neutrinos to leptons of charge -1, where the definite mass leptons are taken to define the weak current basis for their charge. As we indicated above, Higgs universality invites us to define a weak current basis in which the charged lepton fields are not states of definite mass. Thus, were it not for the difference between the charge-raising and charge-lowering definitions, our simple rearrangement of the columns of the TBM matrix would have been sufficient to approximately reproduce the PMNS matrix, with small corrections provided, as for the quarks, from the block-diagonalization of BSM corrections that separates the τ lepton from its less massive partners and also from whatever matrix (U_ν) that is needed to diagonalize the current neutrinos to their mass eigenstates, which would then be expected to be nearly the unit matrix.

The reversal of the direction of charge changing, however, guarantees that U_ν must be substantially different from the identity. The matrix contributed to the PMNS-matrix by the charged leptons is approximately TBM_ℓ (as defined above) so that the total PMNS matrix is (again approximately)

$$\mathrm{PMNS} = \mathrm{TBM}_\ell U_\nu \,. \tag{28}$$

To satisfy PMNS $\sim \mathrm{TBM}_\ell$ thus requires the relation

$$U_\nu \sim \mathrm{TBM}_\ell \times \mathrm{TBM}_\ell \,. \tag{29}$$

4.2. *Sterile neutrino mass matrix and eigenvalues*

We use the relation in Eq. (29) in the seesaw approximation to determine the character of the sterile neutrino mass matrix necessary to achieve this result. We use Eq. (29) to "undo" (as we did for the charged lepton Dirac mass matrix) the diagonal inverse of the active neutrino Majorana mass matrix parametrized in Eq. (19), and pre- and post-multiply by the undiagonalized form of the neutrino Dirac mass matrix in the same form as Eq. (17) to acquire the undiagonalized sterile mass matrix, M_S, as determined by Eq. (25). On expansion of the result in powers of δ and ϵ, we found the exact result for which, not surprisingly, the leading term is "democratic."

We have obtained the eigenvalues as an approximate expansion through 6th order in the small quantities in order to verify the coefficients to that order in the secular equation. This is necessary as the constant term, the determinant of M_S, is of that order. However, we will only examine the behavior of the leading terms for each of the 3 eigenvalues below. The leading terms of the eigenvalues in order down the diagonal are:

$$M_S(1) \approx \frac{-3}{2f-1}\epsilon^2\delta^2\,, \tag{30}$$

$$M_S(2) \approx \frac{(2fg+3f-g)(2f-1)}{6fg}\,, \tag{31}$$

$$M_S(3) \approx \frac{(2f-1)}{2fg+3f-g}\epsilon^2 \tag{32}$$

again with each one times the overall scale, M_s.

4.3. *Sterile neutrino eigenfunctions*

The components of the sterile neutrino eigenfunctions in the sterile subspace are ultimately of great interest for gaining information about Dark Matter. Here, however, we are interested in the components that include the active neutrino space as these are relevant to additional neutrino oscillations that are not due to the misalignment between active neutrino mass and current eigenstates which are described by the PMNS matrix. We show next that reports of short distance (larger mass-squared differences) neutrino oscillations can be consistently described due to these mixing amplitude components.

For this analysis, examining the components of the off-diagonal 3×3 block in the 6×6 neutrino mass matrix that results from the diagonalization of the sterile sector alone provides a sufficient approximation, as the seesaw mechanism applies only if the scale of the active neutrinos is small compared to that of the steriles. If we define U_S as the 3×3 matrix that diagonalizes M_S as determined above, then the calculation

$$\begin{pmatrix} 1 & 0 \\ 0 & U_S^\dagger \end{pmatrix} \begin{pmatrix} 0 & m_D \\ m_D^\dagger & M_S \end{pmatrix} \begin{pmatrix} 1 & 0 \\ 0 & U_S \end{pmatrix} = \begin{pmatrix} 1 & m_D U_S \\ U_S^\dagger m_D^\dagger & (M_S)_{\text{diagonal}} \end{pmatrix}, \tag{33}$$

where m_D is the undiagonalized Dirac mass matrix for the active neutrino mass (Higgs) coupling to the sterile neutrinos in the current basis, we can see immediately that the entries in the columns of $m_D U_S$ describe, to a first approximation, the amplitude of the mixing of the active current neutrinos to each of the approximate eigenmass sterile neutrinos when divided by the sterile neutrino eigenmass. In particular, the $(1,4)$ and $(3,4)$ entries divided by the (presumably) smallest sterile neutrino eigenmass give the mixing amplitude for electron and muon neutrinos (respectively) to that sterile neutrino. Hence, the product describes the amplitude for short-range neutrino oscillations from, for example, ν_μ to ν_e as studied originally in the LSND and KARMEN experiments,[18] where the oscillation scale length is set by the mass of that sterile neutrino.

Simply by solving the eigenvector equations for M_S using the eigenvalues as determined above, we find the normalized eigenvectors and so the U_S matrix. We find

$$U_S = \begin{bmatrix} \frac{2}{\sqrt{6}} & \frac{1}{\sqrt{3}} & \frac{\sqrt{2}(1+f)}{\sqrt{3}(2f-1)}\delta - \frac{(2fg-3f-g)}{\sqrt{3}(2fg+3f-g)}\epsilon \\ -\frac{1}{\sqrt{6}} + \frac{(1+f)}{\sqrt{2}(2f-1)}\delta & \frac{1}{\sqrt{3}} - \frac{(2fg-3f-g)}{\sqrt{2}(2fg+3f-g)}\epsilon & -\frac{1}{\sqrt{2}} - \frac{(1+f)}{\sqrt{6}(2f-1)}\delta - \frac{(2fg-3f-g)}{\sqrt{3}(2fg+3f-g)}\epsilon \\ -\frac{1}{\sqrt{6}} - \frac{(1+f)}{\sqrt{2}(2f-1)}\delta & \frac{1}{\sqrt{3}} + \frac{(2fg-3f-g)}{\sqrt{2}(2fg+3f-g)}\epsilon & \frac{1}{\sqrt{2}} - \frac{(1+f)}{\sqrt{6}(2f-1)}\delta - \frac{(2fg-3f-g)}{\sqrt{3}(2fg+3f-g)}\epsilon \end{bmatrix}. \tag{34}$$

We show this only to leading orders here, although we have carried the calculation out through fourth order in the small quantities (ϵ and δ) to check unitarity. Using m_D in the form as in Eq. (17), we then obtain the values to leading approximation of each entry. However, for our next calculation, we only need

$$\begin{aligned} m_D U_S(1,1) &= \epsilon\delta m_d \left[\frac{1}{\sqrt{3}} \left(\frac{\sqrt{2}(2f-1) - (1+f)}{(2f-1)} \right) \right], \\ m_D U_S(1,3) &= -\epsilon\delta m_d \left[\frac{(2f-1) + (1+f)(\sqrt{2}+\sqrt{3})}{\sqrt{6}(2f-1)} \right]. \end{aligned} \tag{35}$$

5. Application to Experiments

5.1. *Application to short distance neutrino oscillation experiments*

From Eqs. (35), we can conclude that the ν_μ to ν_e mixing amplitude through the mass $3M\epsilon^2\delta^2/(2f-1)$ sterile neutrino is approximately

$$\begin{aligned} \theta_{e\mu} &= -\left(\frac{\sqrt{2}(2f-1) - (1+f)}{\sqrt{3}(2f-1)} \right) \left[\frac{(2f-1)m_d\epsilon\delta}{3M_s\epsilon^2\delta^2} \right]^2 \\ &\quad \times \left(\frac{(2f-1) + (1+f)(\sqrt{2}+\sqrt{3})}{\sqrt{6}(2f-1)} \right) \\ &= \frac{\sqrt{2}m_d^2}{54\epsilon^2\delta^2 M_s^2} \Big[1 + \sqrt{3} + \sqrt{6} + ((2\sqrt{3}+\sqrt{2}-1)\sqrt{6}-1)f \\ &\quad - (2(\sqrt{6}+1) + (\sqrt{6}-1)\sqrt{3})f^2 \Big] \\ &\approx \frac{m_d^2}{\epsilon^2\delta^2 M_s^2} (0.136 + 0.223f - 0.246f^2). \end{aligned} \tag{36}$$

For $0 < f < 1$, the size of the coefficient of $m_d^2/\epsilon^2\delta^2 M_s^2$ is approximately 0.15 and varies only by about 20%.

From the relations between the scaling parameters above, we must have

$$\frac{m_d^2}{M_s} = \mu_a \tag{37}$$

which means that the scale in Eq. (36) must also be equal to μ_a/M_s. If the scale of μ_a is on the order of 100 meV, then, for the oscillation scale reported by LSND of order 1 eV corresponding to $3M_s\epsilon^2\delta^2/(2f-1)$, one expects the approximate value of $\theta_{e\mu}$ to be

$$\theta_{e\mu} \sim 0.15\frac{(2f-1)\mu_a}{3M_s\epsilon^2\delta^2}\frac{3}{(2f-1)} \sim 0.15 \times 10^{-1}\left(\frac{3}{2f-1}\right) \tag{38}$$

which is consistent with the value of $\sin^2(2\theta)$ reported by LSND, for values of f not too close to 1/2. Recall that in the IO hierarchy, $f \sim 1$ for all allowed values of μ.

This is consistent with the results presented in Collin *et al.*,[15] who present a global fit to the contribution to the oscillation of $\bar{\nu}_\mu$ into $\bar{\nu}_e$ from the coupling to a single, mostly sterile, neutrino. For their best fit parameters, the mass of the lightest sterile is about 1.4 eV/c^2.

5.2. *Astrophysical experimental results*

As discussed above, the most striking result we obtain is that the expected masses of the mostly sterile neutrinos should be in the rough ratios of $(1 :: \epsilon^2 :: \epsilon^2\delta^2)$. Using mean values from the charged fermion families to estimate values for the neutral sector leads us to take $\epsilon \approx 0.03$ and $\delta \approx 0.03$ as well.

To convert this into a value for M_s, we need to know the values of ϵ and δ for the neutrino sector. The estimate above suggests ratios of sterile masses of about 10^3, which means that the fit in Collin *et al.* suggests that the next most massive "sterile neutrino" should be expected to be in the range of (1–10) keV/c^2. Cappelluti *et al.*[16] have reviewed the evidence for a 3.55 keV gamma ray as a signal of a dark matter candidate with a mass of 7.1 keV/c^2 and conclude that it cannot be ruled out. This is well within the (broad) region that our analysis suggests should include a sterile neutrino that should be susceptible to decay into a lighter (active) neutrino and a photon. We note that the decay rate and mixing amplitude estimates of Cappelluti *et al.* depend strongly on the assumption that this state constitutes all dark matter. In the event that there are many constituents of dark matter, those extracted values would change markedly and the results could be consistent with the estimates that can be inferred from the discussion presented here.

6. Conclusions

We have applied the concept of Higgs universality, justified by the studies of quark masses and mixing presented in Ref. 14, to leptons. This requires the introduction of three additional neutral Weyl fields sterile under the weak interaction. The results, coupled with the necessary structure of the 6×6 Weyl mass matrix, are that, in general, the ratios between masses for the mostly sterile eigenstates should scale as the square of the ratios observed for the quarks or the charged leptons. Any deviations from this pattern would have strong implications for the distribution of mass eigenstates for the mostly active neutrinos.

References

1. N. Cabibbo, *Phys. Rev. Lett.* **10**, 531 (1963); M. Gell-Mann and M. Lévy, *Nuovo Cimento* **16**, 705 (1960).
2. T. Goldman, G. J. Stephenson Jr. and B. H. J. McKellar, All fundamental fermions are vile, LA-UR-08-3430, unpublished; T. Goldman, G. J. Stephenson Jr. and B. H. J. McKellar, *J. Phys.: Conf. Ser.* **136**, 042023 (2008).
3. Z.-Z. Xing, H. Zhang and S. Zhou, *Phys. Rev. D* **77**, 113016 2008).
4. Particle Data Group (C. Patrignani *et al.*), *Chin. Phys. C.* **40**, 100001 2016).
5. H. Harari, *Phys. Lett. B* **86**, 83 (1979); M. A. Shupe, *Phys. Lett. B* **86**, 87 (1979); H. Harari and N. Seiberg, *Nucl. Phys. B* **204**, 141 (1982); P. Zenczykowski, *Phys. Lett. B* **660**, 567 (2008).
6. S. Weinberg, *Phys. Rev. D* **13**, 974 (1976); S. Weinberg, *Phys. Rev. D* **19**, 1277 (1979); L. Susskind, *Phys. Rev. D* **20**, 2619 (1979); S. Dimopoulos and L. Susskind, *Nucl. Phys. B* **155**, 237 (1979); B. Holdom, *Phys. Rev. D* **24**, 1441 (1981); F. Sannino, *Acta Phys. Pol. B* **40**, 3533 (2009), arXiv:0911.0931.
7. H. Georgi and S. L. Glashow, *Phys. Rev. Lett.* **32**, 438 (1974); J. C. Pati and A. Salam, *Phys. Rev. D* **10**, 275 (1974); H. Fritzsch and P. Minkowski, *Ann. Phys.* **93**, 193 (1975); H. Georgi and D. Nanopoulos, *Nucl. Phys. B* **159**, 16 (1979); R. N. Mohapatra and B. Sakita, *Phys. Rev. D* **21**, 1062 (1980); F. Wilczek and A. Zee, *Phys. Rev. D* **25**, 553 (1982).
8. M. Gell-Mann, P. Ramond and R. Slansky, *Supergravity*, eds. D. Freedman *et al.* (North-Holland, Amsterdam, 1980).
9. H. Fritzsch, *Phys. Lett. B* **70**, 436 1977).
10. H. Harari, H. Haut and J. Weyers, *Phys. Lett. B* **78**, 459 (1978); Y. Koide, *Phys. Lett. B* **120**, 161 (1983); Y. Koide, *Phys. Rev. D* **28**, 252 (1983); *Phys. Rev. D* **39**, 1391 (1989); M. Tanimoto, *Phys. Rev. D* **41**, 1586 (1990); L. Lavoura, *Phys. Lett. B* **228**, 245 (1989).
11. W. G. Hollik and U. J. Saldana Salazar, *Nucl. Phys. B* **892**, 364 (2015); Z.-Z. Xing, *Int. J. Mod. Phys. A* **29**, 1430067 (2014); P. V. Dong, N. T. K. Ngan and D. V. Soa, *Phys. Rev. D* **90**, 075019 (2014); F. Hartmann and W. Kilian, *Eur. Phys. J. C* **74**, 3055 (2014); A. E. Cárcamo Hernández and I. de Medeiros Varzielas, arXiv:1410.2481; P. Fakay, arXiv:1410.7142; J. Berryman and D. Hernández, arXiv:1502.04140; I. T. Dyatlov, arXiv:1502.01501; E. Nardi, arXiv:1503.01476.
12. P. Kaus and S. Meshkov, *Mod. Phys. Lett. A* **3**, 1251 (1988); *Phys. Rev. D* **42**, 1863 (1990); C. Jarlskog, in *Proc. Int. Symp. on Production and Decay of Heavy Flavors*, Heidelberg, Germany, 1986, eds. K. Schubert and R. Waldi (DESY, Hamburg, 1987), p. 331.
13. H. Georgi and T. Goldman, *Phys. Rev. Lett.* **30**, 514 1973).
14. G. J. Stephenson Jr. and T. Goldman, A modest revision of the Standard Model, arXiv:1503.04211.
15. G. H. Collin, C. A. Argulies, J. M. Conrad and M. H. Shaevitz, *Phys. Rev. Lett.* **117**, 221801 2016).
16. N. Cappelluti, E. Bulbul, A. Foster, P. Natarajan, M. C. Urry, M. W. Bautz, F. Civano, E. Miller and R. K. Smith, arXiv:1701.07932v1 [astro-ph.CO].
17. C. Jarlskog, *Phys. Rev. Lett.* **55**, 1039 1985).
18. LSND Collab. (C. Athanassopoulos *et al.*), *Phys. Rev. Lett.* **77**, 3082 (1996); *ibid.* **81**, 1774 (1998); KARMEN Collab. (B. Armbruster *et al.*), *Nucl. Phys. B* **38**, 235 (1995).

Texture Zero Mass Matrices and CP Violation

Gulsheen Ahuja
Department of Physics, Panjab University,
Chandigarh, India
gulsheen@pu.ac.in

Samandeep Sharma
Department of Physics, GGDSD College,
Chandigarh, India

Within the Standard Model, using the facility of making Weak Basis transformations, attempt has been made to examine the most general mass matrices within the texture zero approach. For the case of quarks, interestingly, one finds a particular set of texture four zero quark mass matrices emerging out to be a unique viable option for the description of quark mixing data as well as for accommodation of CP violation. Similarly, general lepton mass matrices, essentially considered as texture zero mass matrices, yield interesting bounds on the CP violating Jarlskog's rephasing invariant parameter in the leptonic sector.

1. Introduction

Despite enormous progress at the experimental front, understanding the pattern of fermion masses and mixings still remains an unfinished task. At present, for the case of quarks, one has a fairly good idea of the quark masses as well as the mixing angles,[1] both interestingly exhibiting a clear cut hierarchy. In the case of neutrinos, the recent refinements of the reactor mixing angle s_{13},[2,3] along with the precision measurement of the solar and atmospheric mixing angles s_{12} and s_{23} and the neutrino mass-squared differences,[4] have given a new impetus to the neutrino oscillation phenomenology. This has further been strengthened by a recent constraint on the sum of absolute neutrino masses provided by the Planck experiment.[5]

The recent observation of the nonzero value of s_{13} has not only restored parallelism between quark and lepton mixing, but also, it has triggered an interest in the exploration of CP violation in the leptonic sector, a phenomenon considered as an indispensable ingredient that could generate the excess of the baryon number of universe. At present, several ongoing and future experiments, including long-baseline facilities, superbeams, and neutrino factories are being planned to pursue the task of establishing the possibility of existence of CP violation in the leptonic sector. It is expected that in the near future these would provide us with useful clues, nevertheless, it is desirable to look for alternative manifestations of CP violation.

In the absence of a fundamental theory of flavor physics, to decipher the mystery of fermion masses and flavor mixings, an interesting idea being explored is the texture zero mass matrices. Texture zero approach is known to provide a satisfactory explanation of quark and lepton mixing phenomenon,[6–9] however, in order to limit the number of free parameters associated with mass matrices, it becomes desirable to exploit certain new principles which tend to reduce this number. Keeping this in mind, in order to arrive at a viable framework for explaining the fermion masses and mixings, we begin with the most general mass matrices and carry out their analysis within the texture zero approach, along with incorporating the concept of Weak Basis (WB) transformations,[10–13] a brief outline of these has been presented in Sec. 2. Further, we have examined these texture zero mass matrices with a focus on obtaining constraints regarding CP violation, the results of the analyses for the case of quarks and leptons have been presented in Secs. 3 and 4, respectively. Finally, in Sec. 5, we summarize our conclusions.

2. Texture Zero Mass Matrices and WB Transformations

In order to achieve insight into origin of fermion masses and flavor mixings, efforts have been made on both the experimental as well as phenomenological front. Experimental attempts have resulted into continuous refinements in the quark as well as lepton mixing data. On the phenomenological front, one can initiate via either the "top-down" or the "bottom-up" approach. In the present work, we follow the bottom-up approach which essentially starts with the phenomenological mass matrices at the weak scale.

To begin with, it may be noted that in the Standard Model (SM), the fermion mass matrices, having their origin in the Higgs fermion couplings, are completely arbitrary. Therefore, the number of free parameters available with a general mass matrix is larger than the number of physical observables. For example, in the quark sector, in the two 3×3 general complex mass matrices in the up and down sector M_U and M_D respectively, there are 36 real free parameters as compared to the 10 physical observables, namely the 6 quark masses, 3 mixing angles and 1 CP violating phase. A similar situation prevails for the leptonic sector as well.

In order to obtain valuable clues for developing an understanding of fermion mixing phenomenology, it is desirable that the number of free parameters of the mass matrices is constrained. Without loss of generality, in the SM and its extensions in which righthanded quarks are singlets, fermion mass matrices can be considered Hermitian, therefore, bringing down the number of free parameters from 36 to 18. To reduce these further, in the context of quark mass matrices, Fritzsch[14,15] had proposed the idea of texture zero mass matrices by assuming certain elements of

Hermitian mass matrices to be zero, e.g.

$$M_U = \begin{pmatrix} 0 & A_U & 0 \\ A_U^* & 0 & B_U \\ 0 & B_U^* & C_U \end{pmatrix}, \quad M_D = \begin{pmatrix} 0 & A_D & 0 \\ A_D^* & 0 & B_D \\ 0 & B_D^* & C_D \end{pmatrix}, \tag{1}$$

M_U and M_D refer to the mass matrices in the up and down sector respectively. Such matrices are called texture zero mass matrices with a particular texture structure defined as texture "n" zero if the sum of the number of diagonal zeros and half the number of the symmetrically placed off diagonal zeros is "n". Each of the above matrix is texture three zero type, together these are known as texture six zero Fritzsch mass matrices. On lines of these ansatze, by considering lesser number of texture zeros, several possible Fritzsch-like texture zero mass matrices can be formulated. Also, one can get non-Fritzsch-like mass matrices by shifting the position of C_i $(i = U, D)$ on the diagonal as well as by shifting the position of zeros among the nondiagonal elements. One can thus obtain a large number of possible texture zero mass matrices, e.g. for the case of texture four zero mass matrices, the number of possible matrices is 24 in each sector resulting into $24 \times 24 = 576$ possible combinations.

While carrying out the analysis of texture zero mass matrices, the viability of the formulated mass matrices is ensured by checking the compatibility of the mixing matrices so obtained from these with the low energy data. Over the past few years, both in the quark as well as lepton sector, a large number of analyses[6–9] have been carried out which clearly establish the texture zero approach as a viable framework for explaining the fermion mixing data. However, despite this considerable success, since the number of possible mass matrices is very large, the limitation of this approach is that an exhaustive case-by-case analysis of all possible texture zero mass matrices needs to be carried out. To account for this limitation, an important supplementary approach is the concept of "Weak Basis (WB) transformations," first discussed in the literature by Branco *et al.*[10,11] and by Fritzsch and Xing.[12,13]

Within the SM and some of its extensions, one has the facility of making Weak Basis (WB) transformations W on the quark fields, e.g. $q_L \to Wq_L$, $q_R \to Wq_R$, $q'_L \to Wq'_L$, $q'_R \to Wq'_R$. These are unitary transformations which leave the gauge currents real and diagonal but transform the mass matrices as

$$M_U \to W^\dagger M_U W\,, \quad M_D \to W^\dagger M_D W\,. \tag{2}$$

Without loss of generality, this approach introduces zeros in the quark mass matrices leading to a reduction in the number of parameters defining the mass matrices. Following this, one can arrive at two kinds of structures of the mass matrices, e.g. Branco *et al.*[10,11] give

$$M_q = \begin{pmatrix} 0 & * & 0 \\ * & * & * \\ 0 & * & * \end{pmatrix}, \quad M_{q'} = \begin{pmatrix} 0 & * & * \\ * & * & * \\ * & * & * \end{pmatrix}, \quad q,\, q' = U, D\,, \tag{3}$$

whereas Fritzsch and Xing[12,13] give

$$M_q = \begin{pmatrix} * & * & 0 \\ * & * & * \\ 0 & * & * \end{pmatrix}, \quad q = U, D\,. \tag{4}$$

The mass matrices so obtained can thereafter be considered texture zero mass matrices and same methodology can be used to analyze these. Interestingly, one now has an additional advantage that the large number of possible structures are not all independent. Several of these are related through WB transformations and therefore yield the same structure of the diagonalizing transformations leading to similar mixing matrices, making the number of matrices to be analyzed much less than before. However, there is a limitation too, i.e. this idea does not result in constraining the parameter space of the elements of the mass matrices. To overcome this, one can further impose a condition on the elements of the mass matrices by considering the following hierarchy for these[8]

$$(1,i) \lesssim (2,j) \lesssim (3,3)\,; \quad i = 1,2,3\,, \quad j = 2,3\,. \tag{5}$$

3. Texture Zero Quark Mass Matrices

The following Hermitian mass matrices can be considered as the most general ones

$$M_q = \begin{pmatrix} E_q & A_q & F_q \\ A_q^* & D_q & B_q \\ F_q^* & B_q^* & C_q \end{pmatrix} \quad (q = U, D)\,. \tag{6}$$

Invoking WB transformations, zeros can be introduced in these matrices using a unitary matrix W, leading to

$$M_U = \begin{pmatrix} E_U & A_U & 0 \\ A_U^* & D_U & B_U \\ 0 & B_U^* & C_U \end{pmatrix}, \quad M_D = \begin{pmatrix} E_D & A_D & 0 \\ A_D^* & D_D & B_D \\ 0 & B_D^* & C_D \end{pmatrix}. \tag{7}$$

In the language of texture zero mass matrices, these are texture one zero each, together these are referred as texture two zero mass matrices.

It may be noted that in the matrices M_U and M_D, instead of the zeros being in the $(1,3)$ and $(3,1)$ positions, these could also be introduced in either the $(1,2)$ and $(2,1)$ or $(2,3)$ and $(3,2)$ position. These other structures are related to the above mentioned matrices as we have the facility of subjecting M_U and M_D to another WB transformation that can be the permutation matrix P. These different mass matrices, however, yield the same Cabibbo–Kobayashi–Maskawa (CKM) matrix,[16] therefore while presenting the results of the analysis, it is sufficient to discuss any one of these matrices. In order to further constrain the parameter space available to the elements of these mass matrices, one can consider the hierarchy for the elements of the matrices given in Eq. (5).

As a next step, to check the viability of the mass matrices given in Eq. (7), the compatibility of the CKM matrix reproduced through these needs to be examined with the recent quark mixing data. A detailed analysis of these matrices, carried out in Ref. 17, reveals that using the following quark masses and the mass ratios at the M_Z scale as inputs[18]

$$m_u = 1.38^{+0.42}_{-0.41}\ \text{MeV}\,, \quad m_d = 2.82\pm 0.48\ \text{MeV}\,, \quad m_s = 57^{+18}_{-12}\ \text{MeV}\,,$$
$$m_c = 0.638^{+0.043}_{-0.084}\ \text{GeV}\,, \quad m_b = 2.86^{+0.16}_{-0.06}\ \text{GeV}\,, \quad m_t = 172.1\pm 1.2\ \text{GeV}\,, \tag{8}$$
$$m_u/m_d = 0.553 \pm 0.043\,, \quad m_s/m_d = 18.9 \pm 0.8$$

and imposing the latest values[1] of the three mixing angles as constraints for the construction of the CKM matrix, one arrives at

$$V_{\text{CKM}} = \begin{pmatrix} 0.9739\text{–}0.9745 & 0.2246\text{–}0.2259 & 0.00337\text{–}0.00365 \\ 0.2224\text{–}0.2259 & 0.9730\text{–}0.9990 & 0.0408\text{–}0.0422 \\ 0.0076\text{–}0.0101 & 0.0408\text{–}0.0422 & 0.9990\text{–}0.9999 \end{pmatrix}, \tag{9}$$

this being fully compatible with the one given by Particle Data Group (PDG).[1] In order to examine whether these mass matrices can accommodate CP violation in the quark sector, in the present work we have made an attempt to reproduce the CP violating Jarlskog's rephasing invariant parameter J. One obtains a range of J from $(2.494\text{–}3.365)\times 10^{-5}$, this again being compatible with its latest experimental value $(3.04^{+0.21}_{-0.20}) \times 10^{-5}$.[1]

Further, in Ref. 17 it has also been shown that the number of texture zeros in the matrices M_U and M_D can be increased as the magnitude of their $(1,1)$ element is quite small in comparison with the other nonzero elements. Therefore, without loss of parameter space, ignoring the $(1,1)$ element of the matrices, we get

$$M_U = \begin{pmatrix} 0 & A_U & 0 \\ A_U^* & D_U & B_U \\ 0 & B_U^* & C_U \end{pmatrix}, \quad M_D = \begin{pmatrix} 0 & A_D & 0 \\ A_D^* & D_D & B_D \\ 0 & B_D^* & C_D \end{pmatrix}, \tag{10}$$

indicating a transition from texture two zero to texture four zero mass matrices. A similar analysis of these matrices yields the corresponding CKM matrix in agreement with the latest quark mixing matrix[1] and also being fully compatible with the CKM matrix given in Eq. (9). The range of the parameter J also is again compatible with its latest experimental range, justifying our conclusion that the $(1,1)$ elements are essentially redundant.

It may be noted that along with the texture four zero mass matrices considered above, there are several other possible texture four zero structures also, all together being 24.[9] One finds that these 24 matrices are all not independent, in particular, several of these are related through WB transformations and therefore yield the same structure of the diagonalizing transformations leading to similar mixing matrices. These can, therefore, be classified into 4 categories, with each category containing 6 matrices each. Carrying out a numerical analysis of all these matrices reveals that out of 24 only 6 matrices, the one given in Eq. (10) and 5 others related to it are actually able to explain the quark mixing data. The remaining 18 mass matrices are not viable as these are not able to reproduce compatible CKM matrices. Also, the quark mixing data rules out texture five and six zero quark mass matrices.[9] Therefore, one can conclude that the above mentioned texture four zero quark mass matrices, similar to the original Fritzsch ansatze, except for their $(2,2)$ element being nonzero, and its permutations can be considered as a unique viable option compatible not only with the recent quark mixing data but also being able to well accommodate CP violation.

4. Texture Zero Lepton Mass Matrices

The quark lepton universality,[19] required by most of the Grand Unified Theories, makes it desirable to carry out an analysis corresponding to the above analysis for quarks in the leptonic sector also. In the case of neutrinos, smallness of their masses is at present best described in terms of the seesaw mechanism[20] expressed as

$$M_\nu = -M_{\nu D}^T M_R^{-1} M_{\nu D}\,, \tag{11}$$

with M_ν, $M_{\nu D}$ and M_R corresponding to the light Majorana neutrino mass matrix, the Dirac neutrino mass matrix and the heavy right-handed Majorana neutrino mass matrix respectively.

In the case of leptons, most of the attempts following the texture zero approach have been made by considering the mass matrices to be in the flavor basis,[21] wherein the chargod lepton mass matrix M_l is considered to be diagonal while a texture is imposed on the Majorana neutrino matrix M_ν. Along with these, some attempts have also been carried out in the nonflavor basis,[22] also followed in the present work, wherein it is usual to impose texture on both, the matrix M_l and the Dirac neutrino mass matrix $M_{\nu D}$. For examining the viability of the mass matrices, the Pontecorvo–Maki–Nakagawa–Sakata (PMNS) matrix[23] is then obtained using the charged lepton mass matrix M_l and the Majorana neutrino matrix M_ν, the latter can be obtained through the seesaw relation, given in Eq. (11) by using $M_{\nu D}$ and the right-handed Majorana neutrino mass matrix M_R.

Within the framework of SM, similar to the case of quarks, the lepton mass matrices are completely arbitrary and are expressed in terms of complex 3×3 matrices. Again, using the facility of WB transformations, wherein it is possible to make a unitary transformation, one can transform the general lepton mass matrices to

$$M_l = \begin{pmatrix} E_l & A_l & 0 \\ A_l^* & D_l & B_l \\ 0 & B_l^* & C_l \end{pmatrix}, \quad M_{\nu D} = \begin{pmatrix} E_{\nu D} & A_{\nu D} & 0 \\ A_{\nu D}^* & D_{\nu D} & B_{\nu D} \\ 0 & B_{\nu D}^* & C_{\nu D} \end{pmatrix}. \tag{12}$$

Table 1. Current data for neutrino mixing parameters from the latest global fits.[25]

Parameter	3σ range
$\Delta m^2_{\rm sol}$ $[10^{-5}\ {\rm eV}^2]$	(7.020–8.09)
$\Delta m^2_{\rm atm}$ $[10^{-3}\ {\rm eV}^2]$	(2.317–2.607) (NO); (2.590–2.307) (IO)
$\sin^2\theta_{13}$ $[10^{-2}]$	(1.86–2.50) (NO); (1.88–2.51) (IO)
$\sin^2\theta_{12}$ $[10^{-1}]$	(2.70–3.44)
$\sin^2\theta_{23}$ $[10^{-1}]$	(3.82–6.43) (NO); (3.89–6.44) (IO)

These texture two zero lepton mass matrices appear similar in structure to the corresponding quark mass matrices, obtained earlier, however, one must keep in mind that the mixing patterns and mass orderings are quite different in the two cases. Therefore, a similar analysis carried out on the lines of the analysis for the case of quarks now reveals that there exist a large number of viable texture zero lepton mass matrices. Therefore, in the present work, we have presented the results pertaining to only the matrices mentioned in Eq. (12). A detailed analysis of these mass matrices has also been carried out in Ref. 24, however, with an emphasis on examining the implications of these matrices for the effective Majorana mass m_{ee}. In the work presented here, for the normal and inverted ordering of neutrino masses, after examining the viability of these mass matrices, we have investigated their implications for CP violation in the leptonic sector.

For the purpose of calculations, we have used the input values of the neutrino mixing parameters derived from the latest global fits,[25] summarized in Table 1.

For the inverted neutrino mass ordering, the mass matrices mentioned in Eq. (12) yield the following magnitudes of the corresponding PMNS matrix elements[24]

$$U^{\rm IO}_{\rm PMNS} = \begin{pmatrix} 0.034-0.859 & 0.0867-0.593 & 0.135-0.996 \\ 0.250-0.971 & 0.068-0.812 & 0.043-0.808 \\ 0.103-0.621 & 0.395-0.822 & 0.088-0.810 \end{pmatrix}. \qquad (13)$$

Recently, Garcia *et al.*[25] have given the following 3σ C.L. ranges of the PMNS matrix elements,

$$U_{\rm PMNS} = \begin{pmatrix} 0.801-0.845 & 0.514-0.580 & 0.137-0.158 \\ 0.225-0.517 & 0.441-0.699 & 0.614-0.793 \\ 0.246-0.529 & 0.464-0.713 & 0.590-0.776 \end{pmatrix}. \qquad (14)$$

It can be easily seen that the 3σ C.L. ranges of the PMNS matrix elements given by Garcia *et al.* are inclusive in the ranges of the PMNS matrix elements found here, thereby ensuring the viability of texture two zero mass matrices for the inverted

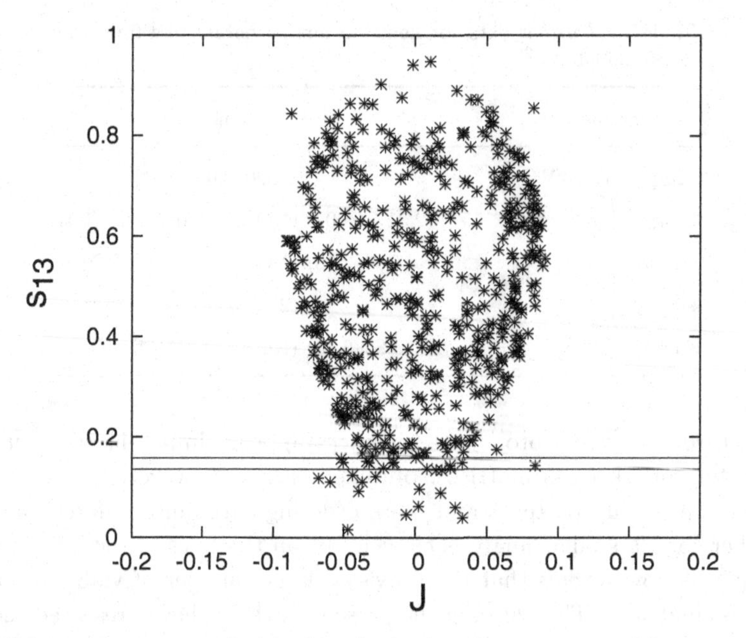

Fig. 1. Mixing angle s_{13} versus Jarlskog's rephasing invariant parameter J for texture two zero leptonic mass matrices (IO).

hierarchy case. Further, analogous to the case of quarks, we have made an attempt to obtain constraints for the CP violating Jarlskog's rephasing invariant parameter in the leptonic sector. To this end, the parameter J versus the mixing angle s_{13} plot has been presented in Fig. 1. The horizontal lines in the graph depict the 3σ C.L. range of the mixing angle s_{13}. A general look at the graph reveals that one obtains a range of J from -0.05–0.05.

For normal ordering of neutrino masses, one obtains[24]

$$U_{\rm PMNS}^{\rm NO} = \begin{pmatrix} 0.444\text{–}0.993 & 0.123\text{–}0.837 & 0.004\text{–}0.288 \\ 0.061\text{–}0.816 & 0.410\text{–}0.941 & 0.047\text{–}0.872 \\ 0.012\text{–}0.848 & 0.049\text{–}0.779 & 0.460\text{–}0.992 \end{pmatrix}, \tag{15}$$

revealing that the 3σ C.L. ranges of the PMNS matrix elements given by Garcia *et al.* are again inclusive in the ranges of the PMNS matrix elements found for this case. This, therefore, ensures the viability of texture two zero mass matrices for the normal neutrino mass ordering also. Further, the parameter J versus the mixing angle s_{13} plot, presented in Fig. 2 gives a range of -0.03–0.03 for J. These observations, therefore, lead one to conclude that the texture two zero leptonic mass matrices are not only compatible with the recent leptonic mixing data but also provide interesting bounds for the Jarlskog's rephasing invariant parameter.

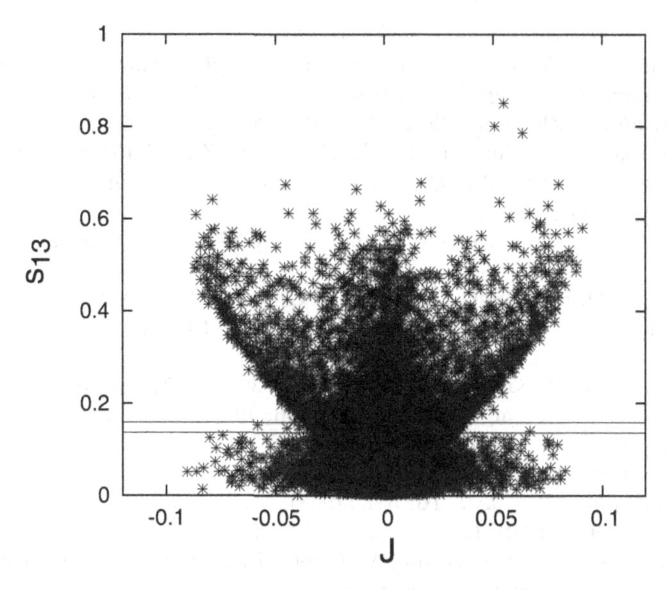

Fig. 2. Mixing angle s_{13} versus Jarlskog's rephasing invariant parameter J for texture two zero leptonic mass matrices (NO).

5. Summary and Conclusions

To summarize, within the Standard Model, starting with the most general mass matrices, we have used the facility of making Weak Basis transformations and have carried out their analysis within the texture zero approach. For the case of quarks, interestingly, our analysis reveals that a particular set of texture four zero quark mass matrices can be considered to be a unique viable option for the description of quark mixing data as well as for accommodation of CP violation. Similarly, general lepton mass matrices can also essentially be considered as texture zero mass matrices. Using these, we obtain interesting bounds on the Jarlskog's rephasing invariant parameter in the leptonic sector for NO and IO cases of texture two zero leptonic mass matrices.

Acknowledgments

G. Ahuja would like to thank M. Gupta for useful discussions. G. Ahuja would also like to acknowledge DST, Government of India (Grant No. SR/FTP/PS-017/2012) for financial support. S. Sharma would like to acknowledge the Principal, GGDSD College, Chandigarh.

References

1. Particle Data Group (C. Patrignani *et al.*), Chin. Phys. C **40**, 100001 (2016).
2. DAYA-BAY Collab. (F. P. An *et al.*), Phys. Rev. Lett. **108**, 171803 (2012).
3. RENO Collab. (J. K. Ahn *et al.*), Phys. Rev. Lett. **108**, 191802 (2012).

4. F. Capozzi *et al.*, Nucl. Phys. B **908**, 218 (2016).
5. Planck Collab. (P. A. R. Ade *et al.*), Astron. Astrophys. **594**, A13 (2016).
6. H. Fritzsch and Z. Z. Xing, Nucl. Phys. B **556**, 49 (1999), and references therein.
7. Z. Z. Xing and H. Zhang, J. Phys. G **30**, 129 (2004), and references therein.
8. N. G. Deshpande, M. Gupta and P. B. Pal, *Phys. Rev. D* **45**, 953 (1992); P. S. Gill and M. Gupta, *Pramana* **45**, 333 (1995); M. Gupta and G. Ahuja, *Int. J. Mod. Phys. A* **26**, 2973 (2011).
9. M. Gupta and G. Ahuja, Int. J. Mod. Phys. A **27**, 1230033 (2012), and references therein.
10. G. C. Branco, D. Emmanuel-Costa and R. G. Felipe, Phys. Lett. B **477**, 147 (2000).
11. G. C. Branco, D. Emmanuel-Costa, R. G. Felipe and H. Serodio, Phys. Lett. B **670**, 340 (2009).
12. H. Fritzsch and Z. Z. Xing, Phys. Lett. B **413**, 396 (1997), and references therein.
13. H. Fritzsch and Z. Z. Xing, Nucl. Phys. B **556**, 49 (1999), and references therein.
14. H. Fritzsch, Phys. Lett. B **70**, 436 (1977).
15. H. Fritzsch, Phys. Lett. B **73**, 317 (1978).
16. N. Cabibbo, *Phys. Rev. Lett.* **10**, 531 (1963); M. Kobayashi and T. Maskawa, *Prog. Theor. Phys.* **49**, 652 (1973).
17. S. Sharma, P. Fakay, G. Ahuja and M. Gupta, Phys. Rev. D **91**, 053004 (2015).
18. Z. Z. Xing, H. Zhang and S. Zhou, Phys. Rev. D **86**, 013013 (2012).
19. M. A. Schmidt and A. Yu. Smirnov, Phys. Rev. D **74**, 113003 (2006).
20. H. Fritzsch, M. Gell-Mann and P. Minkowski, *Phys. Lett. B* **59**, 256 (1975); P. Minkowski, *Phys. Lett. B* **67**, 421 (1977); T. Yanagida, *Proceedings of the Workshop on Unified Theories and Baryon Number in the Universe*, Tsukuba, 1979, eds. A. Sawada and A. Sugamoto (KEK, Tsukuba, 1979), Report No. 79-18; M. Gell-Mann, P. Ramond and R. Slansky, *Supergravity*, eds. F. van Nieuwenhuizen and D. Freedman (North-Holland, Amsterdam, 1979), p. 315; S. L. Glashow, *Quarks and Leptons*, eds. M. Lévy *et al.* (Plenum, New York, 1980), p. 707; R. N. Mohapatra and G. Senjanovic, *Phys. Rev. Lett.* **44**, 912 (1980); J. Schechter and J. W. F. Valle, *Phys. Rev. D* **22**, 2227 (1980).
21. P. H. Frampton, S. L. Glashow and D. Marfatia, *Phys. Lett. B* **536**, 79 (2002); Z. Z. Xing, *Phys. Lett. B* **530**, 159 (2002); B. R. Desai, D. P. Roy and A. R. Vaucher, *Mod. Phys. Lett. A* **18**, 1355 (2003); Z. Z. Xing, *Int. J. Mod. Phys. A* **19**, 1 (2004); A. Merle and W. Rodejohann, *Phys. Rev. D* **73**, 073012 (2006); S. Dev, S. Kumar, S. Verma, S. Gupta and R. R. Gautam, *Phys. Rev. D* **81**, 053010 (2010), and references therein.
22. M. Fukugita, M. Tanimoto and T. Yanagida, *Prog. Theor. Phys.* **89**, 263 (1993); M. Fukugita, M. Tanimoto and T. Yanagida, *Phys. Lett. B* **562**, 273 (2003), arXiv: hep-ph/0303177; M. Randhawa, G. Ahuja and M. Gupta, *Phys. Rev. D* **65**, 093016 (2002); G. Ahuja, S. Kumar, M. Randhawa, M. Gupta and S. Dev, *Phys. Rev. D* **76**, 013006 (2007); G. Ahuja, M. Gupta, M. Randhawa and R. Verma, *Phys. Rev. D* **79**, 093006 (2009); M. Fukugita *et al.*, *Phys. Lett. B* **716**, 294 (2012), arXiv:1204.2389; P. Fakay, S. Sharma, R. Verma, G. Ahuja and M. Gupta, *Phys. Lett. B* **720**, 366 (2013).
23. B. Pontecorvo, *Zh. Eksp. Theor. Fiz.* (*JETP*) **33**, 549 (1957); *ibid.* **34**, 247 (1958); *ibid.* **53**, 1771 (1967); Z. Maki, M. Nakagawa and S. Sakata, *Prog. Theor. Phys.* **28**, 870 (1962).
24. S. Sharma, G. Ahuja and M. Gupta, Phys. Rev. D **94**, 113004 (2016).
25. M. C. Gonzalez-Garcia, M. Maltoni and T. Schwetz, Nucl. Phys. B **908**, 199 (2016).

CP Invariants in Flavor Physics

Manmohan Gupta[a,*], Gulsheen Ahuja[a], Madan Singh[b], Dheeraj Shukla[a]

[a] *Department of Physics, Panjab University, Chandigarh - 160 014, India*
[b] *Department of Physics, NIT Kurukshetra, Kurukshetra, India*
**mmgupta@pu.ac.in*

Issue of CP invariance in the flavor sector has been revisited briefly. Apart from defining the CP invariants in the quark sector as well as in the leptonic sector, we have briefly attempted to find Dirac like leptonic CP violating phase δ_l using analogy of the CP violating phase in the quark sector. Also briefly reviewed in leptonic sector is the importance of CP invariants for texture two zero mass matrices in the flavor basis.

1. Introduction

Study of CP symmetry in various physical phenomenon has been a volatile field of research since the discovery of CP violation in weak interaction. The importance of CP violation has revealed several windows to look into the realm of fundamental interactions. Signatures of the "Direct" CP violation can be traced in the Standard Model (SM) of particle physics if a complex phase appears in the Cabibbo-Kobayashi-Maskawa (CKM) matrix[1] describing quark mixing and Pontecorvo-Maki-Nakagawa-Sakata (PMNS) matrix[2] describing the neutrino mixing. The recent observation of nonzero rather 'large' value of the reactor mixing angle in the case of neutrino oscillations has led to good deal of efforts in finding CP violation in the leptonic sector. Although CP violation in the leptonic sector could be much more complicated than that observed in quark sector, yet there are several features which are common in the two cases.

In the case of quarks, it is very well understood that CP violation in the CKM paradigm is related to the well known Jarlskog's rephasing invariant parameter J[3] which is also related to any of the unitarity triangles. It would be interesting to discuss the corresponding rephasing invariant parameters in the case of leptons. In particular, it would be interesting to find out whether by using the analogy of quarks, one can get any idea regarding the corresponding quantity J_l in the leptonic sector.

To understand CP violation in the context of CKM and the PMNS matrices, one has to formulate appropriate mass matrices. In this context, it is well known that texture zero approach[4–6] has been successful in formulating texture specific mass matrices which are able to describe quark mixing as well as neutrino oscillation data. It is also known that the CP violating phases in the quark as well as the lepton sector, if these exist, are related to the corresponding phases of the mass matrices. In view of the well established relationship of J with the corresponding mass matrices, it becomes desirable to explore the corresponding rephasing invariant quantities having relationship with the lepton mass matrices.

2. CP Invariants in Quark and Lepton Sectors

The flavor dynamics of the standard electroweak model is determined by the structure of the quark and lepton mass matrices, however, these remain undetermined in the SM. Furthermore, the mass matrices are not unique. In fact there is an infinite number of mass matrices, these being related to each other by some unitary transformations, leading to the same physics. For instance, the pair of mass matrices (M, N) and $(X\,M\,X^{\dagger}, X\,N\,X^{\dagger})$ are physically indistinguishable, where X is an arbitrary unitary matrix. It is desirable to define quantities which are free from this redundancy. Another complication is that the number of measurable quantities are less than the number of free parameters related to the mass matrices. Several approaches have been adopted to tackle this problem, however, in the present work, we have followed the texture zero approach, details regarding this can be seen in Ref. 6. Presently we would talk about such CP invariants or CP-violating invariants and some of their implications firstly in the quark sector and then in the leptonic sector.

2.1. *CP invariants in quark sector*

In the SM, the quark mass matrices M_u and M_d are complex in general, these can be diagonalized by bilinear unitary transformations

$$\begin{aligned} U_L\,M_u\,U_R^{\dagger} &= M \equiv diag(m_u, m_c, m_t), \\ U_L\,M_d\,U_R^{\dagger} &= M' \equiv diag(m_d, m_s, m_b). \end{aligned} \tag{1}$$

It is easy to check that

$$U_L\,M\,U_R^{\dagger}U_R\,M^{\dagger}\,U_L^{\dagger} = U_L\,MM^{\dagger}\,U_L^{\dagger} = D^2 = diag(m_u^2, m_c^2, m_t^2).$$

The unitary CKM matrix in case of the quarks can be written as $V_{CKM} \equiv U_{uL}\,U_{dL}'^{\dagger}$. It can be shown that there are total 14 conditions for the CP violation in quark sector[3], i.e.,

$$\begin{aligned} & m_u \neq m_c,\; m_c \neq m_t,\;\; m_t \neq m_u,\; m_d \neq m_s,\; m_s \neq m_b,\; m_b \neq m_d, \\ & \theta_{ij} \neq 0, \frac{\pi}{2};\quad i,j = 1,2,3, \\ & \text{Sin}\delta \neq 0, \text{implying CP violating phase in the quark sector } \delta \neq 0, \pi. \end{aligned} \tag{2}$$

All the above conditions can be collectively expressed through the relation

$$[M_u\,M_u^{\dagger},\; M_d\,M_d^{\dagger}] \equiv i\,C, \tag{3}$$

where

$$\begin{aligned} det\,C = \;& -2J\left(m_t^2 - m_c^2\right)\left(m_c^2 - m_u^2\right)\left(m_u^2 - m_t^2\right) \\ & \left(m_b^2 - m_s^2\right)\left(m_s^2 - m_d^2\right)\left(m_d^2 - m_b^2\right) \neq 0 \end{aligned} \tag{4}$$

and J stands for the Jarlskog's rephasing invariant parameter. In case M_u and M_d are Hermitian quark mass matrices, then we have $[M_u,\, M_d] = i\,C$ where

$$det\,C = \; -2J\,(m_t - m_c)\,(m_c - m_u)\,(m_u - m_t) \\ (m_b - m_s)\,(m_s - m_d)\,(m_d - m_b) \neq 0. \tag{5}$$

2.2. *CP invariants in lepton sector*

In the SM, the charged leptons and quarks get their masses from the Spontaneous Symmetry Breaking, however, neutrinos are massless in SM. Therefore, generation of masses for the neutrinos is not straight-forward. In case of neutrinos, one may have either the Dirac masses or the more general Dirac-Majorana masses. The Dirac-Majorana Lagrangian mass terms are

$$\mathcal{L}^{(\mathcal{D}+\mathcal{M})} = -\overline{l_L}\,M^{(\mathcal{D})}\,l_R - \frac{1}{2}\overline{\nu_R^C}\,M_R^{(\mathcal{M})}\,\nu_R + h.c. \tag{6}$$

where $l_{L/R} = (e,\mu,\tau)_{L/R}$ stands for the SM charged leptonic left/right handed fields and $\nu_{L/R} = (e_\nu,\mu_\nu,\tau_\nu)_{L/R}$ stands for the flavor left/right handed neutrinos. $M^{\mathcal{D}/\mathcal{M}}$ stand for the Dirac/Majorana mass matrices, which are, in general, complex in nature.

Without getting into the details, the lepton mass matrix M_l and the neutrino mass matrix M_ν, the latter obtained through seesaw mechanism encode all the information about lepton masses and mixing. However, like the quark case, there is a redundancy in these matrices stemming from the fact that one has the freedom to make a unitary Weak Basis (WB) transformation[7,8], e.g.,

$$\nu_L = W_L\,\nu_L', \quad l_L = W_L\,l_L', \qquad l_R = W_R\,l_R', \tag{7}$$

under which all gauge currents remain real and diagonal. The mass matrices M_l and M_ν transform in following manner

$$M_l' = W_L^\dagger\,M_l\,W_R, \qquad M_\nu' = W_L^T\,M_\nu\,W_R, \tag{8}$$

One can use this freedom to go to a basis where M_l is real and diagonal. The neutrino mass matrix can be diagonalized by the unitary transformations

$$d_\nu = (U^\nu)^T\,M_\nu\,U^\nu, \tag{9}$$

where $U = U^\nu$ is the Pontecorvo-Maki-Nakagawa-Sakata (PMNS) leptonic mixing matrix. In this representation, the charged current becomes

$$L_W = \frac{g}{\sqrt{2}}\,\overline{l_L}\,\gamma_\mu\,U\,\nu_L\,W^\mu + h.c., \tag{10}$$

It has been shown by Branco et al.[9], a necessary and sufficient condition for low energy CP invariance in the leptonic sector is that the following three WB invariants are identically zero

$$\begin{aligned} I_1 &= Im\, det\, [H_\nu,\, H_l], \\ I_2 &= Im\, tr\, [H_l\, M_\nu\, M_\nu^*\, M_\nu\, H_l^*\, M_\nu^*,\, M_\nu^*], \\ I_3 &= Im\, det\, [M_\nu^*\, H_l\, M_\nu,\, H_l^*], \end{aligned} \tag{11}$$

where, M_l and M_ν are the mass matrices for the charged leptons and the neutrinos, respectively, and $H_l = M_l^\dagger M_l$ and $H_\nu = M_\nu^\dagger M_\nu$. The invariant I_1 is similar to J in the quark sector and is also referred to as J_l. It describes CP violation in the leptonic sector and is sensitive to the Dirac type CP violating phase representing CP violation in the Lepton Number Conserving (LNC) processes. The invariants I_2 and I_3 were proposed by Branco, Lavoura and Rebelo as the WB invariant measures of Majorana type CP violation. The invariant I_3 was shown to have the special feature of being sensitive to Majorana type CP violating phase even in the limit of the exactly degenerate Majorana neutrinos. The CP violation in the lepton number conserving (LNC) processes is contained in Jarlskog CP invariant J_l which can be calculated from the WB invariant I_1 using the relation

$$I_1 = -2J_l\,(m_e^2 - m_\mu^2)\,(m_\mu^2 - m_\tau^2)\,(m_\tau^2 - m_e^2)\,(m_1^2 - m_2^2)\,(m_2^2 - m_3^2)\,(m_3^2 - m_1^2)\,. \tag{12}$$

3. Unitarity Triangles, J and CP Violating Phase

In the context of fermion mixing phenomena, the Pontecorvo-Maki-Nakagawa-Sakata (PMNS) [2] and the Cabibbo-Kobayashi-Maskawa (CKM) [1] matrices have similar parametric structure. Also, regarding the three mixing angles corresponding to the quark and leptonic sector, it is interesting to note that in both the cases the mixing angle s_{13} is smaller as compared to the other two. These similarities of features suggest likelihood of CP violation in the leptonic sector, which, keeping in mind that the value of leptonic s_{13}, observed in the last few years, is not so 'small', could in fact be considerably large. This possibility, in turn, can have deep phenomenological implications. As is well known, the two CP violating Majorana phases do not play any role in the case of neutrino oscillations, therefore any hint regarding the value of Dirac-like CP violating phase in the leptonic sector δ_l will go a long way in the formulation of proposals on observation of CP violation in the Long BaseLine (LBL) experiments. In the absence of any hints from the data regarding leptonic CP violation, keeping in mind the parallelism between the neutrino mixing and the quark mixing, an analysis of the quark mixing phenomena could provide some viable clues regarding this issue in the leptonic sector.

To this end, following Ref. 10, we first determine the Jarlskog's rephasing invariant parameter J, an important parameter as all CP violating effects are proportional to it, in the quark sector. The parameter J can be evaluated from any of the six unitarity triangles in the quark sector whose sides can be defined by magnitudes of the elements of the Cabibbo-Kobayashi-Maskawa (CKM) [1] matrix. For ready reference as well as to facilitate discussion of results, we present below the PDG parameterization[11] of the CKM matrix, e.g.,

$$V_{CKM} = \begin{pmatrix} c_{12}\, c_{13} & s_{12}\, c_{13} & s_{13}\, e^{-i\delta} \\ -s_{12}\, c_{23} - c_{12}\, s_{23}\, s_{13}\, e^{i\delta} & c_{12}\, c_{23} - s_{12}\, s_{23}\, s_{13}\, e^{i\delta} & s_{23}\, c_{13} \\ s_{12}\, s_{23} - c_{12}\, c_{23}\, s_{13}\, e^{i\delta} & -c_{12}\, s_{23} - s_{12}\, c_{23}\, s_{13}\, e^{i\delta} & c_{23}\, c_{13} \end{pmatrix}. \quad (13)$$

The magnitudes of the CKM matrix elements arising from the latest global fits are given by

$$V_{CKM} = \begin{pmatrix} 0.97434 \pm 0.00012 & 0.22506 \pm 0.0005 & 0.00357 \pm 0.00015 \\ 0.22492 \pm 0.00050 & 0.97351 \pm 0.00013 & 0.0411 \pm 0.0013 \\ 0.00875 \pm 0.00033 & 0.0403 \pm 0.0013 & 0.99915 \pm 0.00005 \end{pmatrix}. \quad (14)$$

Making use of the above matrix, one can construct any of the six unitarity triangles in the quark sector. The usually constructed is the '*db*' unitarity triangle, expressed through the relation

$$V_{ud}V_{ub}{}^{*} + V_{cd}V_{cb}{}^{*} + V_{td}V_{tb}{}^{*} = 0\,. \quad (15)$$

Further, the Jarlskog's rephasing invariant parameter J, equal to twice the area of the unitarity triangle, can then be easily constructed using the magnitudes of the elements of the CKM matrix. Consequently, in the PDG representation, the CP violating phase δ can also be evaluated by using the relation between the parameter J and phase δ, e.g., $J = s_{12}s_{23}s_{13}c_{12}c_{23}c_{13}^2 \sin\delta$. Following this procedure, we have obtained a histogram of J, shown in Figure 1a, from which one can find

$$J = (3.36 \pm 0.38) \times 10^{-5}, \quad (16)$$

the corresponding histogram of δ, shown in Figure 1b, yields

$$\delta = 62.60^{\circ} \pm 10.98^{\circ}. \quad (17)$$

Interestingly, we find that the above mentioned J and δ values are compatible with those given by PDG 2016[11].

The above success motivates one to carry out a similar calculation in the case of leptons. We begin with the neutrino mixing matrix[2],

$$\begin{pmatrix} \nu_e \\ \nu_\mu \\ \nu_\tau \end{pmatrix} = \begin{pmatrix} U_{e1} & U_{e2} & U_{e3} \\ U_{\mu 1} & U_{\mu 2} & U_{\mu 3} \\ U_{\tau 1} & U_{\tau 2} & U_{\tau 3} \end{pmatrix} \begin{pmatrix} \nu_1 \\ \nu_2 \\ \nu_3 \end{pmatrix}, \quad (18)$$

where ν_e, ν_μ, ν_τ are the flavor eigenstates and ν_1, ν_2, ν_3 are the mass eigenstates. Following PDG representation, involving three angles and the Dirac-like CP violating phase δ_l as well as the two Majorana phases α_1, α_2, the PMNS matrix U can be written as

$$U = \begin{pmatrix} c_{12}c_{13} & s_{12}c_{13} & s_{13}e^{-i\delta_l} \\ -s_{12}c_{23} - c_{12}s_{23}s_{13}e^{i\delta_l} & c_{12}c_{23} - s_{12}s_{23}s_{13}e^{i\delta_l} & s_{23}c_{13} \\ s_{12}s_{23} - c_{12}c_{23}s_{13}e^{i\delta_l} & -c_{12}s_{23} - s_{12}c_{23}s_{13}e^{i\delta_l} & c_{23}c_{13} \end{pmatrix} \begin{pmatrix} e^{i\rho} & 0 & 0 \\ 0 & e^{i\sigma} & 0 \\ 0 & 0 & 1 \end{pmatrix}. \quad (19)$$

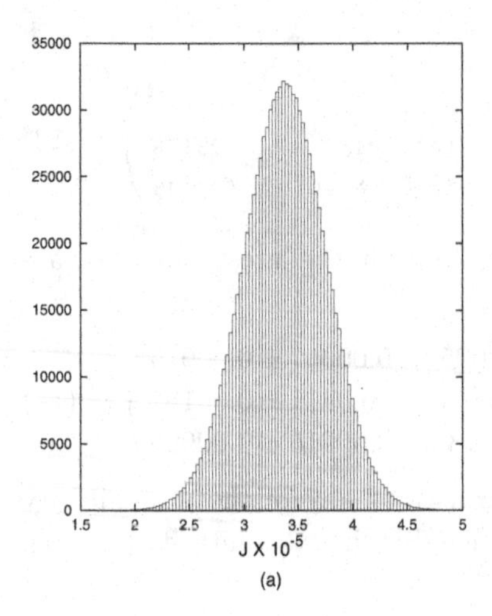

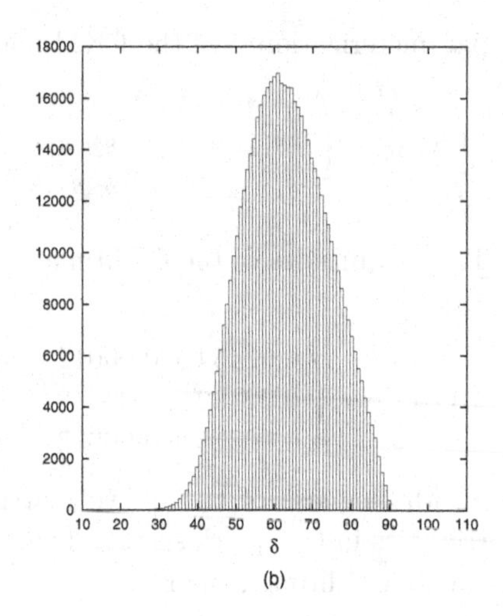

Fig. 1. Histogram of J and δ for 'db' triangle in the case of quarks.

The Majorana phases α_1 and α_2 do not play any role in neutrino oscillations and henceforth would be dropped from the discussion. Recently, by carrying out global fits of the neutrino oscillation data, Garcia et al.[12] have constructed the following magnitudes of the PMNS matrix elements

$$U = \begin{pmatrix} 0.801 - 0.845 & 0.514 - 0.580 & 0.137 - 0.158 \\ 0.225 - 0.517 & 0.441 - 0.699 & 0.614 - 0.793 \\ 0.246 - 0.529 & 0.464 - 0.713 & 0.590 - 0.776 \end{pmatrix}. \tag{20}$$

Analogous to the quark case, using the magnitudes of the PMNS matrix elements, one can construct the unitarity triangles in the lepton sector and consequently evaluation of J_l and δ_l can be carried out. The unitarity of PMNS matrix implies nine relations, three in terms of normalization conditions and other six can be defined as

$$\sum_{i=1,2,3} U_{\alpha i} U^*_{\beta i} = \delta_{\alpha\beta}; \qquad (\alpha \neq \beta)$$

$$\sum_{\alpha=e,\,\mu,\tau} U_{\alpha i} U^*_{\alpha j} = \delta_{ij}; \qquad (i \neq j) \tag{21}$$

where Latin indices run over the eigenmass states i.e. $i, j = 1, 2, 3$ and the Greek indices run over the flavor eigenstates (e, μ, τ). The six non-diagonal relations can

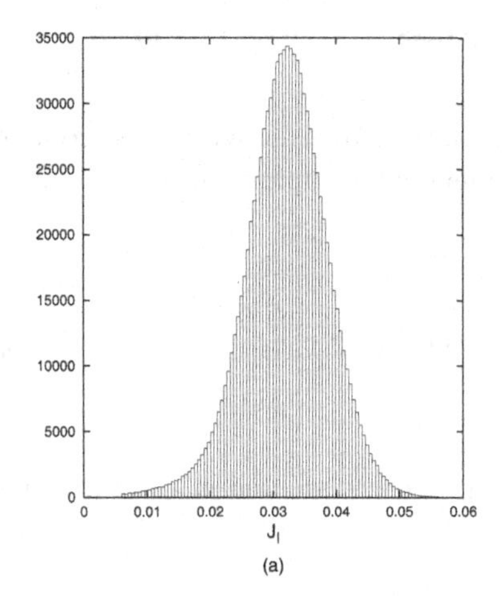

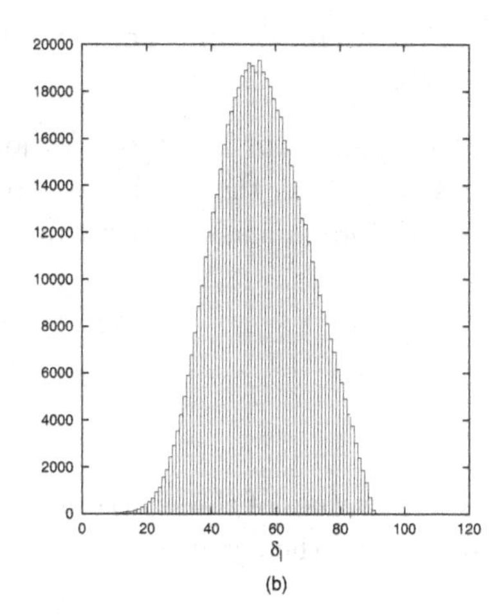

Fig. 2. Histogram of J_l and δ_l for '$\nu_1.\nu_3$' triangle in the case of neutrinos.

be expressed through six independent expressions in the complex plane

$$\begin{array}{ll} e \cdot \mu & U_{e1}\,U^*_{\mu 1} + U_{e2}\,U^*_{\mu 2} + U_{e3}\,U^*_{\mu 3} = 0, \\ e \cdot \tau & U_{e1}\,U^*_{\tau 1} + U_{e2}\,U^*_{\tau 2} + U_{e3}\,U^*_{\tau 3} = 0, \\ \mu \cdot \tau & U_{\mu 1}\,U^*_{\tau 1} + U_{\mu 2}\,U^*_{\tau 2} + U_{\mu 3}\,U^*_{\tau 3} = 0, \\ \nu_1 \cdot \nu_2 & U_{e1}\,U^*_{e2} + U_{\mu 1}\,U^*_{\mu 2} + U_{\tau 1}\,U^*_{\tau 2} = 0, \\ \nu_1 \cdot \nu_3 & U_{e1}\,U^*_{e3} + U_{\mu 1}\,U^*_{\mu 3} + U_{\tau 1}\,U^*_{\tau 3} = 0, \\ \nu_2 \cdot \nu_3 & U_{e2}\,U^*_{e3} + U_{\mu 2}\,U^*_{\mu 3} + U_{\tau 2}\,U^*_{\tau 3} = 0. \end{array} \tag{22}$$

Analogous to the 'db' triangle in the quark sector, we have considered the '$\nu_1.\nu_3$' unitarity triangle, expressed as

$$U_{e1}U^*_{e3} + U_{\mu 1}U^*_{\mu 3} + U_{\tau 1}U^*_{\tau 3} = 0\,. \tag{23}$$

Following the same procedure as in the quark case, using the matrix given in equation (20), we obtain the Jarlskog's rephasing invariant parameter in the leptonic sector J_l as

$$J_l = 0.0318 \pm 0.0065, \tag{24}$$

corresponding distribution has been plotted in figure 2a. Also, we find the corresponding phase δ_l from the histogram given in figure 2b as

$$\delta_l = 54.98^o \pm 13.81^o. \tag{25}$$

Interestingly, the above value of δ_l is coming out to be 'large', however, it should be noted that the quadrant of this δ_l is not fixed as we are considering the magnitude of J_l. One should also note that the recent global analysis give a slight hint of δ_l being near 270°. In this context, our analysis is a very preliminary one, a detailed and comprehensive analysis would be published elsewhere.

4. CP Invariants and Texture Two Zero Mass Matrices in Flavor Basis

In general, in flavor basis, the Majorana mass matrix M_ν is a symmetric complex matrix, therefore, there exist at most six independent complex entries. The form of this Majorana mass matrix can be written as

$$M_\nu = \begin{pmatrix} M_{ee} & M_{e\mu} & M_{e\tau} \\ M_{\mu e} & M_{\mu\mu} & M_{\mu\tau} \\ M_{\tau e} & M_{\tau\mu} & M_{\tau\tau} \end{pmatrix} . \tag{26}$$

If the eigenvalues of the neutrino mass matrix M_ν are m_1, m_2 and m_3, the invariant I_1 in terms of the eigenvalues and its elements is

$$I_1 = 2\left(m_e^2 - m_\mu^2\right)\left(m_\mu^2 - m_\tau^2\right)\left(m_\tau^2 - m_e^2\right) Im\left(M_{ee}\, A_{ee} + M_{\mu\mu}\, A_{\mu\mu} + M_{\tau\tau}\, A_{\tau\tau}\right). \tag{27}$$

The coefficients $A_{ee}, A_{\mu\mu}$ and $A_{\tau\tau}$ are given by

$$\begin{aligned} A_{ee} = {} & M_{\mu\tau}\, M_{e\mu}^*\, M_{e\tau}^* \left(|M_{\mu\mu}|^2 - |M_{\tau\tau}|^2 - |M_{e\mu}|^2 + |M_{e\tau}|^2\right) \\ & + M_{\mu\mu}\left(M_{e\mu}^*\right)^2 \left(|M_{e\tau}|^2 - |M_{\mu\tau}|^2\right) + M_{\mu\mu}^* \left(M_{e\tau}^*\right)^2 \left(M_{\mu\tau}\right)^2 , \\ A_{\mu\mu} = {} & M_{e\tau}\, M_{\mu\tau}^*\, M_{e\mu}^* \left(|M_{\tau\tau}|^2 - |M_{ee}|^2 - |M_{\mu\tau}|^2 + |M_{e\mu}|^2\right) \\ & + M_{\tau\tau}\left(M_{\mu\tau}^*\right)^2 \left(|M_{e\mu}|^2 - |M_{e\tau}|^2\right) + M_{\tau\tau}^* \left(M_{e\mu}^*\right)^2 \left(M_{e\tau}\right)^2 , \\ A_{\tau\tau} = {} & M_{e\mu}\, M_{e\tau}^*\, M_{\mu\tau}^* \left(|M_{ee}|^2 - |M_{\mu\mu}|^2 - |M_{e\tau}|^2 + |M_{\mu\tau}|^2\right) \\ & + M_{ee}\left(M_{e\tau}^*\right)^2 \left(|M_{\mu\tau}|^2 - |M_{e\mu}|^2\right) + M_{ee}^* \left(M_{\mu\tau}^*\right)^2 \left(M_{e\mu}\right)^2 . \end{aligned} \tag{28}$$

Therefore, the Jarlskog CP invariant parameter can be written as

$$J_l = \frac{Im\left(M_{ee}\, A_{ee} + M_{\mu\mu}\, A_{\mu\mu} + M_{\tau\tau}\, A_{\tau\tau}\right)}{\left(m_1^2 - m_2^2\right)\left(m_2^2 - m_3^2\right)\left(m_3^2 - m_1^2\right)} . \tag{29}$$

Following Ref. 13, the expressions for the invariants I_2 and I_3 can be given as

$$\begin{aligned} I_2 = {} & Im\Bigg(M_{ee}\, M_{\mu\mu} \left(M_{e\mu}^*\right)^2 (m_e^2 - m_\mu^2)^2 + M_{\mu\mu} \left(M_{\mu\tau}^*\right)^2 M_{\tau\tau}\, (m_\mu^2 - m_\tau^2)^2 \\ & + M_{\tau\tau} \left(M_{e\tau}^*\right)^2 M_{ee}\, (m_\tau^2 - m_e^2)^2 + 2\, M_{ee}\, M_{e\mu}^*\, M_{e\tau}^*\, M_{\mu\tau}\, (m_e^2 - m_\mu^2)(m_e^2 - m_\tau^2) \\ & + 2\, M_{\mu\mu}\, M_{\mu\tau}^*\, M_{e\mu}^*\, M_{e\tau} (m_\mu^2 - m_\tau^2)(m_\mu^2 - m_e^2)\Bigg), \\ I_3 = {} & 2\left(m_e^2 - m_\mu^2\right)\left(m_\mu^2 - m_\tau^2\right)\left(m_\tau^2 - m_e^2\right) \\ & Im\left(m_e^2\, M_{ee}\, B_{ee} + m_\mu^2\, M_{\mu\mu}\, B_{\mu\mu} + m_\tau^2\, M_{\tau\tau}\, B_{\tau\tau}\right), \end{aligned} \tag{30}$$

where the coefficients $B_{ee}, B_{\mu\mu}$ and $B_{\tau\tau}$ are given as

$$
\begin{aligned}
B_{ee} = {} & M_{\mu\tau}\, M^*_{e\mu}\, M^*_{e\tau}\, (m_\mu^4\, |M_{\mu\mu}|^2 - m_\tau^4\, |M_{\tau\tau}|^2 - m_e^2\, m_\mu^2\, |M_{e\mu}|^2 + m_e^2\, m_\tau^2\, |M_{e\tau}|^2) \\
& + m_\mu^2\, M_{\mu\mu}\, (M^*_{e\mu})^2\, (m_e^2\, |M_{e\tau}|^2 - m_\mu^2\, |M_{\mu\tau}|^2) + m_\mu^2\, m_\tau^2\, M^*_{\mu\mu}\, (M^*_{e\tau})^2\, M^2_{\mu\tau}, \\
B_{\mu\mu} = {} & M_{e\tau}\, M^*_{\mu\tau}\, M^*_{e\mu}\, (m_\tau^4\, |M_{\tau\tau}|^2 - m_e^4\, |M_{ee}|^2 - m_\mu^2\, m_\tau^2\, |M_{\mu\tau}|^2 + m_e^2\, m_\tau^2\, |M_{e\tau}|^2) \\
& + m_\tau^2\, M_{\tau\tau}\, (M^*_{\mu\tau})^2\, (m_\mu^2\, |M_{e\mu}|^2 - m_\tau^2\, |M_{e\tau}|^2) + m_e^2\, m_\tau^2\, M^*_{\tau\tau}\, (M^*_{e\mu})^2\, M^2_{e\tau}, \\
B_{\tau\tau} = {} & M_{e\mu}\, M^*_{e\tau}\, M^*_{\mu\tau}\, (m_e^4\, |M_{ee}|^2 - m_\mu^4\, |M_{\mu\mu}|^2 - m_e^2\, m_\tau^2\, |M_{e\tau}|^2 + m_\mu^2\, m_\tau^2\, |M_{\mu\tau}|^2) \\
& + m_e^2\, M_{ee}\, (M^*_{e\tau})^2\, (m_\tau^2\, |M_{\mu\tau}|^2 - m_e^2\, |M_{e\mu}|^2) + m_e^2\, m_\mu^2\, M^*_{ee}\, (M^*_{\mu\tau})^2\, M^2_{e\mu}.
\end{aligned}
\tag{31}
$$

The symmetric matrix M_ν has more number of unknowns than the data, therefore, attempts have been made to explore the compatibility of M_ν with the neutrino oscillation data by imposing additional constraints. In this context, Frampton, Glashow and Marfatia (FGM)[14] have ruled out all Majorana neutrino mass matrices with three or more texture zeros. In the case of texture two zero mass matrices, they have found the possibilities listed below.

$$
A_1 : \begin{pmatrix} 0 & 0 & \times \\ 0 & \times & \times \\ \times & \times & \times \end{pmatrix}, \qquad A_2 : \begin{pmatrix} 0 & \times & 0 \\ \times & \times & \times \\ 0 & \times & \times \end{pmatrix} \tag{32}
$$

$$
B_1 : \begin{pmatrix} \times & \times & 0 \\ \times & \times & \times \\ 0 & \times & 0 \end{pmatrix}, \quad B_2 : \begin{pmatrix} \times & 0 & \times \\ 0 & \times & \times \\ \times & \times & 0 \end{pmatrix}, \quad B_3 : \begin{pmatrix} \times & 0 & \times \\ 0 & 0 & \times \\ \times & \times & \times \end{pmatrix}, \quad B_4 : \begin{pmatrix} \times & \times & 0 \\ \times & \times & \times \\ 0 & \times & 0 \end{pmatrix} \tag{33}
$$

$$
C : \begin{pmatrix} \times & \times & \times \\ \times & 0 & \times \\ \times & \times & 0 \end{pmatrix} \tag{34}
$$

$$
D_1 : \begin{pmatrix} \times & \times & \times \\ \times & 0 & 0 \\ \times & 0 & \times \end{pmatrix}, \qquad D_2 : \begin{pmatrix} \times & \times & \times \\ \times & \times & 0 \\ \times & 0 & 0 \end{pmatrix} \tag{35}
$$

$$
E_1 : \begin{pmatrix} 0 & \times & \times \\ \times & 0 & \times \\ \times & \times & \times \end{pmatrix}, \quad E_2 : \begin{pmatrix} 0 & \times & \times \\ \times & \times & \times \\ \times & \times & 0 \end{pmatrix}, \quad E_3 : \begin{pmatrix} 0 & \times & \times \\ \times & \times & 0 \\ \times & 0 & \times \end{pmatrix} \tag{36}
$$

$$
F_1 : \begin{pmatrix} \times & 0 & 0 \\ 0 & \times & \times \\ 0 & \times & \times \end{pmatrix}, \quad E_2 : \begin{pmatrix} \times & 0 & \times \\ 0 & \times & 0 \\ \times & 0 & \times \end{pmatrix}, \quad F_3 : \begin{pmatrix} \times & \times & 0 \\ \times & \times & 0 \\ 0 & 0 & \times \end{pmatrix}. \tag{37}
$$

Here the symbol '×' stands for non-zero entries of the matrices. Out of these fifteen possibilities, they had found only seven possibilities ($A_{1,2}, B_{1,2,3,4}, C$) to be viable with the experimental data available at that time.

Table 1. 1σ, 2σ and 3σ CL ranges of neutrino oscillation parameters. NO (IO) refers to normal (inverted) neutrino mass ordering[15].

Parameter	Best Fit	1σ	2σ	3σ
δm^2 [$10^{-5} eV^2$]	7.60	7.42 - 7.79	7.26 - 7.99	7.11 - 8.18
$\|\Delta m^2_{31}\|$ [$10^{-3} eV^2$] (NO)	2.48	2.41 - 2.53	2.35 - 2.59	2.30 - 2.65
$\|\Delta m^2_{31}\|$ [$10^{-3} eV^2$] (IO)	2.38	2.32 - 2.43	2.26 - 2.48	2.20 - 2.54
θ_{12}	34.6°	33.6° - 35.6°	32.7° - 36.7°	31.8° - 37.8°
θ_{23} (NO)	48.9°	41.7° - 50.7°	40.0° - 52.1°	38.8° - 53.3°
θ_{23} (IO)	49.2°	46.9° - 50.7°	41.3° - 52.0°	39.4° - 53.1°
θ_{13} (NO)	8.6°	8.4° - 8.9°	8.2° - 9.1°	7.9° - 9.3°
θ_{13} (IO)	8.7°	8.5° - 8.9°	8.2° - 9.1°	8.0° - 9.4°
δ (NO)	254°	182° - 353°	0° - 360°	0° - 360°
δ (IO)	266°	210° - 322°	0° - 16° $\oplus$ 155° - 360°	0° - 360°

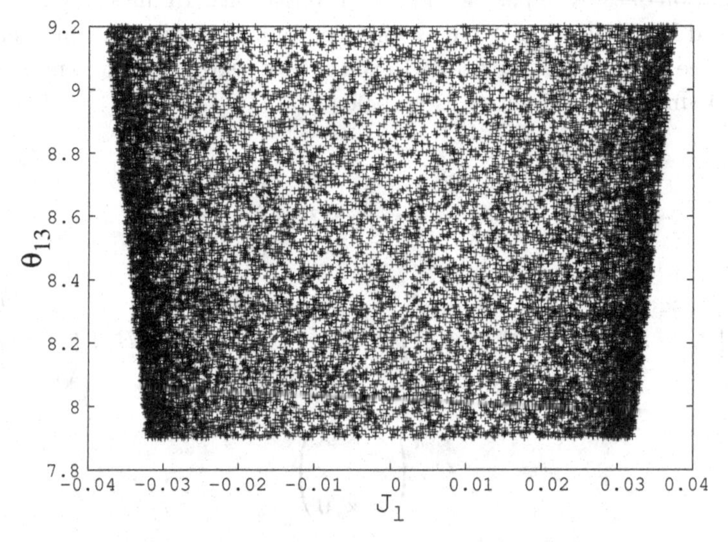

Fig. 3. Plot of parameter J_l versus the mixing angle s_{13} corresponding to matrix A_1 for NO case.

In the present work, we have re-analyzed the above mentioned fifteen possibilities with the most recent data available[15], given in Table 1. For the case of texture two zero mass matrices, out of 15 possible structures, the matrices belonging to classes D, E and F remain ruled out even with the latest data. Our calculations also show that a neutrino mass matrix with three or more zeros gives CP invariance in LNC processes. However, it can give CP violation through the WB invariant I_2 in LNV processes if the three zeros are situated along a row of the mass matrix. The neutrino mass matrices with two texture zeros can, in general, be CP-violating if they are complex and if their phases are not fine tuned.

In the present work, we have discussed the CP invariants for the seven viable texture two zero mass matrices viz. $A_{1,2}, B_{1,2,3,4}$ and C. While the details would

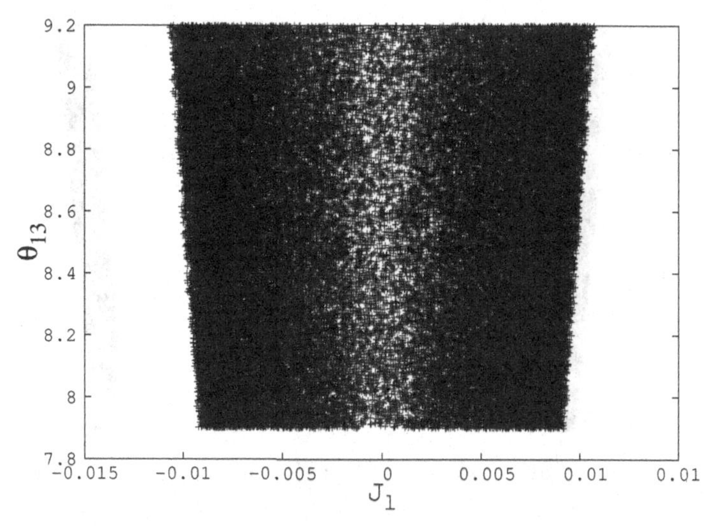

Fig. 4. Plot of parameter J_l versus the mixing angle s_{13} corresponding to matrix B_1 for IO case.

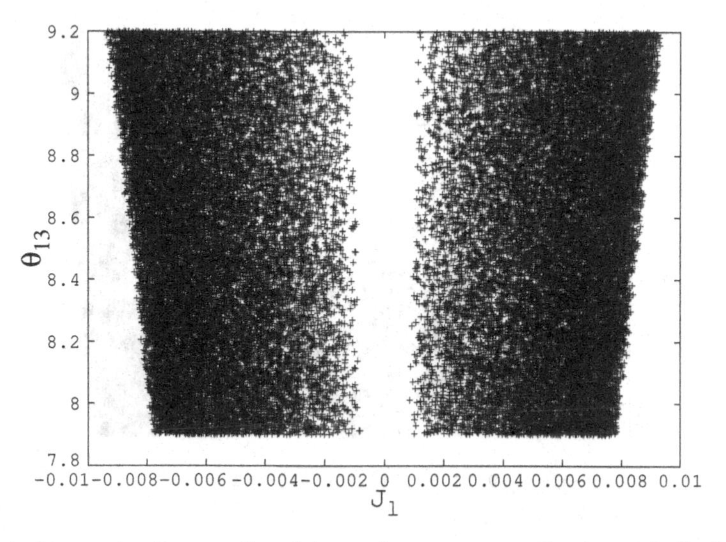

Fig. 5. Plot of parameter J_l versus the mixing angle s_{13} corresponding to matrix B_1 for NO case.

be presented elsewhere, we very briefly discuss some of the important aspects of our calculations related to the CP violation invariants. It may be noted that for the classes A, B and C, analysis has been carried out for both NO and IO. The analysis shows that the matrices of class $A_{1,2}$ are ruled out for IO, whereas for NO, as shown in Figure 3 wherein, corresponding to matrix A_1, we have plotted the Jarlskog's rephasing invariant parameter in the leptonic sector J_l versus the mixing angle s_{13}, one obtains the parameter J_l from -0.03 to 0.03. Similar results are obtained for matrix A_2 also.

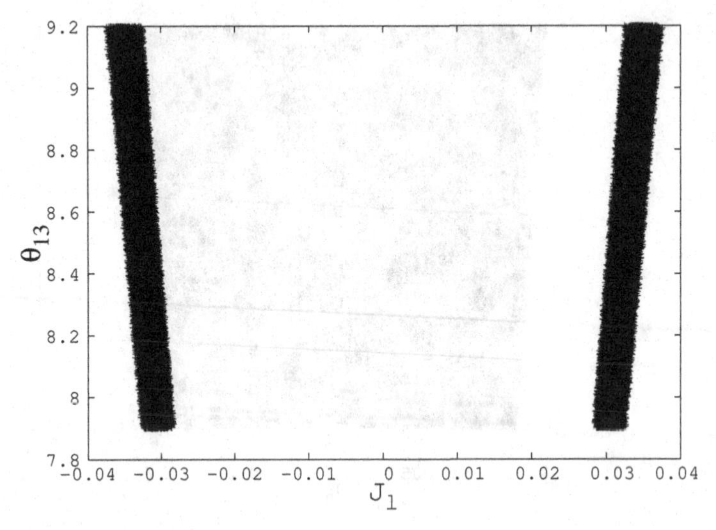

Fig. 6. Plot of parameter J_l versus the mixing angle s_{13} corresponding to matrix B_2 for NO case.

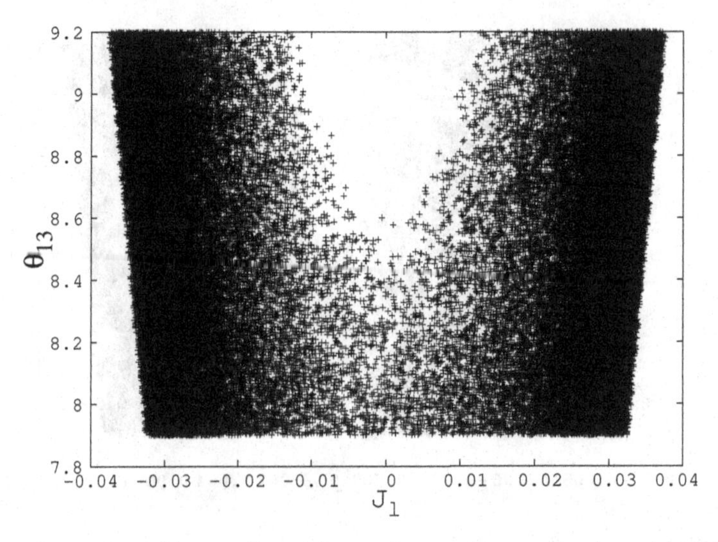

Fig. 7. Plot of parameter J_l versus the mixing angle s_{13} corresponding to matrix C for IO case.

For the class B matrices, all these are viable for both NO and IO. For the case of IO, the matrices $B_{1,2,3,4}$ all yield J_l from -0.01 to 0.01, plot corresponding to matrix B_1 has been presented in Figure 4, the other plots being similar have not been shown here. For the case of NO, corresponding to matrix B_1, the Jarlskog's rephasing invariant parameter in the leptonic sector J_l versus the mixing angle s_{13} plot, shown in Figure 5, gives some constraints on J_l which now is from -0.008 to -0.02 and then from 0.02-0.008. For NO, corresponding to the matrices $B_{2,3,4}$, the constraints on J_l get sharpened and one obtains J_l being around -0.03 and

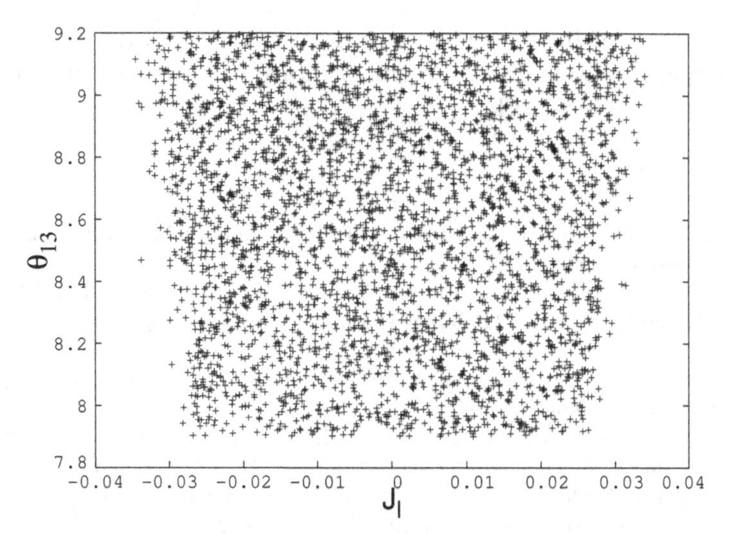

Fig. 8. Plot of parameter J_l versus the mixing angle s_{13} corresponding to matrix C for NO case.

+0.03, plot corresponding to matrix B_2 has been presented in Figure 6. The matrix belonging to class C is again viable for both IO and NO. The Jarlskog's rephasing invariant parameter in the leptonic sector J_l versus the mixing angle s_{13} plots, shown in Figures 7 and 8, for IO and NO respectively, yield the parameter J_l from -0.03 to 0.03.

5. Summary and Conclusions

To summarize, we briefly discuss the issue of CP invariants/ non invariants in the context of quark mixing as well as neutrino oscillation phenomena. To begin with, we have defined the CP invariants in the leptonic sector using the analogy from quark sector. Corresponding to the well known Jarlskog's rephasing invariant parameter J, characterizing all CP violation in the quark sector, a similar invariant J_l (I_1) can be defined in the lepton sector for Dirac neutrinos. In case neutrinos are Majorana particles, two additional CP invariants I_2 and I_3 can be defined. Using the well known existence of CP violation in the quark sector, in an analogous way, we have attempted to calculate probable value of J_l as well as the corresponding CP violating phase δ_l. It is well known that for the case of Majorana neutrinos, texture two zero symmetric mass matrices in the flavor basis are able to accommodate the present neutrino oscillation data. Apart from updating the analysis of texture two zero mass matrices, we have discussed the CP invariant J_l corresponding to viable texture two zero mass matrices.

Acknowledgments

M.G. and D.S. would like to acknowledge CSIR, Government of India (Grant No:03:(1313)14/EMR-II) for financial support. G.A. would also like to acknowledge

DST, Government of India (Grant No: SR/FTP/PS-017/2012) for financial support. M.S. and D.S. would like to thank Chairman, Department of Physics, P.U., for providing facilities to work.

References

1. N. Cabibbo, Phys. Rev. Lett. **10**, 531 (1963); M. Kobayashi and T. Maskawa, Prog. Theor. Phys. **49**, 652 (1973).
2. B. Pontecorvo, Zh. Eksp. Teor. Fiz. (JETP) **33**, 549 (1957); ibid. **34**, 247 (1958); ibid. **53**, 1717 (1967); Z. Maki, M. Nakagawa and S. Sakata, Prog. Theor. Phys. **28**, 870 (1962).
3. C. Jarlskog, in CP violation, Ed. C. Jarlskog, World Scientific Publishing Co. Pte. Ltd., 1989 and references therein.
4. H. Fritzsch, Phys. Lett. **B 70**, 436 (1977); H. Fritzsch, Phys. Lett. **B 73**, 317 (1978); H. Fritzsch and Z. Z. Xing, Prog. Part. Nucl. Phys. **45**, 1 (2000) and references therein; Z. Z. Xing, Int. J. Mod. Phys. **A 19**, 1 (2004) and references therein
5. M. Gupta and G. Ahuja, Int. J. Mod. Phys. **A 26**, 2973 (2011), N. G. Deshpande, M. Gupta and Palash B. Pal, Phys. Rev. **D 45**, 953 (1992); P. S. Gill and M. Gupta, Pramana **45**, 333 (1995).
6. M. Gupta and G. Ahuja, Int. J. Mod. Phys. **A 27**, 1230033 (2012) and references therein.
7. G. C. Branco, D. Emmanuel-Costa and R. G. Felipe, Phys. Lett. **B 477**, 147 (2000); G. C. Branco, D. Emmanuel-Costa, R. G. Felipe and H. Serodio, Phys. Lett. **B 670**, 340 (2009).
8. H. Fritzsch and Z. Z. Xing, Nucl. Phys. **B 556**, 49 (1999) and references therein.
9. G. C. Branco, M. N. Rebelo and J. I. Silva Marcos, Phys. Rev. Lett. **82**, 683 (1999).
10. G. Ahuja and M. Gupta, Phys. Rev. **D 77**, 057301 (2008).
11. C. Patrignani *et al.* Particle Data Group, Chin. Phys. **C 40**, 100001 (2016).
12. M. C. Gonzalez-Garcia, M. Maltoni and T. Schwetz, Nucl. Phys. **B 908**, 199 (2016).
13. S. Dev, S. Kumar and S. Verma, Phys. Rev. **D 79**, 033011 (2009).
14. P. H. Frampton, S. L. Glashow and D. Marfatia, Phys. Lett. **B 536**, 79 (2002).
15. D. V. Forero, M. Trtola and J. W. F. Valle, Phys. Rev. **D 90**, 093006 (2014).

Fermions Masses and Texture Zero

A. Aziz Bhatti

CHEP University of the Punjab Lahore Pakistan
aziz.chep@pu.edu.pk
www.pu.edu.pk

The Standard model of particle physics is a very successful theory of strong weak and electromagnetic interactions. This theory is perturbative at sufficiently high energies and renormalizable, it describes these interactions at quantum level. However it has a number of limitations, one being the fact that it has 28 free parameters assuming massive neutrinos. Within the Standard model these parameters can not be explained, however they can be accommodated in the standard theory. Particularly the masses of the fermions are not predicted by the theory. The existence of the neutrino masses can be regarded as the first glimpse of the physics beyond the standard model. we have described the quark and lepton masses and mixings in context of four zero texture (FZT). In the four zero texture case the fermion masses and mixing can be related. We have made some predictions using tribimaximal mixing, the near tribimaximal (TBM) mixing. Our results show that under the TBM the neutrinos have normal, but weak hierarchy. Under near tribimaximal mixing, we find that the neutrino masses in general increase, if the value of solar angle increases from its TBM value and vice versa. It appears that the neutrinos become more and more degenerate for solar angle values higher than TBM value and hierarchical for lower values of solar angle.

Keywords: Fermions masses; texture zero; neutrinos mass.

1. Introduction

The Standard Model (SM) of electroweak and strong interactions is extremely successful theory of elementary particles[1,2]. The high precision measurement tests[4,5] performed at LEP, SLC, Tevatron etc. has clearly established that the standard model of particle physics is the correct effective theory of the strong and electroweak interactions at the present energies.The quantum corrections and the structure of the $SU(3)_c \times SU(2)_L \times U(1)_Y$ local gauge symmetry has been probed thoroughly. The theory has precisely predicted the measured values of the couplings of the electroweak gauge bosons with the leptons and quarks. The $SU(3)_c$ gauge symmetric sector of the theory has also been tested at LEP, however the scalar sector of the Standard Model(SM) has yet to be tested satisfactorily.

However there are some unpleasant features of SM. Excluding neutrinos there are 19 free parameters in the standard model[3]. Out of these 19, 3 gauge couplings, the Higgs quartic coupling λ and Higgs mass squared μ^2 are flavor universal. The others are flavor parameters. They include 6 quark masses, 3 charged lepton masses, 4 quark mixing parameters and a strong CP phase. The existence of the neutrino mass can be regarded as the first glimpse of the physics beyond the standard model. Including small neutrino masses and mixings, 9 more parameters, 3 neutrino masses,

3 mixing angles and 3 phases have to be introduced. One may ask, why are there so many free parameters? Are the mixing parameters and mass ratios related to each other? Answers to these questions necessarily leads one to the physics beyond the Standard model as the origin of these parameters is still unknown. Within the Standard model these parameters cannot be explained however they can be accommodated in the standard theory.

In the Standard Model the masses of the fermions arise through Yukawa couplings of the fermions fields with the Higgs doublet. As a fermion mass term must involve a left handed and a right handed field, the neutrinos are massless in the Standard Model, since they do not have right handed components.

2. Lepton Masses

The Yukawa lagrangian is given by[6]

$$\mathcal{L}_{H,L} = - \sum_{\alpha,\beta=e,\mu,\tau} Y'^{\ell}_{\alpha\beta} \overline{L}'_{\alpha L} \Phi \ell'_{\beta R} + h.c \tag{1}$$

The products $\overline{L}'_{\alpha L}\ell'_{\beta R}$ are isospin doublets and they have hypercharge $Y = -1$. The Higgs doublet has hypercharge $Y = +1$. Therefore the above mentioned lagrangian is invariant under the $SU(2)_L \times U(1)_Y$ gauge symmetry. After symmetry breaking the Higgs-Lepton Yukawa lagrangian is

$$\mathcal{L}_{H,L} = - \left(\frac{v+H}{\sqrt{2}}\right) \sum_{\alpha,\beta=e,\mu,\tau} Y'^{\ell}_{\alpha\beta} \overline{\ell}'_{\alpha L} \ell'_{\beta R} + h.c \tag{2}$$

In the above expression the term proportional to v (VEV) is the mass term for the charged fermion. Because of the fact that Y'^{ℓ} is usually a non diagonal complex 3×3 matrix. e', μ', τ'do not have definite mass. Charged lepton fields can be obtained, if we diagonalize the Yukawa matrix Y'^{ℓ}. In order to do this lets define an array of charged lepton fields

$$\ell'_L = \begin{pmatrix} e'_L \\ \mu'_L \\ \tau'_L \end{pmatrix}., \qquad \ell'_R = \begin{pmatrix} e'_R \\ \mu'_R \\ \tau'_R \end{pmatrix}$$

The Yukawa lagrangian is

$$\mathcal{L}_{H,L} = - \left(\frac{v+H}{\sqrt{2}}\right) \overline{\ell}'_L Y'^{\ell} \ell'_R + h.c$$

We can diagonalize the Y'^{ℓ} through a biunitary transformation

$$V_L^{\ell\dagger} Y'^{\ell} V_R^{\ell} = Y^{\ell}, \quad \text{with} \quad Y^{\ell}_{\alpha\beta} = y^{\ell}_{\alpha} \delta_{\alpha\beta} \quad (\alpha, \beta = e, \mu, \tau)$$

(V_L^ℓ are V_R^ℓ are appropriate 3×3 matrices). Redefining the charged lepton fields

$$\ell_L = V_L^{\ell\dagger} \ell'_L = \begin{pmatrix} e_L \\ \mu_L \\ \tau_L \end{pmatrix} , \qquad \ell_R = V_R^{\ell\dagger} \ell'_R = \begin{pmatrix} e_R \\ \mu_R \\ \tau_R \end{pmatrix}$$

where ℓ_L,ℓ_R are charged lepton fields with definite mass, the Yukawa lagrangian is given by

$$\mathcal{L}_{H,L} = -\left(\frac{v+H}{\sqrt{2}}\right) \bar{\ell}_L Y^\ell \ell_R + h.c$$

$$\mathcal{L}_{H,L} = -\sum_{\alpha=e,\mu,\tau} \frac{y_\alpha^\ell v}{\sqrt{2}} \bar{\ell}_\alpha \ell_\alpha - \sum_{\alpha=e,\mu,\tau} \frac{y_\alpha^\ell}{\sqrt{2}} \bar{\ell}_\alpha \ell_\alpha H$$

($\ell_\alpha = \ell_{\alpha L} + \ell_{\alpha R}$ are charged lepton fields with definite mass.)

$$\ell_e = e., \qquad \ell_\mu = \mu \, , \qquad \ell_\tau = \tau$$

The mass term for the charged leptons is given by the first term of the lagrangian

$$m_\alpha = \frac{y_\alpha^\ell v}{\sqrt{2}} \qquad \text{with} \qquad (\alpha = e, \mu, \tau) \tag{3}$$

Since y_e^ℓ,y_μ^ℓ,y_τ^ℓ are free parameters in the standard model, the charged lepton masses are not predicted in the SM. If we define

$$\nu_L = V_L^{\ell\dagger} \nu'_L = V_L^{\ell\dagger} \begin{pmatrix} \nu'_{eL} \\ \nu'_{\mu L} \\ \nu'_{\tau L} \end{pmatrix} = \begin{pmatrix} \nu_{eL} \\ \nu_{\mu L} \\ \nu_{\tau L} \end{pmatrix}$$

Then the leptonic charged current can be written as

$$j^\rho_{W,L} = 2\overline{\nu}_L \gamma^\rho \ell_L = 2 \sum_{\alpha,\beta=e,\mu,\tau} \overline{\nu}_{\alpha L} \gamma^\rho \ell_{\alpha L}$$

The neutrino fields ν_{eL}, $\nu_{\mu L}$, $\nu_{\tau L}$ still stay massless as they are linear combinations of the massless (primed) fields. They are called flavor neutrino fields as they only couple to the corresponding charged lepton fields. In the standard model the neutrino flavor eigenstates are also the neutrino mass eigenstates, however in theories beyond the standard model such as $SO(10)$, right handed neutrino field also exist, and neutrinos are massive particles. Therefore in these theories neutrino flavor eigenstates are not in general mass eigenstates.

3. Quark Masses

As the mass term for the fermions possesses both the left handed and the right handed fields of the corresponding fermion, we can have two types of such products

$$\overline{Q}'_{\alpha L} q'^{D}_{\beta R}\,, \qquad \text{where} \qquad \alpha = 1,2,3 \qquad \text{and} \qquad \beta = d,s,b$$
$$\overline{Q}'_{\alpha L} q'^{U}_{\beta R}\,, \qquad \text{where} \qquad \alpha = 1,2,3 \qquad \text{and} \qquad \beta = u,c,t$$

The first term $\overline{Q}'_{\alpha L} q'^{D}_{\beta R}$ has hypercharge -1 and can be combined to the Higgs doublet with hypercharge $+1$ to form an $SU(2)_L \times U(1)_Y$ gauge invariant term

$$-\sum_{\alpha=1,2,3}\ \sum_{\alpha,\beta=d,s,b} Y'^{D}_{\alpha\beta} \overline{Q}'_{\alpha L} \Phi q'^{D}_{\beta R}$$

where Y'^{D} is a 3×3 complex Yukawa couplings matrix, given by (mass term for the down type of quarks). In the unitary gauge we can write the mass term as

$$-\left(\frac{v+H}{\sqrt{2}}\right) \sum_{\alpha,\beta=d,s,b} Y'^{D}_{\alpha\beta} \overline{q}'^{D}_{\alpha L} q'^{D}_{\beta R}$$

where $Y'^{D}_{d,s,b} = Y'^{D}_{1,2,3}$. The term proportional to v belongs to the mass term for the down type quarks. For up type quarks, as $\overline{Q}'_{\alpha L} q'^{U}_{\beta R}$ has hypercharge $+1$, a Higgs doublet with hypercharge -1 is needed in order to have a $SU(2)_L \times U(1)_Y$ gauge invariant term. Therefore we define

$$\widetilde{\Phi} = \iota \tau_2 \Phi^*$$

and write the gauge invariant term as

$$-\sum_{\alpha=1,2,3}\ \sum_{\alpha,\beta=u,c,t} Y'^{U}_{\alpha\beta} \overline{Q}'_{\alpha L} \widetilde{\Phi} q'^{U}_{\beta R}$$

In the unitary gauge we have

$$\widetilde{\Phi} = \frac{1}{\sqrt{2}} \begin{pmatrix} v + H(x) \\ 0 \end{pmatrix}$$

The mass term for the up type of quarks can be written as

$$-\left(\frac{v+H}{\sqrt{2}}\right) \sum_{\alpha,\beta=u,c,t} Y'^{U}_{\alpha\beta} \overline{q}'^{U}_{\alpha L} q'^{U}_{\beta R}$$

Here $Y'^{U}_{d,s,b} = Y'^{U}_{1,2,3}$. The term proportional to v belongs to the masses of the up type of quarks. In the unitary gauge the quark Yukawa lagrangian can be written as

$$\mathcal{L}_{H,Q} = -\left(\frac{v+H}{\sqrt{2}}\right) \left[\sum_{\alpha,\beta=d,s,b} Y'^{D}_{\alpha\beta} \overline{q}'^{D}_{\alpha L} q'^{D}_{\beta R} + \sum_{\alpha,\beta=u,c,t} Y'^{U}_{\alpha\beta} \overline{q}'^{U}_{\alpha L} q'^{U}_{\beta R} \right] \tag{4}$$

The complex Yukawa coupling matrices $Y'^{D}_{\alpha\beta}$, $Y'^{D}_{\alpha\beta}$ are non diagonal in general. In order to get mass terms for the quarks, we must diagonalize the Yukawa matrices. We define

$$q_L^{\prime U} = \begin{pmatrix} u'_L \\ c'_L \\ t'_L \end{pmatrix}, \qquad q_R^{\prime U} = \begin{pmatrix} u'_R \\ c'_R \\ t'_R \end{pmatrix}, \qquad q_L^{\prime D} = \begin{pmatrix} d'_L \\ s'_L \\ b'_L \end{pmatrix}, \qquad q_R^{\prime D} = \begin{pmatrix} d'_R \\ s'_R \\ b'_R \end{pmatrix}$$

This allows us to write the quark Yukawa lagrangian in matrix form as

$$\mathcal{L}_{H,Q} = -\left(\frac{v+H}{\sqrt{2}}\right)\left[\overline{q}_L^{\prime D} Y^{\prime D} q_R^{\prime D} + \overline{q}_L^{\prime U} Y^{\prime U} q_R^{\prime U}\right] + h.c \tag{5}$$

The Yukawa matrices can be diagonalized by a biunitary transformation:

$$\begin{aligned} V_L^{D\dagger} Y^{\prime D} V_R^D &= Y^D, \quad \text{with} \quad Y_{\alpha\beta}^D = y_\alpha^D \delta_{\alpha\beta} \ \ (\alpha, \beta = d, s, b) \\ V_L^{U\dagger} Y^{\prime U} V_R^U &= Y^U, \quad \text{with} \quad Y_{\alpha\beta}^U = y_\alpha^U \delta_{\alpha\beta} \ \ (\alpha, \beta = d, s, b) \end{aligned}$$

(V_L^D, V_R^D, V_L^U, V_R^U are four suitable 3×3 matrices). Now we define unprimed fields in terms of primed fields:

$$\begin{aligned} q_L^U = V_L^{U\dagger} q_L^{\prime U} &= \begin{pmatrix} u_L \\ c_L \\ t_L \end{pmatrix}, \qquad q_R^U = V_R^{U\dagger} q_R^{\prime U} = \begin{pmatrix} u_R \\ c_R \\ t_R \end{pmatrix} \\ q_L^D = V_L^{D\dagger} q_L^{\prime D} &= \begin{pmatrix} d_L \\ s_L \\ b_L \end{pmatrix}, \qquad q_R^D = V_R^{D\dagger} q_R^{\prime D} = \begin{pmatrix} d_R \\ s_R \\ b_R \end{pmatrix} \end{aligned} \tag{6}$$

The Yukawa lagrangian becomes:

$$\begin{aligned} \mathcal{L}_{H,Q} &= -\left(\frac{v+H}{\sqrt{2}}\right)\left[\overline{q}_L^D Y^D q_R^D + \overline{q}_L^U Y^U q_R^U\right] + h.c \\ &= -\sum_{\alpha=d,s,b} \frac{y_\alpha^D v}{\sqrt{2}} \overline{q}_\alpha^D q_\alpha^D - \sum_{\alpha=u,c,t} \frac{y_\alpha^U v}{\sqrt{2}} \overline{q}_\alpha^U q_\alpha^U \\ &\quad - \sum_{\alpha=d,s,b} \frac{y_\alpha^D}{\sqrt{2}} \overline{q}_\alpha^D q_\alpha^D H - \sum_{\alpha=u,c,t} \frac{y_\alpha^U}{\sqrt{2}} \overline{q}_\alpha^U q_\alpha^U H \end{aligned}$$

where the unprimed quarks fields

$$q_\alpha^D = q_{\alpha L}^D + q_{\alpha R}^D, \qquad q_\alpha^U = q_{\alpha L}^U + q_{\alpha R}^U$$

have a definite mass. Their masses are given by:

$$\begin{aligned} m_\alpha &= \frac{y_\alpha^D v}{\sqrt{2}} \qquad \text{with} \qquad (\alpha = d, s, b) \\ m_\alpha &= \frac{y_\alpha^U v}{\sqrt{2}} \qquad \text{with} \qquad (\alpha = u, c, t) \end{aligned} \tag{7}$$

Similar to the leptons case the values of y_d^D, y_s^D, y_b^D, y_u^U, y_c^U, y_t^U are unknown parameters of the SM. Therefore the quark masses cannot be calculated. Due to mismatch between the (massive) unprimed quark fields and primed quark fields, we are led to a particular phenomenon called quark mixing. The quark weak current can be written in matrix form

$$j^{\rho}_{W,Q} = 2\overline{q}'^{U}_{L}\gamma^{\rho}q'^{D}_{L}$$

Now expressing unprimed quark fields in terms of primed quark fields (6), we can write the quark weak current as

$$j^{\rho}_{W,Q} = 2\overline{q}^{U}_{L}V^{U\dagger}_{L}\gamma^{\rho}V^{D}_{L}q^{D}_{L} = 2\overline{q}^{U}_{L}\gamma^{\rho}V^{U\dagger}_{L}V^{D}_{L}q^{D}_{L}$$

The charged current depends on the combination of $V^{U\dagger}_{L}$,V^{D}_{L} called quark mixing matrix or Cabibbo-Kobayashi-Maskawa (CKM) matrix, carrying the physical effects of quark mixing.

$$V = V^{U\dagger}_{L}V^{D}_{L} \tag{8}$$

4. Standard Model and Neutrinos Masses

In the Standard Model the left-handed neutrinos form the electroweak doublets with the charged leptons. They do not posses an electric charge and are colorless. In this theory right handed neutrinos are not included. Due to the absence of the right handed neutrinos in this theory neutrinos are massless at the tree level. Due to the presence of the exact $B - L$ symmetry in this theory, neutrinos remain massless in all orders of the perturbation theory as well as when non perturbative effects are taken into account.[7] It is therefore natural to assume that the nonzero neutrino masses must be associated to the right-handed neutrinos and with the breaking of the $B - L$ symmetry. If the condition of explicit renormalizability of the theory is abandoned, neutrino masses can be obtained even if the SM particles are the only light degrees of freedom. The renormalizable operator generates the Majorana neutrino mass after the electroweak symmetry breaking.[8]

$$\frac{\lambda_{ij}}{M}\left(L_i H\right)^T\left(L_j H\right) \quad , \qquad i,j = e,\mu,\tau$$

$$m_{ij} = \frac{\lambda_{ij}\left\langle H\right\rangle^2}{M}$$

(λ_{ij}, H, M are the dimensionless coupling, the Higgs doublet and the cut off scale respectively). In order to get the correct masses for the neutrinos, some new physics scale below the M_{Pl} must exit.

4.1. *Fermion masses beyond the standard model*

In the Standard Model the mixing of the quark flavors arises after diagonalizing the up and down type quark mass matrices. Both mass matrices cannot be diagonalized by unitary transformations which commute with the charged weak generators. This diagonalization-mismatch gives rise to the phenomenon of flavor mixing, whose dynamical origin is unknown. However it is implied that the mechanism responsible for the generation of the quark masses is also responsible for quark mixing.[9] In many models which go beyond the Standard electroweak model, based of flavor symmetries, the flavor mixing angles are the functions of the mass eigenvalues.[10,11] A hierarchy exists between both the observed values of mass spectrum of quarks and the observed values of flavor mixing parameters. This hierarchical structure can be understood as a result of a specific pattern of chiral symmetries whose breaking would cause the hierarchical tower of masses to appear step by step.[12,13] Such a chiral evolution of the quark mass matrices leads to specific way of describing the flavor mixing. Here we describe a parameterization of flavor mixing which is unique in the sense that it incorporates the chiral evolution of the mass matrices in a natural way.

We assume that the quark mass eigenvalues are the dynamical entities whose values can be changed to study certain symmetry limits (as is done in QCD). Without loss of generality we can take the quark mass matrices as Hermitian matrices.[14,15] A general hermitian mass matrix can be written as

$$M_q = \begin{pmatrix} E_q & D_q & F_q \\ D_q^* & C_q & B_q \\ F_q^* & B_q^* & A_q \end{pmatrix}$$

$(A >> C, |B| >> E, |D|, |F|)$

With a common unitary transformation of the up and the down quark fields, it is always possible to arrange the mass matrices M_u, M_d in such way that $F_q = F_u = F_d = 0$

$$M_q = \begin{pmatrix} E_q & D_q & 0 \\ D_q^* & C_q & B_q \\ 0 & B_q^* & A_q \end{pmatrix} \tag{9}$$

The basis in which the mass matrices take the form (9) is a basis in which up and down quark mass matrices exhibit two texture zero. In this basis the $(1,3)$ and $(3,1)$ elements of the mass matrices in M_u, M_d are zero. Therefore no direct mixing of the heavy quark t (*or* b) and the light quark u (*or* d) is present.[16] In the hierarchy limit of the quark masses the mass matrix of the type (9), can be diagonalized by a rotation matrix having just two angles[10].

$$V = \begin{pmatrix} s_u s_d c + c_u c_d e^{-i\varphi} & s_u c_d c - c_u s_d e^{-i\varphi} & s_u s \\ c_u s_d c - s_u c_d e^{-i\varphi} & c_u c_d c + s_u s_d e^{-i\varphi} & c_u s \\ -s_d s & -c_d s & c \end{pmatrix} \tag{10}$$

($c_u = \cos\theta_u, s_u = \sin\theta_u$ etc.) As all the three angles $\theta_u, \theta_d, \theta$ can be arranged to lie in the first quadrant, s_u, s_d, c all are positive definite. The CP violating phase φ can lie in the range $0 - 2\pi$. CP violation is present in the weak interactions if $\varphi \neq 0, \pi, 2\pi$.

The parameterization (10) has many advantages[9]. Some of them are as follows

(1) The three mixing angles θ_u, θ_d and θ have precise physical meaning. The angle θ describes the mixing between the second and the third families in the limit $m_u << m_c << m_t$, generated by B_u and B_d in (9) which can be named as "heavy quark mixing". The angle θ_u describes the mixing between $u - c$ channel by term D_u in (9) and is called "$u-$channel mixing". Similarly the "$d-$channel mixing" is mixing given by angle θ_d which describes the mixing between $d - s$ channel by the term D_d in (9) in the limit $m_d << m_s << m_b$.

(2) A simple relation exists between the three mixing angles and some observable quantities in the B-meson Physics. From (10) we can get the following simple relations

$$\sin\theta = |V_{cb}| \sqrt{1 + \left|\frac{V_{ub}}{V_{cb}}\right|^2}$$

$$\tan\theta_u = \left|\frac{V_{ub}}{V_{cb}}\right|, \qquad \tan\theta_d = \left|\frac{V_{td}}{V_{ts}}\right|$$

Another advantage which the parameterization (10) has over the standard parameterization is that the renormalization group evolution of the V is to a very high degree of accuracy associated with only angle θ. This can be verified easily if one keeps only the Yukawa coupling of t, b and neglect the possible threshold effects in renormalization group equations of the Yukawa matrices[17]. Thus if the underlying scale is changed from weak scale ($\sim 10^2$ GeV) to grand unified scale ($\sim 10^{16}$ GeV), the value of θ changes where as the values of θ_u, θ_d and φ remain independent of this variation. So heavy quark mixing is subjected to renormalization group equation effects where as $u-$and $d-$ channels and CP- violation phase are not.

We can predict the flavor mixing angles in terms of quark masses in the following way. We take $E_q = 0$ in the mass matrix (9)

$$M_q = \begin{pmatrix} 0 & D_q & 0 \\ D_q^* & C_q & B_q \\ 0 & B_q^* & A_q \end{pmatrix}$$

The physical constraints are as follows; in the flavor basis in which the (1,3) and (3,1) elements of $M_{u,d}$ vanish the (1,1) element also vanish. The vanishing of (1,1) elements can be viewed as a result of some underlying discrete or continuous flavor symmetry [18]. The prediction about the mixing angle obtained from the above texture is almost independent from renormalization-group effects. Therefore there is no need to specify any energy scale at which the above texture is realized.

5. Flavor Mixing Angles and Masses

For a quark mass matrix of the type (9) the following relations hold [18]

$$
\begin{aligned}
X_u &\equiv \left| \frac{|D_u|}{\lambda_1 - E_u} \frac{|D_d| \left(\lambda_3^d - A_d\right)}{|B_d| \left(\lambda_3^d - E_d\right)} + \frac{\lambda_3^d - A_d}{|B_d|} e^{i\varphi_1} + \frac{|B_u|}{\lambda_1^u - A_u} e^{i(\varphi_1 + \varphi_2)} \right| \\
Y_u &\equiv \left| \frac{|D_u|}{\lambda_2 - E_u} \frac{|D_d| \left(\lambda_3^d - A_d\right)}{|B_d| \left(\lambda_3^d - E_d\right)} + \frac{\lambda_3^d - A_d}{|B_d|} e^{i\varphi_1} + \frac{|B_u|}{\lambda_2^u - A_u} e^{i(\varphi_1 + \varphi_2)} \right|
\end{aligned}
\tag{11}
$$

(λ_i, $i = 1, 2, 3$ represent the quark mass eigenvalues and $\varphi_{1,2}$ are related to the phases of $B_{u,d}$, $D_{u,d}$)

Similar relations for down type quarks can be obtained by changing the subscripts in (11) $u \leftrightarrow d$.

$$
\tan\theta_u = \frac{O_{21}^u}{O_{22}^u} \frac{X_u}{Y_u}, \qquad \tan\theta_d = \frac{O_{21}^d}{O_{22}^d} \frac{X_d}{Y_d}
$$

(O_{ij}^q, $ij = 21, 22$, $q = u, d$, are the elements of the matrices used to diagonalize.(9)).

We can define $\frac{|B_q|}{C_q} \equiv r_q$ having magnitude $\vartheta(1)$ with $C_q \neq 0$ for each quark sector. The parameters $A_q, |B_q|, C_q$ and $|D_q|$ can be expressed in terms of quark mass eigenvalues and r_q. Using relations (11) and (12) we can find two mixing angles in terms of the quark masses given by

$$
\tan\theta_u = \sqrt{\frac{m_u}{m_c}} \left(1 + \Delta_u\right), \qquad \tan\theta_d = \sqrt{\frac{m_d}{m_s}} \left(1 + \Delta_d\right)
$$

$$
\Delta_u = \sqrt{\frac{m_c m_d}{m_u m_s} \frac{m_s}{m_b}} \cos\varphi_1, \quad \Delta_d = 0
$$

In leading order the above relations give

$$
\tan\theta_u = \sqrt{\frac{m_u}{m_c}} \tag{12}
$$

$$
\tan\theta_d = \sqrt{\frac{m_d}{m_s}} \tag{13}
$$

6. Predicting the Neutrino Mass

Our understanding of the neutrinos has changed in past few years. Now we know that neutrinos produced in a well defined flavor eigenstate, after traveling some macroscopic distance, appear as a different eigenstate. The simplest answer to this phenomenon is that neutrinos, like all other fermions are massive particles. Their mass eigenstates are different than their flavor eigenstates. This scenario opens up many of the new possibilities which does not exist for the massless neutrinos, such as, massive neutrinos can have nonzero magnetic dipole moments, they can decay to lighter neutrinos and can have a Majorana mass term.[19]

At least three light physical neutrinos with left handed flavor eigenstates are required to explain the current solar and atmospheric neutrino oscillation data. These three neutrino flavor eigenstates are related to the three mass eigenstates by a mixing matrix called lepton mixing matrix or PMNS matrix V.[20]

$$\begin{pmatrix} \nu_e \\ \nu_\mu \\ \nu_\tau \end{pmatrix} = \begin{pmatrix} V_{e1} & V_{e2} & V_{e3} \\ V_{\mu 1} & V_{\mu 2} & V_{\mu 3} \\ V_{\tau 1} & V_{\tau 2} & V_{\tau 3} \end{pmatrix} \begin{pmatrix} \nu_1 \\ \nu_2 \\ \nu_3 \end{pmatrix}$$

Assuming that the neutrino masses are Majorana masses V can be parameterized by the three mixing angles and three complex phases. The standard parameterization of V is similar to the parameterization of quarks and is given as

$$V_{PMNS} = \begin{pmatrix} c_{12}c_{13} & s_{13}c_{12} & s_{13}e^{-\iota\delta_{13}} \\ -s_{12}c_{23} - c_{12}s_{23}s_{13}e^{\iota\delta_{13}} & c_{12}c_{23} - c_{12}s_{23}s_{13}e^{\iota\delta_{13}} & s_{23}c_{13} \\ s_{12}s_{12} - c_{12}c_{23}s_{13}e^{\iota\delta_{13}} & -c_{12}s_{23} - s_{12}c_{23}s_{13}e^{\iota\delta_{13}} & c_{23}c_{13} \end{pmatrix}$$

($c_{ab} = \cos\theta_{ab}$ and $s_{ab} = \sin\theta_{ab}$) with solar angle $= \theta_{12}$, atmospheric angle $= \theta_{23}$, reactor angle $= \theta_{13}$.

The discussion in the previous section for quarks can be extended for the leptons in a straight forward way.[21] If neutrinos are Majorana particles, then the lepton mixing matrix is given by $V = UP$, where

$$U_{PMNS} = \begin{pmatrix} c_l & s_l & 0 \\ -s_l & c_l & 0 \\ 0 & 0 & 1 \end{pmatrix} \begin{pmatrix} e^{-i\varphi} & 0 & 0 \\ 0 & c & s \\ 0 & -s & c \end{pmatrix} \begin{pmatrix} c_\nu & -s_\nu & 0 \\ s_\nu & c_\nu & 0 \\ 0 & 0 & 1 \end{pmatrix} \tag{14}$$

$$U_{PMNS} = \begin{pmatrix} s_l s_\nu c + c_l c_\nu e^{-i\varphi} & s_l c_\nu c - c_l s_\nu e^{-i\varphi} & s_l s \\ c_l s_\nu c - s_l c_\nu e^{-i\varphi} & c_l c_\nu c + s_l s_\nu e^{-i\varphi} & c_l s \\ -s_\nu s & -c_\nu s & c \end{pmatrix} \tag{15}$$

($c_l = \cos\theta_l$, $c_\nu = \cos\theta_\nu$, $s = \sin\theta$). The Majorana phase matrix is given as $P = \{e^{i\rho}, e^{i\sigma}, 1\}$.

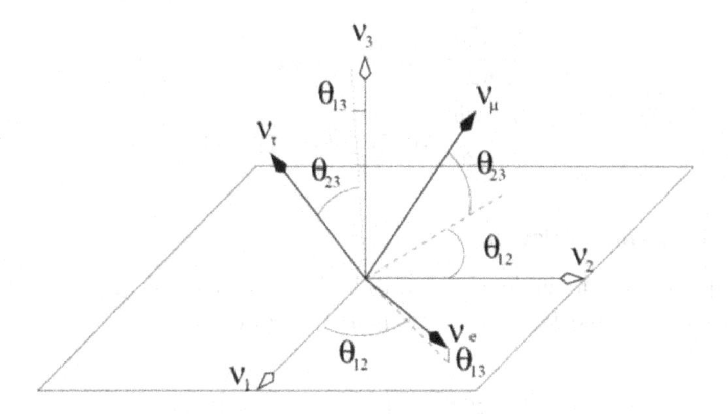

Fig. 1. The relation between the neutrino flavor-mass eigenstates.

These mixing angles have direct physical interpretations angle θ describes the mixing between the second and the third family, θ_l describes the mixing between charged lepton sector ($e - \mu$ mixing), and the angle θ_ν describes the solar neutrino mixing($\nu_e - \nu_\mu$ mixing). In the approximation that solar and atmospheric neutrino oscillations nearly decouple, the solar angle θ_{12}, the atmospheric angle θ_{23} and the CHOOZ angle θ_{13} can be expressed in terms of angles θ_l,θ_ν and θ.

$$\theta_{12} \approx \theta_\nu, \quad \theta_{23} \approx \theta, \quad \theta_{13} \approx \theta_l \sin\theta \tag{16}$$

Neutrino oscillation data can be described in a very convenient way by using this parameterization. In order to relate the neutrino masses with the neutrino mixing angles (similar to the case of quarks), we speculate that the neutrino has the normal mass hierarchy. By using the equation (12), we can write

$$\tan\theta_l = \sqrt{\frac{m_e}{m_\mu}}, \qquad \tan\theta_\nu = \sqrt{\frac{m_1}{m_2}} \tag{17}$$

Using the relations (17) and the fact that neutrino oscillations are associated with the mass squared differences $\Delta m_{21}^2 = m_2^2 - m_1^2$ and $\Delta m_{32}^2 = m_3^2 - m_2^2$, we arrive at the following relations:

$$m_1^2 = \frac{\sin^4\theta_\nu}{\cos 2\theta_\nu}\Delta m_{21}^2 \tag{18}$$

$$m_2^2 = \frac{\cos^4\theta_\nu}{\cos 2\theta_\nu}\Delta m_{21}^2 \tag{19}$$

$$m_3^2 = \frac{\cos^4\theta_\nu}{\cos 2\theta_\nu}\Delta m_{21}^2 + \Delta m_{32}^2$$
$$= \frac{\cos^4\theta_\nu}{\cos 2\theta_\nu}\Delta m_{21}^2 + (\Delta m_{31}^2 - \Delta m_{21}^2)\,. \tag{20}$$

6.1. *Tribimaximal mixing*

It was conjectured independently by Cabibo [22,23] and Wolfenstein [24], that the mixing matrix linking charged leptons to the neutrino could be given by

$$U_{l\nu}^{CW} = \begin{pmatrix} 1 & 1 & 1 \\ 1 & \omega & \omega^2 \\ 1 & \omega^2 & \omega \end{pmatrix}$$

($\omega = \exp\left(\frac{2\pi i}{3}\right) = -\frac{1}{2} + i\frac{\sqrt{3}}{2}$).

In the conventional notation this can be written as

$$\theta_{12} = \theta_{23} = 45°, \quad \theta_{13} = 35.3°, \quad \delta_{CP} = 90°$$

After the discovery of the neutrino oscillations Harrison, Perkins and Scott proposed in 2002 the tribimaximal mixing matrix (or HPS) mixing matrix [25], which describes the current neutrino oscillation data fairly well and is given by

$$U^{TBM} = U_{l\nu}^{HPS} = \begin{pmatrix} \sqrt{\frac{2}{3}} & \frac{1}{\sqrt{3}} & 0 \\ -\frac{1}{\sqrt{6}} & \frac{1}{\sqrt{3}} & -\frac{1}{\sqrt{2}} \\ -\frac{1}{\sqrt{6}} & \frac{1}{\sqrt{3}} & \frac{1}{\sqrt{2}} \end{pmatrix} \tag{21}$$

The tribimaximal mixing leads to $\tan\theta_{12} = \frac{1}{\sqrt{2}} \Rightarrow \theta_{12} \approx 35.26°$. For the atmospheric neutrino oscillations we have $\tan\theta_{23} = 1 \Rightarrow \theta_{23} = 45°$. For the CHOOZ angle we take $\theta_{13} = 0°$.

From the matrices (15) and (21) we can calculate the numerical values of the angles θ_l, θ_ν and θ

$$\cos\theta = \frac{1}{\sqrt{2}} \Longrightarrow \theta_{23} \approx \theta = 45°$$

$$-\sin\theta_\nu \sin\theta = \frac{1}{\sqrt{6}} \Longrightarrow \theta_{12} \approx \theta_\nu = 35.26°$$

$$\cos\theta_l \sin\theta = -\frac{1}{\sqrt{2}} \Longrightarrow \theta_l = \pi\,, \quad \theta_{13} \approx \theta_l \sin\theta = 2.22°$$

Using the best fit values of the mass splitting parameters $\triangle m_{21}^2 \approx 7.65 \times 10^{-5}$ eV^2, $\triangle m_{31}^2 \approx 2.40 \times 10^{-3}$eV^2 [26] and the mixing angle $\theta_\nu \approx 35°$, we can predict the masses of the neutrinos using relations (19) (19) (20)

$$m_1 = \sqrt{\frac{\sin^4\theta_\nu}{\cos 2\theta_\nu}\triangle m_{21}^2}$$

$$\approx 0.005 \text{ eV}$$

$$m_2 = \sqrt{\frac{\cos^4\theta_\nu}{\cos 2\theta_\nu}\triangle m_{21}^2}$$

$$\approx 0.01 \text{ eV}$$

$$m_3 = \sqrt{\frac{\cos^4\theta_\nu}{\cos 2\theta_\nu}\triangle m_{21}^2 + (\triangle m_{31}^2 - \triangle m_{21}^2)}$$

$$\approx 0.05 \text{ eV}$$

The analysis shows that the neutrinos have normal but weak hierarchy i.e. $m_1 : m_2 : m_3 = 1 : 2 : 10$. Neutrino masses obtained using tribimaximal mixing agree with those obtained by Fritzsch and Xing[21].

7. Triminimal Parameterization

Even if tribimaximal mixing, in the lepton sector is the result of some flavor symmetry, in general there will be deviations from this scheme.[27,28] The triminiml parameterization is a completely general scheme of the PMNS matrix, treating Tribimaximal mixing as the zeroth order basis. Four independenent parameters of the U_{MNS} are given by ϵ_{jk}, $jk = 21, 32, 13$, which are the deviations of the θ_{jk} from their tribimaximal values and CP- violating phase δ. One can obtain the usual tribimaximal mixing by taking $\theta_{jk} = 0$. The triminimal parameterization is given by

$$U_{TMin} = R_{32}\left(\frac{\pi}{3}\right) U_\epsilon\left(\epsilon_{32}\,,\,\epsilon_{13}\,,\,\epsilon_{21},\,\delta\right) R_{21}\left(\sin^{-1}\frac{1}{\sqrt{3}}\right)$$

$$U_\epsilon = R_{32}\left(\epsilon_{32}\right) U_\delta^\dagger R_{13}\left(\epsilon_{13}\right) U_\delta R_{21}\left(\epsilon_{21}\right) \tag{22}$$

$$U_{TMin} = \begin{pmatrix} \sqrt{2} & 0 & 0 \\ 0 & 1 & 1 \\ 0 & -1 & 1 \end{pmatrix} \frac{U_\epsilon}{\sqrt{6}} \begin{pmatrix} \sqrt{2} & 1 & 0 \\ -1 & \sqrt{2} & 0 \\ 0 & -0 & \sqrt{3} \end{pmatrix}$$

The neutrino observables up to second order in ϵ_{jk} can be expressed in terms of the triminimal parameters as

$$
\begin{aligned}
\sin^2\theta_{21} &= \frac{1}{3}\left(\cos\epsilon_{21} + \sqrt{2}\sin\epsilon_{21}\right)^2 \\
&\simeq \frac{1}{3} + \frac{2\sqrt{2}}{3}\epsilon_{21} + \frac{1}{3}\epsilon_{21}^2 \\
\sin^2\theta_{23} &= \frac{1}{2} + \sin\epsilon_{32}\cos\epsilon_{32} \\
&\simeq \frac{1}{2} + \epsilon_{32} \\
U_{e3} &= \sin\epsilon_{13}e^{-i\delta}
\end{aligned}
\tag{23}
$$

(This means $\theta_{23} = \epsilon_{13}$ and $\delta = \delta$)

Using equations (23) and (19–20), one can obtain the neutrino mass relations in terms of the deviations from solar angle ϵ_{12}:

$$
\begin{aligned}
m_1 &\approx \sqrt{\frac{1 + 4\sqrt{2}\,\epsilon_{12}}{3 - 12\sqrt{2}\,\epsilon_{12}}\Delta m_{21}^2} \\
m_2 &\approx \sqrt{\frac{4 - 8\sqrt{2}\,\epsilon_{12}}{3 - 12\sqrt{2}\,\epsilon_{12}}\Delta m_{21}^2} \\
m_3 &\approx \sqrt{\frac{4 - 8\sqrt{2}\,\epsilon_{12}}{3 - 12\sqrt{2}\,\epsilon_{12}}\Delta m_{21}^2 + \Delta m_{32}^2}
\end{aligned}
\tag{24}
$$

In the limit $\epsilon_{12} \to 0$ using the best fit values of the mass squared differences $\Delta m_{21}^2 \approx 7.65 \times 10^{-5}$ eV^2, $\Delta m_{31}^2 \approx 2.40 \times 10^{-3}$eV^2 [26], we obtain the neutrino masses for the tribimaximal mixing (TBM):

$$
m_1 \approx \sqrt{\frac{1}{3}\Delta m_{21}^2} \approx 0.00504 \text{ eV}
$$

$$
m_2 \approx \sqrt{\frac{4}{3}\Delta m_{21}^2} \approx 0.010 \text{ eV}
$$

$$
m_3 \approx \sqrt{\frac{4}{3}\Delta m_{21}^2 + \Delta m_{32}^2} \approx 0.05 \text{ eV}
$$

The behavior of the neutrino masses in the vicinity of tribimaximal mixing (TBM) ($\epsilon_{12} = 0$) is shown by the figure 2.

Four zero texture relates the fermions mass ratios and the mixing angles in the quark and lepton sectors [18,21]. Using tribimaximal mixing predictions about the neutrino mass are made. The model predicts the consistent values for the quarks and charged leptons. For neutrino masses we can use the triminimal mixing

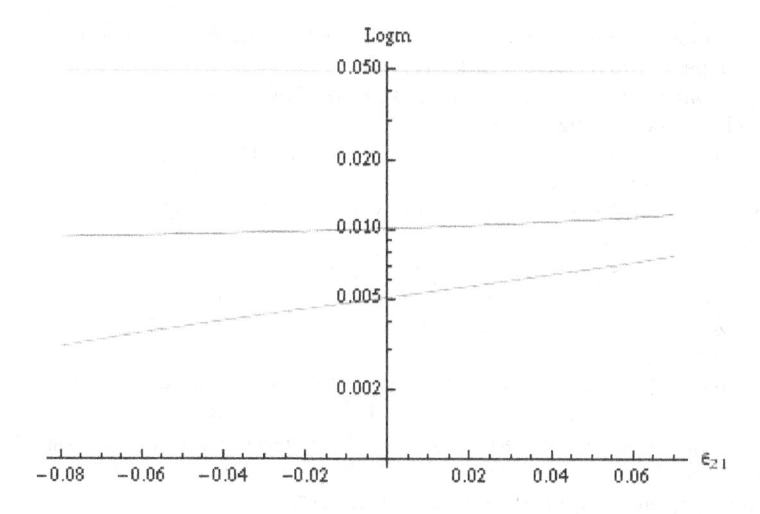

Fig. 2. Behavior of m1, m2,m3 in the vicinity of TBM.

matrix and determine the values of neutrino masses for tribimaximal mixing given by $m_1 \approx 0.005$ eV, $m_1 \approx 0.01$ eV and $m_1 \approx 0.00504$ eV. This shows that neutrinos have weak but normal hierarchy. We have also studied the effect on neutrino masses when the value of solar angle is changed from its TBM value. Our analysis shows that the neutrinos masses m_1, m_2 in general tend to increase as the value of solar angle increases from its TBM value and vice versa. However m_3 seems to be rather stable against the change in solar angle. It appears that the neutrinos become more and more degenerate for solar angle values higher than TBM value and hierarchical for lower values of solar angle.

Acknowledgments

I am grateful to Prof. Dr. Harald Fritzsch for useful discussions. I also highly appreciate the director NEC NTU Singapore for partial travel funding to attend the conference.

References

1. S. Glashow, Nuclear Phys. 22 (1961) 579; S. weinberg, Phys. Rev. Lett. 19 (1967) 1264 A. Salam, in N. Svartholm (Ed), Elementary Particle theory, Almqvist and Wiksells, Stockholm 1969, p. 367.
2. M. Gell-Mmann, Phys. Lett. 8 (1964) 214; G. Zweig, CERN-Report 8182/TH401(1964); H. Fritzsch, M. Gell-Mann and H. Leutwyler, Phys. Lett. B 47 (1973) 365; D. Gross and F. Wilczek, Phys. Rev. Lett. 30 (1973) 1343; H. D. Politzer, Phys. Rev. Lett. 30 (1976) 1346.
3. K. S. Babu, arXiv:org/abs/0910.2948v1.
4. The LEP Collaborations (ALEPH, DELPHI, L3 and OPAL), the LEP Electroweak Working Group and the SLD Heavy Flavour Group, A combination of preliminary

Electroweak measurements and constraints on the Standard Model, hep-ex/0412015; http://lepewwg.web.cern.ch/LEPEWWG.
5. Particle Data Group, K. Hagiwara et al., Phys. Rev. D 66 (2002) 010001; S. Eidelman et al., Phys. Let. B 592 (2004) 1.
6. C. Giunti and C. W. Kim, Fundamentals of Neutrino Physics and Astrophysics, Oxford University Press (2007).
7. R. N. Mohapatra and A. Y. Smirnov, Ann. Rev. Nucl. Part. Sci. 56:569–628, 2006; arXiv:hep-ph/0603118v2.
8. S. Weinberg, Phys. Rev. Lett. 43, 1566 (1979).
9. H. Fritzsch and Z. Z. Xing, Phys. Lett. B 413 (1997) 396–404.
10. H. Fritzsch, Phys. Lett. B 70 (1977) 436; B 73 (1978) 317.; H. Fritzsch, Nucl. Phys. B 155 (1979) 189.
11. S. Weinberg, Transactions of the New York Academy of Sciences, Series II, 38 (1977) 185; F. Wilczck and A. Zee, Phys. Lett. B 70 (1977) 418.
12. H. Fritzsch, Phys. Lett. B 184 (1987) 391; H. Fritzsch, Phys. Lett. B 189 (1987) 191.
13. L. J. Hall and S. Weinberg, Phys. Rev. D 48 (1993) 979.
14. M. Neubert, Int. J. Mod. Phys. A 11 (1996) 4173.
15. R. Forty, talk given at the Second International Conference on B Physics and CP Violation, Honolulu, Hawaii, March 24-27, 1997
16. G. C. Branco, L. Lavoura and F. Mota, Phys. Rev. D 39 (1989) 3443.
17. K. S. Babu and Q. Shafi, Phys. Rev. D 47 (1993) 5004; and references therein.
18. H. Fritzsch and Z. Z. Xing, Nucl. Phys. B 556 (1999) 49.
19. R. N. Mohapatra et al., 2007 Rep. Prog. Phys. 70, 1757–1867.
20. Z. Maki, M. Nakagawa and S. Sakata, Prog. Theo. Phys. 28 (1962) 247; B. W. Lee, S. Pakvasa, R. E. Shrock and H. Sugawara, Phys. Rev. Lett. 38 (1977) 937 [Erratum-ibid. 38 (1977) 1230].
21. A. Aziz Bhatti and Harald Fritzsch, Mod. Phys. Lett. A 29, 1450059 (2014); H. Fritzsch and Z. Z. Xing, Phys. Lett. B 634 (2006) 514.
22. N. Cabibbo, Phys. Lett. B 72, 333 (1978).
23. Ernest Ma, arXiv:0908.1770v1.
24. L. Wolfenstein, Phys. Rev. D 18, 958 (1978).
25. P. F. Harrison, D. H. Perkins and W. G. Scott, Phys. Lett. B 530, 167 (2002).
26. Thomas Schwetz et al 2008 New J. Phys. 10 113011 (10pp).
27. S. Pakvasa, W. Rodejohann and T. J. Weiler, Phys. Rev. Lett. 100, 111801 (2008).
28. F. Plentinger and W. Rodejohann, Phys. Lett. B 625, 264 (2005); S. Antusch and S. F. King, Phys. Lett. B 631, 42 (2005); S. Luo and Z. Z. Xing, Phys. Lett. B 632, 341 (2006); N. Haba, A. Watanabe and K. Yoshioka, Phys. Rev. Lett. 97, 041601 (2006); M. Hirsch, E. Ma, J. C. Romao, J. W. F. Valle and A. Villanova del Moral, Phys. Rev. D 75, 053006 (2007); X. G. He and A. Zee, Phys. Lett. B 645, 427 (2007); A. Dighe, S. Goswami and W. Rodejohann, Phys. Rev. D 75, 073023 (2007); S. Antusch, P. Huber, S. F. King and T. Schwetz, JHEP 0704, 060 (2007); M. Lindner and W. Rodejohann, JHEP0705, 089 (2007). K. A. Hochmuth, S. T. Petcov and W. Rodejohann, Phys. Lett. B 654, 177 (2007).

Radiative Neutrino Mass Generation: Models, Flavour and the LHC*

Raymond R. Volkas

ARC Centre of Excellence for Particle Physics at the Terascale,
School of Physics, The University of Melbourne, Victoria 3010, Australia
raymondv@unimelb.edu.au

Radiative neutrino mass models and the seesaw models are viewed from the unifying framework of standard model effective operators that explicitly violate lepton number by two units ($\Delta L = 2$). After some comments on naturalness and leptogenesis in the minimal type 1 seesaw model, a full list of minimal renormalisable models that produce mass dimension-7, $\Delta L = 2$ operators at low energies is presented. By way of example, phenomenological bounds from Run 1 LHC and lepton flavour violation data are then placed on one of these models. A possible connection between radiative neutrino mass models and the current flavour anomalies in $b \to c$ and $b \to s$ transitions is then described.

Keywords: Neutrino mass; radiative mass generation; seesaw models; LHC constraints; lepton flavour violation; flavour anomalies.

1. Introduction

The origin of neutrino mass is a very important open problem in high-energy physics. While experiments are now measuring the squared-mass differences and leptonic mixing angles with good precision, the possibilities for neutrino mass theories remain extremely broad. From the simple but unsatisfying case of Dirac neutrino masses generated in exactly the same way as the charged fermions, through various types of tree-level seesaw mechanisms for Majorana neutrinos, to the many different radiative models, the viable "theory space" is huge. Because of this, it is useful to have an organising principle through which a systematic search strategy can be constructed. At least for the Majorana case, a useful starting point is the set of standard model (SM) gauge-invariant effective operators that violate lepton number by two units.

*Based on a talk presented at the Conference on Cosmology, Gravitational Wave and Particles, 6–10 February 2017, NTU, Singapore.

These operators exist at odd mass dimension. A full list of operators at mass dimension-5, -7, -9 and -11 that contain fermion fields and the Higgs doublet only has been provided by Babu and Leung.[1] Dimension-7 operators that also contain SM gauge fields are listed in Ref. 2. At dimension-5, there is only one operator up to family replication:

$$O_1 = LLHH\,, \tag{1}$$

where L and H are the left-handed lepton and Higgs doublets, respectively. Replacing H with its vacuum expectation value, $\langle H \rangle = v$, immediately yields the famous seesaw formula

$$m_\nu \sim v^2/M\,, \tag{2}$$

where M is the scale of new physics. In the regime $M \gg v$, the smallness of the neutrino masses relative to the weak scale obtains a natural explanation. By "opening up" the operator O_1 in all possible minimal ways at tree-level, one obtains the underlying, renormalisable type-1, -2 and -3 seesaw models,[3] where the new physics mass scale M is related to, respectively, fermion singlet, Higgs triplet and fermion triplet masses. By starting with the effective operator, one is guaranteed to construct *all* the parent theories (subject to additional e.g. minimality assumptions), so the process is a systematic one.

At mass dimension-7 and higher, almost all of the effective operators contain fields other than neutrinos and Higgs bosons, the exceptions being the O_1 generalisations $LLHH(\bar{H}H)^n$. While each such operator may be opened up at tree-level, the operator itself does not immediately produce a neutrino mass, unlike the case of O_1 and its generalisations. Instead, once needs to construct a neutrino self-energy graph from the opened-up operator together with SM Yukawa interactions. Because these effective operators contain charged-lepton and/or quark external lines, the neutrino self-energy diagram is necessarily at loop order, with the SM charged fermion lines becoming those of virtual particles, joining the virtual exotics used to open up the operator in the first place. Figure 1 illustrates this process for the historically important case of the 2-loop Zee–Babu model.[4] The exotic scalars k^{++} and h^{+} can be searched for at the LHC (see Ref. 5 for the latest ATLAS and CMS bounds on doubly-charged scalars).

One (good) reason for being interested in radiative models is that they are typically more testable than seesaw models. While the type-2 and -3 seesaw models introduce exotics that feel electroweak interactions and can thus be produced at colliders, the seesaw idea in its purest form requires the scale of new physics to be very high. Using $m_\nu \sim 0.1$ eV in Eq. (2) produces $M \sim 10^{14}$ GeV. Of course, there are Yukawa couplings constants in the numerator, so the scale can be made lower, reaching the testable 100 GeV range for Yukawas of order 10^{-6} (which is similar to that of the electron). But, it would hardly be a surprise if the new seesaw physics was quite out of reach of the LHC or any conceivable future machine. Radiative models require a lower scale of new physics, and generally (but not inevitably) the

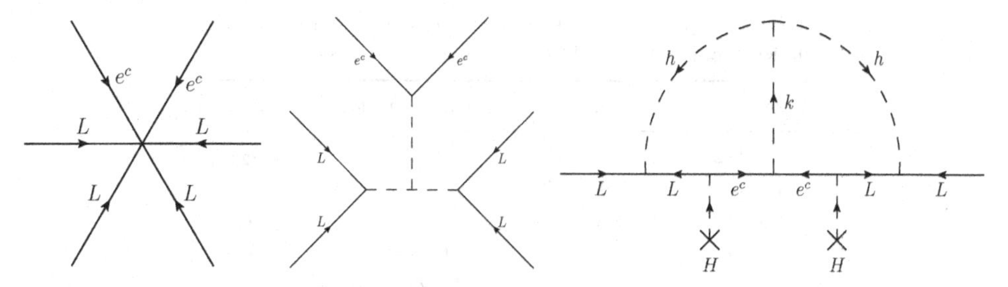

Fig. 1. The $\Delta L = 2$ operator $O_9 = LLLe^cLe^c$ is depicted in the leftmost graph, where L is a lepton doublet and e is the charged-lepton singlet. The middle graph shows one way to open up this operator at tree level. It features a doubly-charged scalar k Yukawa-coupling to e^ce^c and a singly-charged scalar h coupling to LL. Both scalars are isosinglets. Invoking SM Yukawa interactions in addition to the opened-up operator, a self-energy graph for Majorana neutrino mass is obtained in the rightmost plot at 2-loop level.

higher the loop-order at which neutrino masses arise, the lower the scale has to be. Three-loop models are the most likely general class to require TeV-scale new physics, with 4-loop models being marginal at best. (For an interesting 1-loop model that is very testable, see Ref. 6.)

2. Naturalness and Leptogenesis in the Type-1 Seesaw Model

The type-1 seesaw model is a minimal and compelling extension of the SM. Unfortunately, its simplest version is notoriously immune from experimental test, both because the scale of the right-handed Majorana neutrinos is expected to be very high, and because these gauge-singlet neutrinos barely interact with other species. Can one cite any problem with the type-1 seesaw model to motivate more testable theories such as radiative models? There is at least one problem with the minimal model: a potential violation of naturalness due to the destabilisation of the weak scale by the high seesaw scale.[7,8] Computing the 1-loop correction to the μ^2 parameter in the SM Higgs potential and demanding on naturalness grounds that it be less than $(1\text{ TeV})^2$, one obtains the upper bounds[8]

$$M_{N_1} \lesssim 4 \times 10^7 \text{ GeV}\,, \quad M_{N_2} \lesssim 7 \times 10^7 \text{ GeV}\,, \quad M_{N_3} \lesssim 3 \times 10^7 \text{ GeV}\left(\frac{0.05 \text{ eV}}{m_{\text{min}}}\right), \tag{3}$$

where $N_{1,2,3}$ are the heavy neutral fermions in ascending mass order. These upper bounds somewhat compromise the spirit of the seesaw mechanism, requiring Yukawa coupling constants of order 10^{-4} or less. They also are at odds with hierarchical, thermal leptogenesis,[9] which require heavy neutral fermion masses above 5×10^8 GeV.[10] (The N_3 case is not immune, because as the smallest active neutrino mass m_{min} is made smaller, the more decoupled N_3 becomes.) One may avoid this conflict between naturalness and hierarchical, thermal leptogenesis by extending the minimal type-1 seesaw model using, for example, supersymmetry, or two Higgs doublets without supersymmetry.[11]

Table 1. Quantum numbers of exotic scalars and fermions in underlying theories for the mass dimension-7 operators.

Scalar	Scalar	Dirac fermion	Operator
$(1,2,1/2)$	$(1,1,1)$		$O_{2,3,4}$
$(3,2,1/6)$	$(3,1,-1/3)$		$O_{3,8}$
$(3,2,1/6)$	$(3,3,-1/3)$		O_3
$(1,1,1)$		$(1,2,-3/2)$	O_2
$(1,1,1)$		$(3,2,-5/6)$	O_3
$(1,1,1)$		$(3,1,2/3)$	O_3
$(3,2,1/6)$		$(3,2,-5/6)$	O_3
$\mathbf{(3,1,-1/3)}$		$\mathbf{(3,2,-5/6)}$	$\boldsymbol{O_{3,8}}$
$(3,3,-1/3)$		$(3,2,-5/6)$	O_3
$(3,2,1/6)$		$(3,3,2/3)$	O_3
$(1,1,1)$		$(3,2,7/6)$	O_4
$(1,1,1)$		$(3,1,-1/3)$	O_4
$(3,2,1/6)$		$(3,2,7/6)$	O_8
$(3,2,1/6)$		$(1,2,-1/2)$	O_8

3. Neutrino Mass Models from Mass Dimension-7 Operators

We now proceed to examine the mass dimension-7 $\Delta L = 2$ operators,[1]

$$O_2 = LLLe^cH\,, \quad O_3 = LLQd^cH\,, \quad O_4 = LL\bar{Q}\bar{u}^cH\,, \quad O_8 = L\bar{e}^c\bar{u}^cd^cH\,, \qquad (4)$$

to see what underlying, minimal renormalisable models they imply. (There is also the generalised Weinberg operator $LLHH\bar{H}H$, and operators containing gauge fields,[2] which are interesting to consider but beyond the scope of this talk.) The above equation specifies the flavour content of each operator, and it should be noted that the O_3 and O_4 cases encompass two independent operators each, because of different index contraction possibilities.

Our definition of "minimal" for the underlying theories is that the exotic particles integrated out to obtain the effective operators are either exotic scalars or vector-like fermions. (Exotic Majorana fermions turn out to be irrelevant for the mass dimension-7 cases.) The full list of theories was constructed in Ref. 12, building on Ref. 1, and the particle content of each is summarised in Table 1.

The phenomenological bounds on the model in boldface in the table were derived in Ref. 12. The massive, vector-like Dirac fermion transforming as $(3,2,-5/6)$ contains a charge $-1/3$ quark b' in its upper component and a rather exotic charge $-4/3$ quark in its lower component. The latter can easily be arranged to be unstable through the charged-current interaction. The former will mix with the standard charge $-1/3$ quarks. For the purposes of neutrino mass generation, it suffices to consider mixing with the standard b quark. Two mass eigenstates are produced, with the more massive one dominated by a b' admixture, and the less massive one, which must be identified with the physical bottom quark, is of course predominantly a b. CMS has searched for massive b'-like quarks,[13] and from run 1 data the lower bound is about 620 GeV. Additional collider bounds come from leptoquark searches which constrain the $(3, 1, -1/3)$ scalar. Depending on the parameter space of the model and the decay modes considered, run 1 LHC bounds[14] in the range 520–730 GeV were derived.[12] Finally, there are important constraints from the charged-lepton flavour-violating processes $\mu \to e\gamma$, $\mu \to eee$ and μ to e conversion on nuclei; see Ref. 12 for the relevant plots.

4. Flavour Anomalies

The discussion in this section is an updated and expanded version of what was presented at the workshop.

In recent years, some very interesting anomalies have been found in the decays of B-mesons. If confirmed by the up-coming Belle 2 experiment and future analyses by LHCb, then this will be the discovery of new physics at the precision frontier. These anomalies are extremely relevant for radiative neutrino mass models, because, for example, leptoquark scalars may contribute wholly or partially to the new physics, if new physics there be. One set of anomalies concerns semileptonic $b \to s$ transition measurements by the LHCb collaboration. The ratios

$$R_{K^{(*)}} \equiv \frac{\Gamma(B \to K^{(*)}\mu^+\mu^-)}{\Gamma(B \to K^{(*)}e^+e^-)} \tag{5}$$

have been measured to be[15]

$$R_K = 0.745^{+0.090}_{-0.074} \pm 0.036\,, \tag{6}$$

and[16]

$$R_{K^*} = \begin{cases} 0.660^{+0.110}_{-0.070} \pm 0.024 & \text{for } 0.045 \text{ GeV}^2 < q^2 < 1.1 \text{ GeV}^2 \\ 0.685^{+0.113}_{-0.069} \pm 0.047 & \text{for } 1.1 \text{ GeV}^2 < q^2 < 6 \text{ GeV}^2\,, \end{cases} \tag{7}$$

suggesting a violation of lepton flavour universality (LFU), since the SM expectation is very close to unity for all of these ratios. Each of these discrepancies is at about the 2.5σ level, and systematic errors from hadronic effects cancel out in the ratios. There are also interesting anomalies in other branching ratios and some angular observables; see Ref. 17 for a recent survey. The combined significance of these anomalies has been estimated to be anywhere from 4σ to almost 6σ.[19]

Another set of anomalies concerns semileptonic $b \to c$ transitions in the ratios

$$R_{D^{(*)}} \equiv \frac{\Gamma(B \to D^{(*)}\tau\nu)}{\Gamma(B \to D^{(*)}\ell\nu)}\,, \tag{8}$$

where $\ell = e$, μ, again pointing to violation of LFU. In the SM, these ratios differ from unity only because of phase space suppression due to the large mass of the τ relative to μ and e, and the calculations are reliable because hadronic uncertainties, once again, cancel out. The SM values are computed to be 0.299 ± 0.011 for R_D

and 0.252 ± 0.003 for $R_{D^{(*)}}$.[18] Experimental measurements have been performed by Babar and Belle for both R_D and R_{D^*} and LHCb for R_{D^*}.[20] Babar and LHCb are in good agreement with each other and are significantly discrepant from the SM, while Belle's results lie between the SM prediction and the Babar/LHCb figures and are consistent with both. The HFAG have done a combined fit to obtain $0.397 \pm 0.040 \pm 0.028$ and $0.316 \pm 0.016 \pm 0.010$ for R_D and R_{D^*}, respectively, which is about a 4σ discrepancy from the SM.[21] By incorporating the most recent result for R_{D^*} by Belle, we have estimated that the combined average for that quantity is now 0.311 ± 0.016.[22]

It has been suggested that the $(3, 1, -1/3)$ scalar leptoquark that, for example, appears in the radiative neutrino mass models of Refs. 12 and 23 might play some role in resolving the anomalies.[24] We have recently reconsidered this possibility, incorporating all relevant phenomenological bounds, and find that such a new particle can ameliorate both the $b \to c$ anomalies (which it contributes to at tree-level) and the $b \to s$ discrepancies (via a 1-loop effect) to within 2σ of the experimentally favoured values, which is a significant improvement over the SM.[22] When also required to help generate neutrino mass in the model of Ref. 23, the parameter space becomes so restricted that it can explain the $b \to c$ anomalies only. Another scalar leptoquark, with quantum numbers $(3, 3, -1/3)$, has been widely mentioned as of relevance, because one component of it can contribute to the $b \to s$ processes at tree-level.[25] Radiative models featuring both of these particles are possible.[22,26]

5. Final Remarks

Radiative neutrino mass models form a large class of viable and relatively testable theories resolving one of the major problems of fundamental physics: the origin of neutrino masses and mixings. Some of them can also help resolve some, or all, of the flavour anomalies in semileptonic $b \to s$ and $b \to c$ transitions that point to violations of lepton flavour universality. Although they were well beyond the scope of this talk, it is worth mentioning that there are other radiative models that also incorporate dark matter.[27] Our focus was on the generation of Majorana rather than Dirac masses, thus tying these lepton number violating theories to the baryogenesis problem as well, for good or ill (the latter because of the prospect of washout). Curious connections evidently may exist between the three great empirical proofs of the need for new physics: neutrino masses, dark matter and the matter–antimatter asymmetry of the universe, augmented now by the indications, not yet proof, of lepton flavour universality violation in semileptonic B-meson decays.

Acknowledgments

This work was supported in part by the Australian Research Council. I would like to thank collaborators P. Angel, Yi Cai, J. Clarke, J. Gargalionis, N. Rodd and M. A. Schmidt, and H. Fritzsch for the invitation to the workshop.

References

1. K. S. Babu and C. N. Leung, *Nucl. Phys. B* **619**, 667 (2001); See also, A. de Gouvêa and J. Jenkins, *Phys. Rev. D* **77**, 013008 (2008).
2. S. Bhattacharya and J. Wudka, *Phys. Rev. D* **94**, 055022 (2016) [Erratum: *ibid.* **95**, 039904 (2017)].
3. P. Minkowski, *Phys. Lett. B* **67**, 421 (1977); T. Yanagida, in *Proceedings of the Workshop on Unified Theories and Baryon Number in the Universe*, eds. O. Sawada and A. Sugamoto (KEK, Tsukuba, 1979); S. L. Glashow, in *Proceedings of the 1979 Cargèse Summer Institute on Quarks and Leptons* (Plenum Press, New York, 1980); M. Gell-Mann, P. Ramond and R. Slansky, *Supergravity*, eds. P. van Nieuwenhuizen and D. Z. Freedman (North-Holland, Amsterdam, 1979); R. N. Mohapatra and G. Senjanović, *Phys. Rev. Lett.* **44**, 912 (1980); M. Magg and C. Wetterich, *Phys. Lett. B* **94**, 61 (1980); J. Schechter and J. Valle, *Phys. Rev. D* **22**, 2227 (1980); T. Cheng and L.-F. Li, *Phys. Rev. D* **22**, 2860 (1980); C. Wetterich, *Nucl. Phys. B* **187**, 343 (1981); G. Lazarides, Q. Shafi and C. Wetterich, *Nucl. Phys. B* **181**, 287 (1981); R. N. Mohapatra and G. Senjanović, *Phys. Rev. D* **23**, 165 (1981); R. Foot, H. Lew, X.-G. He and G. C. Joshi, *Z. Phys. C* **44**, 441 (1989).
4. A. Zee, *Phys. Lett. B* **161**, 141 (1985); *Nucl. Phys. B* **264**, 99 (1986); K. S. Babu, *Phys. Lett. B* **203**, 132 (1988).
5. ATLAS Collab., ATLAS-CONF-2016-051; CMS Collab., CMS-PAS-HIG-16-036.
6. J. Herrero-García *et al.*, arXiv:1701.05345.
7. F. Vissani, *Phys. Rev. D* **57**, 7027 (1998).
8. J. D. Clarke, R. Foot and R. R. Volkas, *Phys. Rev. D* **91**, 073009 (2015).
9. M. Fukugita and T. Yanagida, *Phys. Lett. B* **174**, 45 (1986).
10. S. Davidson and A. Ibarra, *Phys. Lett. B* **535**, 25 (2002); G. Giudice *et al.*, *Nucl. Phys. B* **685**, 89 (2004).
11. J. D. Clarke, R. Foot and R. R. Volkas, *Phys. Rev. D* **92**, 033006 (2015).
12. Y. Cai, J. D. Clarke, M. A. Schmidt and R. R. Volkas, *J. High Energy Phys.* **1502**, 161 (2015).
13. CMS Collab., Tech. Rep. CMS-PAS-B2G-12-019 (2012); Tech. Rep. CMS-PAS-B2G-12-021 (2013); Tech. Rep. CMS-PAS-B2G-13-003 (2013); S. Chatrchyan *et al.*, *Phys. Lett. B* **729**, 149 (2014).
14. G. Aad *et al.*, *J. High Energy Phys.* **1310**, 189 (2013); *J. High Energy Phys.* **06**, 124 (2014); CMS Collab., Tech. Rep. CMS-PAS-SUS-13-018 (2014); ATLAS Collab., Tech. Rep. ATLAS-CONF-2013-037 (2013); S. Chatrchyan *et al.*, *Eur. Phys. J. C* **73**, 2677 (2013).
15. LHCb Collab. (R. Aaij *et al.*), *Phys. Rev. Lett.* **113**, 151601 (2014).
16. for the LHCb Collab. (S. Bifani), CERN Seminar, 18 April 2017.
17. W. Altmannshofer and D. M. Straub, *Eur. Phys. J. C* **75**, 382 (2015).
18. MILC Collab. (J. A. Bailey *et al.*), *Phys. Rev. D* **92**, 034506 (2015); M. Tanaka and R. Watanabe, *Phys. Rev. D* **87**, 034028 (2013).
19. W. Altmannshofer, P. Stangl and D. M. Straub, arXiv:1704.05435; B. Capdevila, A. Crivellin, S. Descotes-Genon, J. Matias and J. Virto, arXiv:1704.05340.
20. BaBar Collab. (J. P. Lees *et al.*), *Phys. Rev. Lett.* **109**, 101802 (2012); *Phys. Rev. D* **88**, 072012 (2013); Belle Collab. (M. Huschle *et al.*), *Phys. Rev. D* **92**, 072014 (2015); Belle Collab. (Y. Sato *et al.*), *Phys. Rev. D* **94**, 072007 (2016); Belle Collab. (S. Hirose *et al.*), arXiv:1612.00529; LHCb Collab. (R. Aaij *et al.*), *Phys. Rev. Lett.* **115**, 111803 (2015).
21. Heavy Flavor Averaging Group Collab. (Y. Amhis *et al.*), arXiv:1207.1158; See also, M. Freytsis, Z. Ligeti and J. T. Ruderman, *Phys. Rev. D* **92**, 054018 (2015).

22. Y. Cai, J. Gargalionis, M. A. Schmidt and R. R. Volkas, arXiv:1704.05849.
23. P. W. Angel, Y. Cai, N. L. Rodd, M. A. Schmidt and R. R. Volkas, *J. High Energy Phys.* **10**, 118 (2013); See also, P. W. Angel, N. L. Rodd and R. R. Volkas, *Phys. Rev. D* **87**, 073007 (2013).
24. M. Bauer and M. Neubert, *Phys. Rev. Lett.* **116**, 141802 (2016).
25. See, for example, G. Hiller and I. Nišandžić, arXiv:1704.05444; A. Crivellin, D. Müller and T. Ota, arXiv:1703.09226.
26. I. E. Bigaran, J. Gargalionis and R. R. Volkas, work in progress.
27. E. Ma, *Phys. Rev. D* **73**, 077301 (2006); C. Boehm *et al.*, *Phys. Rev. D* **77**, 043516 (2008).

Neutrinoless Double-Beta Decays: New Insights

Z. Z. Xing

Institute of High Energy Physics & School of Physical Sciences, University of Chinese Academy of Sciences, Beijing 100049, China
Center for High Energy Physics, Peking University, Beijing 100080, China
xingzz@ihep.ac.cn

Z. H. Zhao

Department of Physics, Liaoning Normal University, Dalian 116029, China
zhzhao@itp.ac.cn

We give some new insights into the effective Majorana neutrino mass $\langle m \rangle_{ee}$ responsible for the neutrinoless double-beta ($0\nu 2\beta$) decays. We put forward a three-dimensional way of plotting $|\langle m \rangle_{ee}|$ against the lightest neutrino mass and the Majorana phases, which can provide more information as compared with the two-dimensional one. With the help of such graphs we discover a novel threshold of $|\langle m \rangle_{ee}|$ in terms of the neutrino masses and flavor mixing angles: $|\langle m \rangle_{ee}|_* = m_3 \sin^2 \theta_{13}$ in connection with $\tan \theta_{12} = \sqrt{m_1/m_2}$ and $\rho = \pi$, which can be used to signify observability of the future $0\nu 2\beta$-decay experiments. Fortunately, the possibility of $|\langle m \rangle_{ee}| < |\langle m \rangle_{ee}|_*$ turns out to be very small, promising a hopeful prospect for the $0\nu 2\beta$-decay searches.

Keywords: Majorana neutrino; neutrinoless double-beta decay; effective Majorana neutrino mass.

1. Introduction

The experimental discoveries of solar, atmospheric, reactor and accelerator neutrino oscillations[1] establish the fact that neutrinos are massive and mix among different flavors. While the neutrino mixing pattern has been figured out more or less, the (Dirac or Majorana) nature of massive neutrinos remains an open question in particle physics. Actually, neutrinos are currently the only realistic candidate for the Majorana fermions — a new form of matter proposed by Majorana.[2] In particular after the advent of the seesaw mechanism which allows for a natural explanation of the smallness of neutrino masses, neutrinos are widely believed to be really of the Majorana nature. If this is the case, then some even–even nuclei may undergo the neutrinoless double-beta ($0\nu 2\beta$) decays[3] (i.e. $N(A, Z) \rightarrow N(A, Z+2) + 2e^-$

with a two-unit violation of the lepton number). The $0\nu2\beta$ decay thus serves as a unique and feasible process for demonstrating the Majorana nature of massive neutrinos, considering that the other lepton-number-violating processes are more desperately suppressed or more experimentally inaccessible. In this respect a number of ambitious experiments are either underway or in preparation.[4]

In the standard three-neutrino scheme, the $0\nu2\beta$-decay rate is controlled by the size of the effective Majorana neutrino mass term[5,a]

$$\langle m \rangle_{ee} = m_1 |U_{e1}|^2 e^{i\rho} + m_2 |U_{e2}|^2 + m_3 |U_{e3}|^2 e^{i\sigma} , \quad (1)$$

where m_i is the ith neutrino mass (for $i = 1, 2, 3$), U_{ei} stands for the corresponding element of the 3×3 neutrino mixing matrix U,[7] whereas ρ and σ denote the Majorana phases. And $|U_{ei}|$ is commonly parametrized in terms of the mixing angles as follows: $|U_{e1}| = \cos\theta_{12}\cos\theta_{13}$, $|U_{e2}| = \sin\theta_{12}\cos\theta_{13}$ and $|U_{e3}| = \sin\theta_{13}$. Thanks to the various neutrino oscillation experiments,[1] θ_{12}, θ_{13}, $\Delta m^2_{21} \equiv m^2_2 - m^2_1$ and the modulus of $\Delta m^2_{31} \equiv m^2_3 - m^2_1$ have been determined to a good degree of accuracy. But the two phase parameters in Eq. (1), the sign of Δm^2_{31} and the absolute neutrino mass scale remain unknown. That is why one usually chooses to plot $|\langle m \rangle_{ee}|$ against m_1 in the normal mass ordering (NMO) case ($\Delta m^2_{31} > 0$) or m_3 in the inverted mass ordering (IMO) case ($\Delta m^2_{31} < 0$) while allowing ρ and σ to vary from 0 to 2π.[8] In such a so-called Vissani graph, a two-dimensional "well" can appear in the NMO case for 0.4 meV $\lesssim m_1 \lesssim$ 10 meV due to a significant cancellation among the three components of $\langle m \rangle_{ee}$. The bottom of this well stands for the limit of $|\langle m \rangle_{ee}| \to 0$,[9] an unwelcome possibility which is nevertheless allowed by current experimental data. Notice that the dependence of $|\langle m \rangle_{ee}|$ on the Majorana phases has not been explicitly shown in the Vissani graph. In order to present this in a transparent way, one may invoke some three-dimensional presentations by plotting $|\langle m \rangle_{ee}|$ against the lightest neutrino mass and the Majorana phases. With the aid of these three-dimensional graphs, we can particularly study the structure of the well and how possible for $|\langle m \rangle_{ee}|$ to fall into it.

2. Three-Dimensional Graphs

Let us focus on the three-dimensional way of plotting $|\langle m \rangle_{ee}|$ against the neutrino masses and Majorana phases, which can provide more information as compared with the two-dimensional one.[10] Figure 1, the three-dimensional graph of $|\langle m \rangle_{ee}|$ against the lightest neutrino mass m_1 and the relevant Majorana phase ρ in the NMO case, will be taken as a special example and studied in some detail for the following considerations.[12] On the one hand, a combination of recent atmospheric (Super-Kamiokande[13]) and accelerator-based (T2K[14] and NOvA[15]) neutrino oscillation data preliminarily favors the NMO at a 2σ level. If this turns out to be true, it

[a] The phase convention taken here is highly advantageous as ρ (or σ) will automatically disappear when one considers the interesting and experimentally-allowed possibility $m_1 \to 0$ (or $m_3 \to 0$).[6]

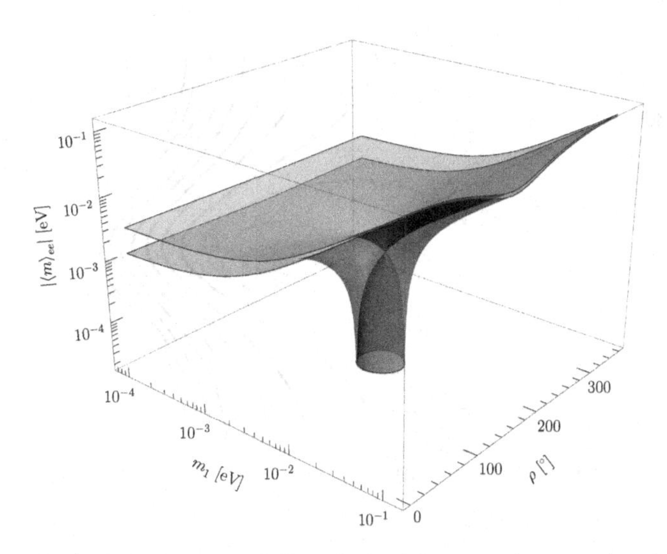

Fig. 1. (color online) Three-dimensional illustration of the upper (in orange) and lower (in blue) bounds of $|\langle m\rangle_{ee}|$ as functions of m_1 and ρ in the NMO case. Here and in the following, the best-fit values $\Delta m^2_{21} = 7.54\times10^{-5}$ eV^2, $\Delta m^2_{31} = 2.47\times10^{-3}$ eV^2, $\sin^2\theta_{12} = 0.308$ and $\sin^2\theta_{13} = 0.0234$ [11] have typically been input.

will be highly desirable to figure out the structure of the well and the possibility for $|\langle m\rangle_{ee}|$ to fall into it so that a discovery or null result of the $0\nu2\beta$-decay searches can be interpreted properly. On the other hand, $|\langle m\rangle_{ee}|$ is more sensitive to ρ than to σ since the latter is associated with the suppression factor $\sin^2\theta_{13}$ in Eq. (1).

The upper ("U") and lower ("L") bounds of $|\langle m\rangle_{ee}|$ presented in Fig. 1 are derived in a straightforward way. Substituting the expression of σ obtained from

$$\frac{\partial|\langle m\rangle_{ee}|}{\partial\sigma} = 0 \Rightarrow \tan\sigma = \frac{m_1\sin\rho}{m_1\cos\rho + m_2\tan^2\theta_{12}}\,, \tag{2}$$

into Eq. (1) simply gives their expressions as

$$|\langle m\rangle_{ee}|_{\rm U,L} = |\overline{m}_{12}\cos^2\theta_{13} \pm m_3\sin^2\theta_{13}|\,, \tag{3}$$

where the sign "+" (or "−") corresponds to "U" (or "L"), and

$$\overline{m}_{12} \equiv \sqrt{m_1^2\cos^4\theta_{12} + \frac{1}{2}m_1 m_2\sin^2 2\theta_{12}\cos\rho + m_2^2\sin^4\theta_{12}}\,. \tag{4}$$

This result can be understood in an intuitive way: for any given values of m_1 and ρ, the maximum of $|\langle m\rangle_{ee}|$ comes out when the sum of the first two components of $\langle m\rangle_{ee}$ has the same phase as the third one (i.e. σ); and the minimum of $|\langle m\rangle_{ee}|$ arises when these two phases differ from each other by $\pm\pi$. As an immediate consequence of Eq. (3), the condition for $|\langle m\rangle_{ee}|_{\rm L} = 0$ which corresponds to the bottom of the well in Fig. 1 reads

$$\overline{m}_{12} = m_3\tan^2\theta_{13}\,. \tag{5}$$

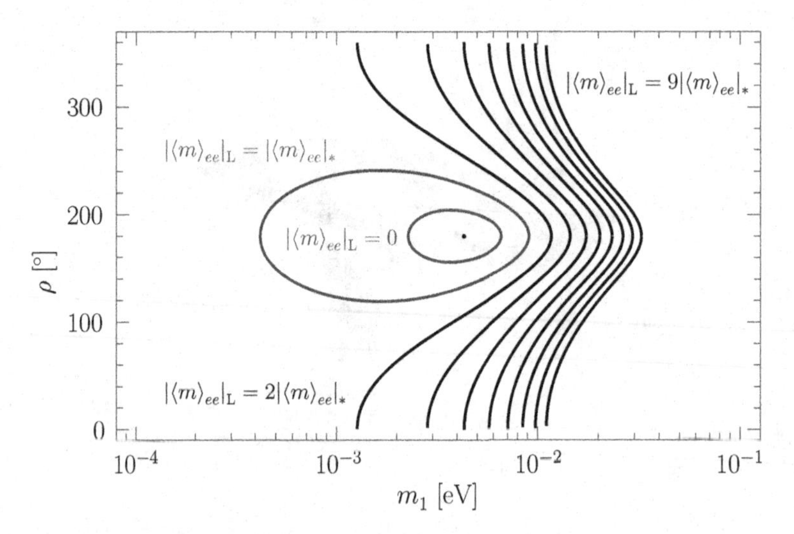

Fig. 2. (color online) The numerical correlation between m_1 and ρ in three typical cases: (a) $|\langle m\rangle_{ee}|_{\rm L} = 0$ (the red curve); (b) $|\langle m\rangle_{ee}|_{\rm L} = |\langle m\rangle_{ee}|_* = m_3 \sin^2\theta_{13}$ (the black dot and the blue curve); and (c) $|\langle m\rangle_{ee}|_{\rm L} = n|\langle m\rangle_{ee}|_*$ with $n \geq 2$ (the black curves).

Given the expressions $m_2 = \sqrt{m_1^2 + \Delta m_{21}^2}$ and $m_3 = \sqrt{m_1^2 + \Delta m_{31}^2}$ in the NMO case, this equation allows us to fix the correlation between m_1 and ρ, and the numerical result is shown by the red curve in Fig. 2. One can see that touching the bottom of the well (i.e. $|\langle m\rangle_{ee}| \to 0$) is not a highly probable event, because it requires m_1 and ρ to lie in the narrow regions 2 meV $\lesssim m_1 \lesssim$ 7 meV and $0.86 \lesssim \rho/\pi \lesssim 1.14$, respectively.[16]

Another salient feature of the well is the "bullet"-like structure of $|\langle m\rangle_{ee}|_{\rm L}$ as shown in Fig. 1. The surface of this bullet is described by

$$|\langle m\rangle_{ee}|_{\rm L} = m_3 \sin^2\theta_{13} - \overline{m}_{12}\cos^2\theta_{13}\,, \tag{6}$$

whose extremum (i.e. the tip of the bullet) is supposed to be located at a point determined by the conditions

$$\begin{aligned} \frac{\partial|\langle m\rangle_{ee}|_{\rm L}}{\partial\rho} &= \frac{m_1 m_2 \sin^2 2\theta_{12}\cos^2\theta_{13}}{4\overline{m}_{12}}\sin\rho = 0\,, \\ \frac{\partial|\langle m\rangle_{ee}|_{\rm L}}{\partial m_1} &= \frac{m_1}{m_3}\sin^2\theta_{13} - \cos^2\theta_{13} \\ &\quad \times \frac{m_2\overline{m}_{12}^2 - \Delta m_{21}^2\sin^2\theta_{12}(m_2\sin^2\theta_{12} + m_1\cos^2\theta_{12}\cos\rho)}{m_1 m_2 \overline{m}_{12}} = 0\,. \end{aligned} \tag{7}$$

The first condition definitely leads us to $\rho = 0$ or π. But Fig. 2 shows that ρ should take a value around π inside the well, so it is appropriate for us to take $\rho = \pi$.

In this case the second condition is simplified to

$$\frac{\partial|\langle m\rangle_{ee}|_{\rm L}}{\partial m_1} = \frac{m_1}{m_3}\sin^2\theta_{13} \pm \left(\cos^2\theta_{12} - \frac{m_1}{m_2}\sin^2\theta_{12}\right)\cos^2\theta_{13} = 0\,, \tag{8}$$

where "$\pm$" correspond to the prerequisites $m_1 < m_2\tan^2\theta_{12}$ and $m_1 > m_2\tan^2\theta_{12}$, respectively. Note that Eq. (8) can never be fulfilled as the second term is much larger than the first one in the NMO case. It instead allows us to reach

$$\begin{aligned} \frac{\partial|\langle m_{ee}|_{\rm L}}{\partial m_1} > 0 \quad \text{for } m_1 < m_2\tan^2\theta_{12}\,, \\ \frac{\partial|\langle m\rangle_{ee}|_{\rm L}}{\partial m_1} < 0 \quad \text{for } m_1 > m_2\tan^2\theta_{12}\,, \end{aligned} \tag{9}$$

which implies that $|\langle m\rangle_{ee}|_{\rm L}$ increases (or decreases) for $m_1 < m_2\tan^2\theta_{12}$ (or $m_1 > m_2\tan^2\theta_{12}$). Hence there must be a local maximum for $|\langle m\rangle_{ee}|_{\rm L}$, denoted as

$$|\langle m\rangle_{ee}|_* = m_3\sin^2\theta_{13} = \sqrt{m_1^2 + \Delta m_{31}^2}\sin^2\theta_{13}\,, \tag{10}$$

at the position fixed by $\rho = \pi$ and

$$m_1 = m_2\tan^2\theta_{12} = \sqrt{m_1^2 + \Delta m_{21}^2}\tan^2\theta_{12}\,. \tag{11}$$

Namely, the local maximum of $|\langle m\rangle_{ee}|_{\rm L}$ inside the well arises from Eq. (6) at $\overline{m}_{12} = 0$. Given the best-fit values of Δm_{21}^2, Δm_{31}^2, $\sin^2\theta_{12}$ and $\sin^2\theta_{13}$, the location of the tip of the bullet turns out to be $(m_1, \rho, |\langle m\rangle_{ee}|_*) \simeq (4\ \text{meV}, 180^\circ, 1\ \text{meV})$. In fact $|\langle m\rangle_{ee}|_{\rm U}$ takes the same value as $|\langle m\rangle_{ee}|_{\rm L}$ at this location as can be seen from Eq. (3), thereby explaining why they connect each other at the tip of the bullet in Fig. 1. In the same way as above, one can verify that this special value is a local minimum of $|\langle m\rangle_{ee}|_{\rm U}$. Consequently, $|\langle m\rangle_{ee}|_* = m_3\sin^2\theta_{13} \simeq 1$ meV can be taken as a threshold value of $|\langle m\rangle_{ee}|$ in the NMO case.

From the experimental point of view, this threshold value may mark an ultimate limit of the reachable sensitivity to $|\langle m\rangle_{ee}|$ in the foreseeable future.[b] At present the most sensitive $0\nu2\beta$-decay experiments can only set an upper limit around 165 meV for $|\langle m\rangle_{ee}|$,[19] which certainly depends on some theoretical uncertainties in calculating the relevant nuclear matrix elements.[20] While the most ambitious next-generation high-sensitivity $0\nu2\beta$-decay experiments (e.g. nEXO[21]) are expected to probe $|\langle m\rangle_{ee}|$ at the level of a few tens of meV,[4] a sensitivity still much larger than the threshold value. So it is fair for us to say that there would be no hope to observe any $0\nu2\beta$-decay signal if $|\langle m\rangle_{ee}|$ were unfortunately around or below the value of $|\langle m\rangle_{ee}|_*$ in the standard three-flavor scheme. Using this special value as a unit, we project the contours of $|\langle m\rangle_{ee}|_{\rm L}$ corresponding to $n|\langle m\rangle_{ee}|_*$ (for $n = 0, 1, 2, \ldots$) onto

[b] In Ref. 17 a purely statistical analysis of the possibility of $|\langle m\rangle_{ee}| \lesssim 1$ meV has been done to see to what extent the Majorana phases ρ and σ can be constrained for a given value of m_1, and in Ref. 18 the conditions for $|\langle m\rangle_{ee}| > 1$ meV are analyzed in the special case of $m_1 \to 0$ or $\theta_{13} \to 0$.

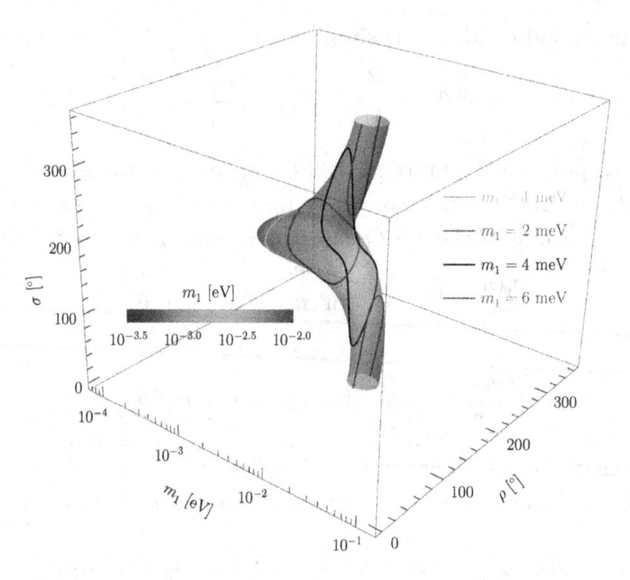

Fig. 3. (color online) The parameter space of m_1, ρ and σ allowed for $|\langle m \rangle_{ee}| < |\langle m \rangle_{ee}|_*$ to hold. The intersecting surfaces for $m_1 = 1, 2, 4$ and 6 meV on the ρ-σ plane are explicitly shown in the figure.

the m_1-ρ plane in Fig. 2. The purpose is to reveal how steep the slope of $|\langle m \rangle_{ee}|_{\rm L}$ around the well is, such that one can get a ballpark feeling for the possibility of falling into the well. It is especially interesting to compare between the contours of the well at its bottom with $|\langle m \rangle_{ee}|_{\rm L} = 0$ (the red curve) and at its threshold height with $|\langle m \rangle_{ee}|_{\rm L} = |\langle m \rangle_{ee}|_*$ (the blue curve and the black point). They clearly show how the well becomes narrower when the value of $|\langle m \rangle_{ee}|_{\rm L}$ goes down. The profile of $|\langle m \rangle_{ee}|_{\rm L}$ will be partially open on the $m_1 \to 0$ side and thus lose its "well" feature as $|\langle m \rangle_{ee}|_{\rm L} \geq 2|\langle m \rangle_{ee}|_*$ is taken into account. Now that $|\langle m \rangle_{ee}|_{\rm L} > |\langle m \rangle_{ee}|_*$ always holds outside the blue curve, we argue that the parameter space of $|\langle m \rangle_{ee}|_{\rm L} \leq |\langle m \rangle_{ee}|_*$ (i.e. 0.4 meV $\lesssim m_1 \lesssim$ 10 meV and $0.66 \lesssim \rho/\pi \lesssim 1.34$) is a simple measure of the chance for $|\langle m \rangle_{ee}|$ to fall into the well and become completely unobservable.

In the above discussions we have not explicitly shown the dependence of $|\langle m \rangle_{ee}|$ on σ, which plays the role of separating $|\langle m \rangle_{ee}|_{\rm U}$ from $|\langle m \rangle_{ee}|_{\rm L}$. To have a knowledge of the preferred values of σ in the $|\langle m \rangle_{ee}| < |\langle m \rangle_{ee}|_*$ case, we present the resulting three-dimensional parameter space of m_1, ρ and σ in Fig. 3. As a reference, the intersecting surfaces on the ρ-σ plane corresponding to $m_1 = 1, 2, 4$ and 6 meV are also specified. It turns out that a large part of the range of σ is allowed when the first two components of $\langle m \rangle_{ee}$ in Eq. (1) essentially cancel each other out (e.g. the black intersecting surface corresponding to $m_1 = 4$ meV). But when the value of m_1 decreases, the value of σ should approach π, such as the green intersecting surface corresponding to $m_1 = 1$ meV. In this case the second component of $\langle m \rangle_{ee}$ can be cancelled by the other two components to the maximal level. For a similar reason, the value of σ should approach 0 or 2π when the value of m_1 increases (e.g. the blue

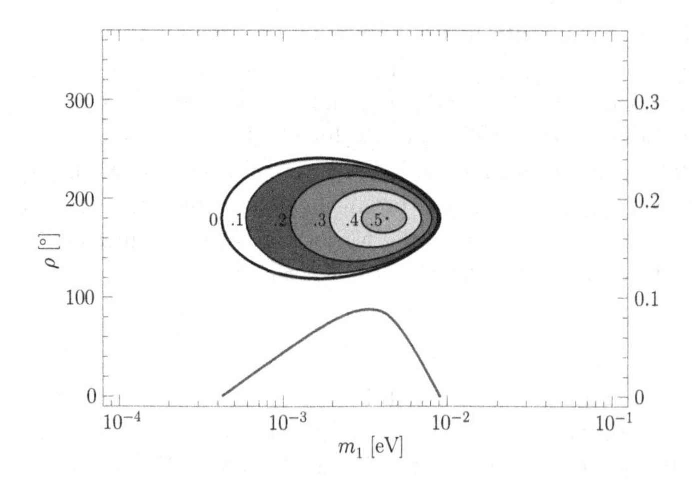

Fig. 4. (color online) The contours (labelled by 0, 0.1 and so on) illustrate the ratio of the allowed range of σ in the $|\langle m\rangle_{ee}| < |\langle m\rangle_{ee}|_*$ case to the whole range 360° for any given values of m_1 and ρ, while the red line measures the ratio of the allowed range of (ρ, σ) also in this case to the whole range 360° × 360° for any given value of m_1.

intersecting surface corresponding to $m_1 = 6$ meV). To quantitatively illustrate how possible a randomly selected value of σ may lead $|\langle m\rangle_{ee}|$ to have a value smaller than $|\langle m\rangle_{ee}|_*$ for some given values of m_1 and ρ, in Fig. 4 we display the ratio of the allowed range of σ in such a case to the whole range 360° in terms of some contours (labelled by 0, 0.1 and so on) on the m_1-ρ plane. This ratio takes a value close to 0.5 when (m_1, ρ) are around (4 meV, 180°) (i.e. the location of the tip of the well), and eventually vanishes for (m_1, ρ) beyond the contour labelled by 0 (i.e. the blue curve in Fig. 2). Furthermore, the red line in Fig. 4 measures the ratio (whose values should be inferred from the coordinate labels on the right-hand side) of the range of (ρ, σ) allowing for $|\langle m\rangle_{ee}| < |\langle m\rangle_{ee}|_*$ to the whole range 360° × 360° for a given value of m_1. One finds that such a ratio only takes finite but small values in the range 0.4 meV $\lesssim m_1 \lesssim$ 10 meV. Based on the above results we conclude that the possibility of $|\langle m\rangle_{ee}| < |\langle m\rangle_{ee}|_*$ involves significant cancellations among its three components and thus is really small.

3. Summary

We have achieved some new insights into the effective Majorana neutrino mass $\langle m\rangle_{ee}$ of the $0\nu2\beta$ decays in the NMO case — a case which seems to be more likely than the IMO one according to today's preliminary experimental data. Considering that $|\langle m\rangle_{ee}|$ depends not only on the unknown neutrino mass m_1 but also on the free Majorana phases ρ and σ, we put forward a novel three-dimensional presentation of $|\langle m\rangle_{ee}|$ against m_1 and ρ, which reveals an intriguing "well" structure in the NMO case. As touching the bottom of the well poses a disappointing possibility of $|\langle m\rangle_{ee}| \to 0$ for the $0\nu2\beta$-decay searches, the structure of the well and the possibility of falling into it are studied in particular. An interesting threshold of $|\langle m\rangle_{ee}|$ is found: $|\langle m\rangle_{ee}|_* = m_3 \sin^2\theta_{13}$ in connection with $\tan\theta_{12} = \sqrt{m_1/m_2}$ and $\rho = \pi$, which links the local minimum and maximum of $|\langle m\rangle_{ee}|$. From the experimental point of view, this special value can be used to signify observability of the future $0\nu 2\beta$-decay experiments. The numerical results show that the possibility of $|\langle m\rangle_{ee}| < |\langle m\rangle_{ee}|_*$ is very small, promising a hopeful prospect for detecting the $0\nu 2\beta$ decays and thus identifying the Majorana nature of massive neutrinos in a foreseeable future even if the NMO turns out to be true.

Acknowledgments

Z.Z.X. would like to thank Harald Fritzsch and Kok Khoo Phua for organizing this conference together and for warm hospitality at the IAS of NTU during this conference. He is also grateful to Janet Chan, Phil Chan, Chee-Hok Lim, Louis Lim, Chris Ong and Chi Xiong for the hospitality to him and his family in Singapore. This work is supported in part by the National Natural Science Foundation of China under Grant No. 11135009 (Z.Z.X.) and Grant No. 11605081 (Z.H.Z.).

References

1. Particle Data Group (C. Patrignani *et al.*), Chin. Phys. C **40**, 100001 (2016).
2. E. Majorana, Nuovo Cimento **14**, 171 (1937).
3. W. H. Furry, Phys. Rev. **15**, 1184 (1939).
4. For a recent review with extensive references, see S. M. Bilenky and C. Giunti, *Int. J. Mod. Phys. A* **30**, 0001 (2015); S. Dell'Oro, S. Marcocci, M. Viel and F. Vissani, *Adv. High Energy Phys.* **2016**, 2162659 (2016); J. D. Vergados, H. Ejiri and F. Simkovic, *Int. J. Mod. Phys. E* **25**, 1630007 (2016).
5. S. M. Bilenky, J. Hosek and S. T. Petcov, *Phys. Lett. B* **94**, 495 (1980); J. Schechter and J. W. F. Valle, *Phys. Rev. D* **22**, 2227 (1980); M. Doi, T. Kotani, H. Nishiura, K. Okuda and E. Takasugi, *Phys. Lett. B* **102**, 323 (1981).
6. Z. Z. Xing and Y. L. Zhou, *Chin. Phys. C* **39**, 011001 (2015); *Mod. Phys. Lett. A* **30**, 1530019 (2015); *Adv. Ser. Direct. High Energy Phys.* **25**, 157 (2015).
7. Z. Maki, M. Nakagawa and S. Sakata, *Progr. Theor. Phys.* **28**, 870 (1962); B. Pontecorvo, *Sov. Phys. JETP* **26**, 984 (1968).
8. F. Vissani, JHEP **06**, 022 (1999).
9. W. Rodejohann, *Nucl. Phys. B* **597**, 110 (2001); Z. Z. Xing, *Phys. Rev. D* **68**, 053002 (2003); W. Rodejohann, *Int. J. Mod. Phys. E* **20**, 1833 (2011); S. Dell'oro, S. Marcocci and F. Vissani, *Phys. Rev. D* **90**, 033005 (2014).
10. Z. Z. Xing, Z. H. Zhao and Y. L. Zhou, Eur. Phys. J. C **75**, 423 (2015).
11. F. Capozzi *et al.*, *Phys. Rev. D* **89**, 093018 (2014); *Nucl. Phys. B* **908**, 218 (2016); D. V. Forero, M. Tortola and J. W. F. Valle, *Phys. Rev. D* **90**, 093006 (2014); M. C. Gonzalez-Garcia, M. Maltoni and T. Schwetz, *JHEP* **1411**, 052 (2014).
12. Z. Z. Xing and Z. H. Zhao, Eur. Phys. J. C **77**, 192 (2017).
13. B. Rebel, talk given at the *XIV Int. Conf. on Topics in Astroparticle and Underground Physics*, September 2015, Torino, Italy.
14. K. Abe *et al.*, Phys. Rev. Lett. **112**, 181801 (2014).
15. C. Kachulis, talk given at the *EPS Conf. on High Energy Physics*, July 2015, Vienna, Austria.

16. G. Benato, Eur. Phys. J. C **75**, 563 (2015).
17. S. F. Ge and M. Lindner, Phys. Rev. D **95**, 033003 (2017).
18. S. Pascoli and S. T. Petcov, Phys. Rev. D **77**, 113003 (2008).
19. KamLAND-Zen Collab. (A. Gando *et al.*), Phys. Rev. Lett. **117**, 082503 (2016).
20. J. Engel and J. Menendez, Rep. Prog. Phys. **80**, 046301 (2017).
21. Y. Lin, talk given at the APR15 meeting of APS, 2015.

Fermions and Bosons in the Expanding Universe by the Spin-charge-family Theory

N. S. Mankoč Borštnik

Department of Physics, University of Ljubljana,
SI-1000 Ljubljana, Slovenia
norma.mankoc@fmf.uni-lj.si
www.fmf.uni-lj.si

The *spin-charge-family* theory, which is a kind of the Kaluza-Klein theories in $d = (13+1)$ — but with the two kinds of the spin connection fields, the gauge fields of the two Clifford algebra objects, S^{ab} and $\tilde{S}^{ab}$ — explains all the assumptions of the *standard model*: The origin of the charges of fermions appearing in one family, the origin and properties of the vector gauge fields of these charges, the origin and properties of the families of fermions, the origin of the scalar fields observed as the Higgs's scalar and the Yukawa couplings. The theory explains several other phenomena like: The origin of the dark matter, of the matter-antimatter asymmetry, the "miraculous" triangle anomaly cancellation in the *standard model* and others. Since the theory starts at $d = (13 + 1)$ the question arises how and at which d had our universe started and how it came down to $d = (13 + 1)$ and further to $d = (3 + 1)$. In this short contribution some answers to these questions are presented.

Keywords: Unifying theories; beyond the standard model; Kaluza-Klein-like theories; vector and scalar gauge fields and their origin; fermions, their families in their properties in the expanding universe.

1. Introduction

Both standard models, the *standard model* of elementary fermion and boson fields and the *standard cosmological model*, have quite a lot of assumptions, guessed from the properties of observables. Although in the history physics was and still is (in particular when many degrees of freedom are concerned) relying on small theoretical steps, confirmed by experiments, there are also a few decisive steps, without which no real further progress would be possible. Among such steps there are certainly the *general theory of relativity* and the *standard model* of elementary fermion and boson fields. Both theories enabled much better understanding of our universe and its elementary fields — fermions and bosons.

With more and more accurate experiments is becoming increasingly clear that a new decisive step is again needed in the theory of elementary fields as well as in cosmology.

Both theories rely on observed facts built into innovative mathematical models. However, the assumptions remain unexplained.

Among the non understood assumptions of the *standard model* of the elementary fields of fermions and bosons are: i. The origin of massless family members with their charges related to spins. ii. The origin of families of fermions. iii. The origin of the massless vector gauge fields of the observed charges. iv. The origin of masses of family members and heavy bosons. v. The origin of the Higgs's scalar and the Yukawa couplings. vi. The origin of matter-antimatter asymmetry. vii. The origin of the dark matter. viii. The origin of the electroweak phase transition scale. ix. The origin of the colour phase transition scale. And others.

Among the non understood assumptions of the *cosmological model* are: *a.* The differences in the origin of the gravity, of the vector gauge fields and the (Higgs's) scalars. *b.* The origin of the dark matter, of the matter-antimatter asymmetry of the (ordinary) matter. *c.* The appearance of fermions. *d.* The origin of the inflation of the universe. *e.* While it is known how to quantize vector gauge fields, the quantization of gravity is still an open problem.

The L(arge) H(adron) C(collider) and other accelerators and measuring apparatus produce a huge amount of data, the analyzes of which should help to explain the assumptions of both standard models. But it looks like so far that the proposed models, relying more or less on small extensions of the standard models, can not offer much help. The situation in elementary particle physics is reminiscent of the situation in the nuclear physics before the *standard model* of the elementary fields was proposed, opening new insight into physics of elementary fermion and boson fields.

The deeper into the history of our universe we are succeeding to look by the observations and experiments the more both standard models are becoming entangled, dependent on each other, calling for the next step which would offer the explanation for most of the above mentioned non understood assumptions of both standard models.

The *spin-charge-family* theory[1–8] does answer open questions of the *standard model* of the elementary fields and also several of cosmology.

The *spin-charge-family* theory[1–4] is promising to be the right next step beyond the *standard model* of elementary fermion and boson fields by offering the explanation for all the assumptions of this model. By offering the explanation also for the dark matter and matter-antimatter asymmetry the theory makes a new step also in cosmology, in particular since it starts at $d \geq 5$ with spinors and gravitational fields only — like the Kaluza-Klein theories (but with the two kinds of the spin connection fields, which are the gauge fields of the two kinds of the Clifford algebra objects). Although there are still several open problems waiting to be solved, common to most of proposed theories — like how do the boundary conditions influence the breaking of the starting symmetry of space-time and how to quantize gravity in any d, while we know how to quantize at least the vector gauge fields in $d = (3+1)$ — the *spin-charge-family* theory is making several predictions (not just stimulated by the current experiments what most of predictions do).

The *spin-charge-family* theory (Refs. 1–13, 15 and the references therein) starts in $d = (13+1)$: **i.** with the simple action for spinors, Eq. (1), which carry two kinds of spins, **i.a.** the Dirac one described by γ^a and manifesting at low energies in $d = (3+1)$ as spins and all the charges of the observed fermions of one family, Table 1, **i.b.** the second one named[15] (by the author of this paper) $\tilde{\gamma}^a$ ($\{\tilde{\gamma}^a, \gamma^b\}_+ = 0$, Eq. (2)), and manifesting at low energies the family quantum numbers of the observed fermions. **ii.** Spinors interact in $d = (13+1)$ with the gravitational field only, **ii.a.** the vielbeins and **ii.b.** the two kinds of the spin connection fields (Refs. 1, 4, and the references therein). Spin connection fields — ω_{stm} ($(s,t) \geq 5$, $m = (0,1,2,3,4)$), Eq. (1) — are the gauge fields of S^{st}, Eq. (7), and manifest at low energies in $d = (3+1)$ as the vector gauge fields (the colour, weak and hyper vector gauge fields are directly or indirectly observed vector gauge fields). Spin connections $\omega_{sts'}$ ($(s,t) \geq 5, s' = (7,8)$) manifest as scalar gauge fields, contributing to the Higgs's scalar and the Yukawa couplings together with the scalar spin connection gauge fields — $\tilde{\omega}_{abs'}$ ($(a,b) = (m,s,t)$, $s' = (7,8)$), Eq. (1) — which are the gauge fields of $\tilde{S}^{ab}$, Eq. (7)[1–4]. Correspondingly these (several) scalar gauge fields determine after the electroweak break masses of the families of all the family members and of the heavy bosons (Refs. 1–4, and the references therein).

Scalar fields $\omega_{sts'}$ ($(s,t) \geq 5, s' = (9, \cdots, 14)$), Ref. 4 (and the references therein), cause transitions from anti-leptons to quarks and anti-quarks into quarks and back. In the presence of the condensate of two right handed neutrinos[2,4] the matter-antimatter symmetry breaks.

2. Short Presentation of the *spin-charge-family* Theory

The *spin-charge-family* theory[2–4,7–10] assumes a simple action, Eq. (1), in an even dimensional space ($d = 2n$, $d > 5$). d is chosen to be $(13+1)$, what makes the simple starting action in d to manifest in $d = (3+1)$ in the low energy regime all the observed degrees of freedom, explaining all the assumptions of the *standard model* as well as other observed phenomena. Fermions interact with the vielbeins $f^{\alpha}{}_a$ and the two kinds of the spin-connection fields — $\omega_{ab\alpha}$ and $\tilde{\omega}_{ab\alpha}$ — the gauge fields of $S^{ab} = \frac{i}{4}(\gamma^a \gamma^b - \gamma^b \gamma^a)$ and $\tilde{S}^{ab} = \frac{i}{4}(\tilde{\gamma}^a \tilde{\gamma}^b - \tilde{\gamma}^b \tilde{\gamma}^a)$, respectively, where:

$$\begin{aligned} \mathcal{A} = &\int d^dx \; E \; \tfrac{1}{2} (\bar{\psi} \gamma^a p_{0a} \psi) + h.c. + \\ &\int d^dx \; E \; (\alpha \, R + \tilde{\alpha} \, \tilde{R}), \end{aligned} \tag{1}$$

here $p_{0a} = f^{\alpha}{}_a \, p_{0\alpha} + \frac{1}{2E} \{p_{\alpha}, E f^{\alpha}{}_a\}_-$, $p_{0\alpha} = p_{\alpha} - \frac{1}{2} S^{ab} \omega_{ab\alpha} - \frac{1}{2} \tilde{S}^{ab} \tilde{\omega}_{ab\alpha}$ [a],

$$R = \frac{1}{2} \{ f^{\alpha [a} f^{\beta b]} \, (\omega_{ab\alpha,\beta} - \omega_{ca\alpha} \, \omega^c{}_{b\beta}) \} + h.c.,$$

$$\tilde{R} = \frac{1}{2} \{ f^{\alpha [a} f^{\beta b]} \, (\tilde{\omega}_{ab\alpha,\beta} - \tilde{\omega}_{ca\alpha} \, \tilde{\omega}^c{}_{b\beta}) \} + h.c..$$

The action introduces two kinds of the Clifford algebra objects, γ^a and $\tilde{\gamma}^a$,

$$\{\gamma^a, \gamma^b\}_+ = 2\eta^{ab} = \{\tilde{\gamma}^a, \tilde{\gamma}^b\}_+ \, . \tag{2}$$

[a] Whenever two indexes are equal the summation over these two is meant.

$f^{\alpha}{}_{a}$ are vielbeins inverted to $e^{a}{}_{\alpha}$, Latin letters $(a, b, ..)$ denote flat indices, Greek letters $(\alpha, \beta, ..)$ are Einstein indices, $(m, n, ..)$ and $(\mu, \nu, ..)$ denote the corresponding indices in $(0, 1, 2, 3)$, $(s, t, ..)$ and $(\sigma, \tau, ..)$ denote the corresponding indices in $d \geq 5$:

$$e^{a}{}_{\alpha} f^{\beta}{}_{a} = \delta^{\beta}_{\alpha}\,, \quad e^{a}{}_{\alpha} f^{\alpha}{}_{b} = \delta^{a}_{b}\,, \tag{3}$$

$E = \det(e^{a}{}_{\alpha})$ [b].

The action $\mathcal{A}$ offers the explanation for the origin and all the properties of the observed fermions (of the family members and families), of the observed vector gauge fields, of the Higgs's scalar and of the Yukawa couplings, explaining the origin of the matter-antimatter asymmetry, the appearance of the dark matter and predicts the new scalars and the new (fourth) family coupled to the observed three to be measured at the LHC (Refs. 2, 4 and the references therein).

The *standard model* groups of spins and charges are the subgroups of the $SO(13, 1)$ group with the generator of the infinitesimal transformations expressible with S^{ab} — for spins

$$\vec{N}_{\pm}(= \vec{N}_{(L,R)}) := \tfrac{1}{2}(S^{23} \pm i S^{01}, S^{31} \pm i S^{02}, S^{12} \pm i S^{03})\,, \tag{4}$$

— for the weak charge, $SU(2)_{I}$, and the second $SU(2)_{II}$, these two groups are the invariant subgroups of $SO(4)$

$$\begin{aligned} \vec{\tau}^{1} &:= \tfrac{1}{2}(S^{58} - S^{67},\, S^{57} + S^{68},\, S^{56} - S^{78})\,, \\ \vec{\tau}^{2} &:= \tfrac{1}{2}(S^{58} + S^{67},\, S^{57} - S^{68},\, S^{56} + S^{78})\,, \end{aligned} \tag{5}$$

— for the colour charge $SU(3)$ and for the "fermion charge" $U(1)_{II}$, these two groups are subgroups of $SO(6)$

$$\begin{aligned} \vec{\tau}^{3} &:= \frac{1}{2}\{S^{9\,12} - S^{10\,11}, S^{9\,11} + S^{10\,12},\, S^{9\,10} - S^{11\,12}, \\ &\quad S^{9\,14} - S^{10\,13},\, S^{9\,13} + S^{10\,14}, S^{11\,14} - S^{12\,13}, \\ &\quad S^{11\,13} + S^{12\,14},\, \frac{1}{\sqrt{3}}(S^{9\,10} + S^{11\,12} - 2S^{13\,14})\}\,, \\ \tau^{4} &:= -\frac{1}{3}(S^{9\,10} + S^{11\,12} + S^{13\,14})\,, \end{aligned} \tag{6}$$

— while the hyper charge Y is $Y = \tau^{23} + \tau^{4}$. The breaks of the symmetries, manifesting in Eqs. (4, 5, 6), are in the *spin-charge-family* theory caused by the condensate and the constant values of the scalar fields carrying the space index $(7, 8)$ (Refs. 3, 4 and the references therein). The space breaks first to $SO(7, 1) \times SU(3) \times U(1)_{II}$ and then further to $SO(3, 1) \times SU(2)_{I} \times U(1)_{I} \times SU(3)$, what explains the connections between the weak and the hyper charges and the handedness of spinors.

The equivalent expressions for the family charges, expressed by $\tilde{S}^{ab}$, follow if in Eqs. (4 - 6) S^{ab} are replaced by $\tilde{S}^{ab}$.

[b] This definition of the vielbein and the inverted vielbein is general, no specification about the curled space is assumed yet, but is valid also in the low energies regions, when the starting symmetry is broken [1].

2.1. *A short inside into the spinor states of the spin-charge-family theory*

I demonstrate in this subsection on two examples how transparently can properties of spinor and anti-spinor states be read from these states[3,13,15], when the states are expressed with $\frac{d}{2}$ nilpotents and projectors, formed as odd and even objects of γ^a's (Eq. (10)) and chosen to be the eigenstates of the Cartan subalgebra (Eq. (8)) of the algebra of the two groups, as in Table 1.

Recognizing that the two Clifford algebra objects (S^{ab}, S^{cd}), or $(\tilde{S}^{ab}, \tilde{S}^{cd})$, fulfilling the algebra,

$$\begin{aligned}
\{S^{ab}, S^{cd}\}_- &= i(\eta^{ad} S^{bc} + \eta^{bc} S^{ad} - \eta^{ac} S^{bd} - \eta^{bd} S^{ac}) ,\\
\{\tilde{S}^{ab}, \tilde{S}^{cd}\}_- &= i(\eta^{ad} \tilde{S}^{bc} + \eta^{bc} \tilde{S}^{ad} - \eta^{ac} \tilde{S}^{bd} - \eta^{bd} \tilde{S}^{ac}) ,\\
\{S^{ab}, \tilde{S}^{cd}\}_- &= 0 ,
\end{aligned} \tag{7}$$

commute, if all the indexes (a, b, c, d) are different, the Cartan subalgebra is in $d = 2n$ selected as follows

$$\begin{aligned}
&S^{03}, S^{12}, S^{56}, \cdots, S^{d-1\, d}, \quad \text{if} \quad d = 2n \geq 4 ,\\
&\tilde{S}^{03}, \tilde{S}^{12}, \tilde{S}^{56}, \cdots, \tilde{S}^{d-1\, d}, \quad \text{if} \quad d = 2n \geq 4 .
\end{aligned} \tag{8}$$

Let us define as well one of the Casimirs of the Lorentz group — the handedness Γ $(\{\Gamma, S^{ab}\}_- = 0)$ in $d = 2n$ [c]

$$\Gamma^{(d)} : = (i)^{d/2} \prod_a (\sqrt{\eta^{aa}} \gamma^a), \quad \text{if} \quad d = 2n , \tag{9}$$

which can be written also as $\Gamma^{(d)} = i^{d-1} \cdot 2^{\frac{d}{2}}\; S^{03} \cdot S^{12} \cdots S^{(d-1)\, d}$. The product of γ^a's must be taken in the ascending order with respect to the index a: $\gamma^0 \gamma^1 \cdots \gamma^d$. It follows from the Hermiticity properties of γ^a for any choice of the signature η^{aa} that $\Gamma^{(d)\dagger} = \Gamma^{(d)}$, $(\Gamma^{(d)})^2 = I$. One proceeds equivalently for $\tilde{\Gamma}^{(d)}$, substituting γ^a's by $\tilde{\gamma}^a$'s. We also find that for d even the handedness anticommutes with the Clifford algebra objects γ^a $(\{\gamma^a, \Gamma\}_+ = 0)$.

Spinor states can be, as in Table 1, represented as products of nilpotents and projectors defined by γ^a's

$$\stackrel{ab}{(k)}: = \frac{1}{2}(\gamma^a + \frac{\eta^{aa}}{ik} \gamma^b) , \qquad \stackrel{ab}{[k]}:= \frac{1}{2}(1 + \frac{i}{k} \gamma^a \gamma^b) , \tag{10}$$

where $k^2 = \eta^{aa} \eta^{bb}$.

It is easy to check that the nilpotent $\stackrel{ab}{(k)}$ and the projector $\stackrel{ab}{[k]}$ are "eigenstates" of S^{ab} and $\tilde{S}^{ab}$

$$\begin{aligned}
&S^{ab} \stackrel{ab}{(k)} = \frac{1}{2} k \stackrel{ab}{(k)} , \qquad S^{ab} \stackrel{ab}{[k]} = \frac{1}{2} k \stackrel{ab}{[k]} ,\\
&\tilde{S}^{ab} \stackrel{ab}{(k)} = \frac{1}{2} k \stackrel{ab}{(k)} , \qquad \tilde{S}^{ab} \stackrel{ab}{[k]} = -\frac{1}{2} k \stackrel{ab}{[k]} ,
\end{aligned} \tag{11}$$

[c] The reader can find the definition of handedness for d odd in Refs. 4, 13 and the references therein.

where in Eq. (11) the vacuum state $|\psi_0\rangle$ is meant to stay on the right hand sides of projectors and nilpotents. This means that one gets when multiplying nilpotents $\stackrel{ab}{(k)}$ and projectors $\stackrel{ab}{[k]}$ by S^{ab} the same objects back multiplied by the constant $\frac{1}{2}k$, while $\tilde{S}^{ab}$ multiply $\stackrel{ab}{(k)}$ by k and $\stackrel{ab}{[k]}$ by $(-k)$ rather than k.

One can namely see, taking into account Eq. (2), that

$$\gamma^a \stackrel{ab}{(k)} = \eta^{aa} \stackrel{ab}{[-k]}, \; \gamma^b \stackrel{ab}{(k)} = -ik \stackrel{ab}{[-k]}, \; \gamma^a \stackrel{ab}{[k]} = \stackrel{ab}{(-k)}, \; \gamma^b \stackrel{ab}{[k]} = -ik\eta^{aa} \stackrel{ab}{(-k)},$$
$$\tilde{\gamma}^a \stackrel{ab}{(k)} = -i\eta^{aa} \stackrel{ab}{[k]}, \; \tilde{\gamma}^b \stackrel{ab}{(k)} = -k \stackrel{ab}{[k]}, \; \tilde{\gamma}^a \stackrel{ab}{[k]} = \; i \stackrel{ab}{(k)}, \; \tilde{\gamma}^b \stackrel{ab}{[k]} = -k\eta^{aa} \stackrel{ab}{(k)} \,. \quad (12)$$

One recognizes also that γ^a transform $\stackrel{ab}{(k)}$ into $\stackrel{ab}{[-k]}$, never to $\stackrel{ab}{[k]}$, while $\tilde{\gamma}^a$ transform $\stackrel{ab}{(k)}$ into $\stackrel{ab}{[k]}$, never to $\stackrel{ab}{[-k]}$.

In Table 1 [2,3,5] the left handed ($\Gamma^{(13,1)} = -1$, Eq. (9)) massless multiplet of one family (Table 2) of spinors — the members of the fundamental representation of the $SO(13,1)$ group — is presented as products of nilpotents and projectors, Eq. (10). All these states are eigenstates of the Cartan sub-algebra (Eq. (8)). Table 1 manifests the subgroup $SO(7,1)$ of the colour charged quarks and anti-quarks and the colourless leptons and anti-leptons [13,15]. The multiplet contains the left handed ($\Gamma^{(3,1)} = -1$) weak ($SU(2)_I$) charged ($\tau^{13} = \pm\frac{1}{2}$, Eq. (5)), and $SU(2)_{II}$ chargeless ($\tau^{23} = 0$, Eq. (5)) quarks and leptons and the right handed ($\Gamma^{(3,1)} = 1$) weak ($SU(2)_I$) chargeless and $SU(2)_{II}$ charged ($\tau^{23} = \pm\frac{1}{2}$) quarks and leptons, both with the spin S^{12} up and down ($\pm\frac{1}{2}$, respectively). Quarks and leptons (and separately anti-quarks and anti-leptons) have the same $SO(7,1)$ part. They distinguish only in the $SU(3) \times U(1)$ part: Quarks are triplets of three colours ($c^i = (\tau^{33}, \tau^{38}) = [(\frac{1}{2}, \frac{1}{2\sqrt{3}}), (-\frac{1}{2}, \frac{1}{2\sqrt{3}}), (0, -\frac{1}{\sqrt{3}})]$, Eq. (6)) carrying the "fermion charge" ($\tau^4 = \frac{1}{6}$, Eq. (6)). The colourless leptons carry the "fermion charge" ($\tau^4 = -\frac{1}{2}$).

The same multiplet contains also the left handed weak ($SU(2)_I$) chargeless and $SU(2)_{II}$ charged anti-quarks and anti-leptons and the right handed weak ($SU(2)_I$) charged and $SU(2)_{II}$ chargeless anti-quarks and anti-leptons. Anti-quarks are anti-triplets, carrying the "fermion charge" ($\tau^4 = -\frac{1}{6}$). The anti-colourless anti-leptons carry the "fermion charge" ($\tau^4 = \frac{1}{2}$). S^{12} defines the ordinary spin $\pm\frac{1}{2}$. $Y = (\tau^{23} + \tau^4)$ is the hyper charge, the electromagnetic charge is $Q = (\tau^{13} + Y)$. The vacuum state, on which the nilpotents and projectors operate, is not shown.

All these properties of states can be read directly from the table. *Example 1. and 2.* demonstrate how this can be done.

The states of opposite charges (anti-particle states) are reachable from the particle states (besides by S^{ab}) also by the application of the discrete symmetry operator $\mathcal{C}_{\mathcal{N}}\, \mathcal{P}_{\mathcal{N}}$, presented in Refs. 12 and in the footnote of this subsection.

In Table 1 the starting state is chosen to be $\stackrel{03}{(+i)}\stackrel{12}{(+)} | \stackrel{56}{(+)}\stackrel{78}{(+)} || \stackrel{9\,10}{(+)}\;\stackrel{11\,12}{(-)}\;\stackrel{13\,14}{(-)}$. We could make any other choice of products of nilpotents and projectors, let say

Table 1. The left handed ($\Gamma^{(13,1)} = -1$, Eq. (9)) multiplet of spinors — the members of (one family of) the fundamental representation of the $SO(13,1)$ group anti-quarks and the colourless leptons and anti-leptons, with the charges, spin and handedness manifesting in the low energy regime — is presented in the massless basis using the technique[2,3,5], explained in the text and in *Examples 1.,2.*.

i		$\vert^a\psi_i>$, $\Gamma^{(7,1)} = (-1)\,1\,,\,\Gamma^{(6)} = (1)\,-1$ (Anti)octet of (anti)quarks and (anti)leptons	$\Gamma^{(3,1)}$	S^{12}	τ^{13}	τ^{23}	τ^{33}	τ^{38}	τ^4	Y	Q
1	u_R^{c1}	$\stackrel{03}{(+i)}\stackrel{12}{(+)}\vert\stackrel{56}{(+)}\stackrel{78}{(+)}\vert\vert\stackrel{9\,10}{(+)}\;\stackrel{11\,12}{(-)}\;\stackrel{13\,14}{(-)}$	1	$\frac{1}{2}$	0	$\frac{1}{2}$	$\frac{1}{2}$	$\frac{1}{2\sqrt{3}}$	$\frac{1}{6}$	$\frac{2}{3}$	$\frac{2}{3}$
2	u_R^{c1}	$\stackrel{03}{[-i]}\stackrel{12}{[-]}\vert\stackrel{56}{(+)}\stackrel{78}{(+)}\vert\vert\stackrel{9\,10}{(+)}\;\stackrel{11\,12}{(-)}\;\stackrel{13\,14}{(-)}$	1	$-\frac{1}{2}$	0	$\frac{1}{2}$	$\frac{1}{2}$	$\frac{1}{2\sqrt{3}}$	$\frac{1}{6}$	$\frac{2}{3}$	$\frac{2}{3}$
3	d_R^{c1}	$\stackrel{03}{(+i)}\stackrel{12}{(+)}\vert\stackrel{56}{[-]}\stackrel{78}{[-]}\vert\vert\stackrel{9\,10}{(+)}\;\stackrel{11\,12}{(-)}\;\stackrel{13\,14}{(-)}$	1	$\frac{1}{2}$	0	$-\frac{1}{2}$	$\frac{1}{2}$	$\frac{1}{2\sqrt{3}}$	$\frac{1}{6}$	$-\frac{1}{3}$	$-\frac{1}{3}$
4	d_R^{c1}	$\stackrel{03}{[-i]}\stackrel{12}{[-]}\vert\stackrel{56}{[-]}\stackrel{78}{[-]}\vert\vert\stackrel{9\,10}{(+)}\;\stackrel{11\,12}{(-)}\;\stackrel{13\,14}{(-)}$	1	$-\frac{1}{2}$	0	$-\frac{1}{2}$	$\frac{1}{2}$	$\frac{1}{2\sqrt{3}}$	$\frac{1}{6}$	$-\frac{1}{3}$	$-\frac{1}{3}$
5	d_L^{c1}	$\stackrel{03}{[-i]}\stackrel{12}{(+)}\vert\stackrel{56}{[-]}\stackrel{78}{(+)}\vert\vert\stackrel{9\,10}{(+)}\;\stackrel{11\,12}{(-)}\;\stackrel{13\,14}{(-)}$	-1	$\frac{1}{2}$	$-\frac{1}{2}$	0	$\frac{1}{2}$	$\frac{1}{2\sqrt{3}}$	$\frac{1}{6}$	$\frac{1}{6}$	$-\frac{1}{3}$
6	d_L^{c1}	$\stackrel{03}{(+i)}\stackrel{12}{[-]}\vert\stackrel{56}{[-]}\stackrel{78}{(+)}\vert\vert\stackrel{9\,10}{(+)}\;\stackrel{11\,12}{(-)}\;\stackrel{13\,14}{(-)}$	-1	$-\frac{1}{2}$	$-\frac{1}{2}$	0	$\frac{1}{2}$	$\frac{1}{2\sqrt{3}}$	$\frac{1}{6}$	$\frac{1}{6}$	$-\frac{1}{3}$
7	u_L^{c1}	$\stackrel{03}{[-i]}\stackrel{12}{(+)}\vert\stackrel{56}{(+)}\stackrel{78}{[-]}\vert\vert\stackrel{9\,10}{(+)}\;\stackrel{11\,12}{(-)}\;\stackrel{13\,14}{(-)}$	-1	$\frac{1}{2}$	$\frac{1}{2}$	0	$\frac{1}{2}$	$\frac{1}{2\sqrt{3}}$	$\frac{1}{6}$	$\frac{1}{6}$	$\frac{2}{3}$
8	u_L^{c1}	$\stackrel{03}{(+i)}\stackrel{12}{[-]}\vert\stackrel{56}{(+)}\stackrel{78}{[-]}\vert\vert\stackrel{9\,10}{(+)}\;\stackrel{11\,12}{(-)}\;\stackrel{13\,14}{(-)}$	-1	$-\frac{1}{2}$	$\frac{1}{2}$	0	$\frac{1}{2}$	$\frac{1}{2\sqrt{3}}$	$\frac{1}{6}$	$\frac{1}{6}$	$\frac{2}{3}$
9	u_R^{c2}	$\stackrel{03}{(+i)}\stackrel{12}{(+)}\vert\stackrel{56}{(+)}\stackrel{78}{(+)}\vert\vert\stackrel{9\,10}{[-]}\;\stackrel{11\,12}{[+]}\;\stackrel{13\,14}{(-)}$	1	$\frac{1}{2}$	0	$\frac{1}{2}$	$-\frac{1}{2}$	$\frac{1}{2\sqrt{3}}$	$\frac{1}{6}$	$\frac{2}{3}$	$\frac{2}{3}$
10	u_R^{c2}	$\stackrel{03}{[-i]}\stackrel{12}{[-]}\vert\stackrel{56}{(+)}\stackrel{78}{(+)}\vert\vert\stackrel{9\,10}{[-]}\;\stackrel{11\,12}{[+]}\;\stackrel{13\,14}{(-)}$	1	$-\frac{1}{2}$	0	$\frac{1}{2}$	$-\frac{1}{2}$	$\frac{1}{2\sqrt{3}}$	$\frac{1}{6}$	$\frac{2}{3}$	$\frac{2}{3}$
11	d_R^{c2}	$\stackrel{03}{(+i)}\stackrel{12}{(+)}\vert\stackrel{56}{[-]}\stackrel{78}{[-]}\vert\vert\stackrel{9\,10}{[-]}\;\stackrel{11\,12}{[+]}\;\stackrel{13\,14}{(-)}$	1	$\frac{1}{2}$	0	$-\frac{1}{2}$	$-\frac{1}{2}$	$\frac{1}{2\sqrt{3}}$	$\frac{1}{6}$	$-\frac{1}{3}$	$-\frac{1}{3}$
12	d_R^{c2}	$\stackrel{03}{[-i]}\stackrel{12}{[-]}\vert\stackrel{56}{[-]}\stackrel{78}{[-]}\vert\vert\stackrel{9\,10}{[-]}\;\stackrel{11\,12}{[+]}\;\stackrel{13\,14}{(-)}$	1	$-\frac{1}{2}$	0	$-\frac{1}{2}$	$-\frac{1}{2}$	$\frac{1}{2\sqrt{3}}$	$\frac{1}{6}$	$-\frac{1}{3}$	$-\frac{1}{3}$
13	d_L^{c2}	$\stackrel{03}{[-i]}\stackrel{12}{(+)}\vert\stackrel{56}{[-]}\stackrel{78}{(+)}\vert\vert\stackrel{9\,10}{[-]}\;\stackrel{11\,12}{[+]}\;\stackrel{13\,14}{(-)}$	-1	$\frac{1}{2}$	$-\frac{1}{2}$	0	$-\frac{1}{2}$	$\frac{1}{2\sqrt{3}}$	$\frac{1}{6}$	$\frac{1}{6}$	$-\frac{1}{3}$
14	d_L^{c2}	$\stackrel{03}{(+i)}\stackrel{12}{[-]}\vert\stackrel{56}{[-]}\stackrel{78}{(+)}\vert\vert\stackrel{9\,10}{[-]}\;\stackrel{11\,12}{[+]}\;\stackrel{13\,14}{(-)}$	-1	$-\frac{1}{2}$	$-\frac{1}{2}$	0	$-\frac{1}{2}$	$\frac{1}{2\sqrt{3}}$	$\frac{1}{6}$	$\frac{1}{6}$	$-\frac{1}{3}$
15	u_L^{c2}	$\stackrel{03}{[-i]}\stackrel{12}{(+)}\vert\stackrel{56}{(+)}\stackrel{78}{[-]}\vert\vert\stackrel{9\,10}{[-]}\;\stackrel{11\,12}{[+]}\;\stackrel{13\,14}{(-)}$	-1	$\frac{1}{2}$	$\frac{1}{2}$	0	$-\frac{1}{2}$	$\frac{1}{2\sqrt{3}}$	$\frac{1}{6}$	$\frac{1}{6}$	$\frac{2}{3}$
16	u_L^{c2}	$\stackrel{03}{(+i)}\stackrel{12}{[-]}\vert\stackrel{56}{(+)}\stackrel{78}{[-]}\vert\vert\stackrel{9\,10}{[-]}\;\stackrel{11\,12}{[+]}\;\stackrel{13\,14}{(-)}$	-1	$-\frac{1}{2}$	$\frac{1}{2}$	0	$-\frac{1}{2}$	$\frac{1}{2\sqrt{3}}$	$\frac{1}{6}$	$\frac{1}{6}$	$\frac{2}{3}$

i		$\vert^a\psi_i>$, $\Gamma^{(7,1)} = (-1)\,1$, $\Gamma^{(6)} = (1)\,-1$ (Anti)octet of (anti)quarks and (anti)leptons	$\Gamma^{(3,1)}$	S^{12}	τ^{13}	τ^{23}	τ^{33}	τ^{38}	τ^{4}	Y	Q
17	u_R^{c3}	$\stackrel{03}{(+i)}\stackrel{12}{(+)}\vert\stackrel{56}{(+)}\stackrel{78}{(+)}\Vert\stackrel{9\,10}{[-]}\stackrel{11\,12}{(-)}\stackrel{13\,14}{[+]}$	1	$\frac{1}{2}$	0	$\frac{1}{2}$	0	$-\frac{1}{\sqrt{3}}$	$\frac{1}{6}$	$\frac{2}{3}$	$\frac{2}{3}$
18	u_R^{c3}	$\stackrel{03}{[-i]}\stackrel{12}{[-]}\vert\stackrel{56}{(+)}\stackrel{78}{(+)}\Vert\stackrel{9\,10}{[-]}\stackrel{11\,12}{(-)}\stackrel{13\,14}{[+]}$	1	$-\frac{1}{2}$	0	$\frac{1}{2}$	0	$-\frac{1}{\sqrt{3}}$	$\frac{1}{6}$	$\frac{2}{3}$	$\frac{2}{3}$
19	d_R^{c3}	$\stackrel{03}{(+i)}\stackrel{12}{(+)}\vert\stackrel{56}{[-]}\stackrel{78}{[-]}\Vert\stackrel{9\,10}{[-]}\stackrel{11\,12}{(-)}\stackrel{13\,14}{[+]}$	1	$\frac{1}{2}$	0	$-\frac{1}{2}$	0	$-\frac{1}{\sqrt{3}}$	$\frac{1}{6}$	$-\frac{1}{3}$	$-\frac{1}{3}$
20	d_R^{c3}	$\stackrel{03}{[-i]}\stackrel{12}{[-]}\vert\stackrel{56}{[-]}\stackrel{78}{[-]}\Vert\stackrel{9\,10}{[-]}\stackrel{11\,12}{(-)}\stackrel{13\,14}{[+]}$	1	$-\frac{1}{2}$	0	$-\frac{1}{2}$	0	$-\frac{1}{\sqrt{3}}$	$\frac{1}{6}$	$-\frac{1}{3}$	$-\frac{1}{3}$
21	d_L^{c3}	$\stackrel{03}{[-i]}\stackrel{12}{(+)}\vert\stackrel{56}{[-]}\stackrel{78}{(+)}\Vert\stackrel{9\,10}{[-]}\stackrel{11\,12}{(-)}\stackrel{13\,14}{[+]}$	-1	$\frac{1}{2}$	$-\frac{1}{2}$	0	0	$-\frac{1}{\sqrt{3}}$	$\frac{1}{6}$	$\frac{1}{6}$	$-\frac{1}{3}$
22	d_L^{c3}	$\stackrel{03}{(+i)}\stackrel{12}{[-]}\vert\stackrel{56}{[-]}\stackrel{78}{(+)}\Vert\stackrel{9\,10}{[-]}\stackrel{11\,12}{(-)}\stackrel{13\,14}{[+]}$	-1	$-\frac{1}{2}$	$-\frac{1}{2}$	0	0	$-\frac{1}{\sqrt{3}}$	$\frac{1}{6}$	$\frac{1}{6}$	$-\frac{1}{3}$
23	u_L^{c3}	$\stackrel{03}{[-i]}\stackrel{12}{(+)}\vert\stackrel{56}{(+)}\stackrel{78}{[-]}\Vert\stackrel{9\,10}{[-]}\stackrel{11\,12}{(-)}\stackrel{13\,14}{[+]}$	-1	$\frac{1}{2}$	$\frac{1}{2}$	0	0	$-\frac{1}{\sqrt{3}}$	$\frac{1}{6}$	$\frac{1}{6}$	$\frac{2}{3}$
24	u_L^{c3}	$\stackrel{03}{(+i)}\stackrel{12}{[-]}\vert\stackrel{56}{(+)}\stackrel{78}{[-]}\Vert\stackrel{9\,10}{[-]}\stackrel{11\,12}{(-)}\stackrel{13\,14}{[+]}$	-1	$-\frac{1}{2}$	$\frac{1}{2}$	0	0	$-\frac{1}{\sqrt{3}}$	$\frac{1}{6}$	$\frac{1}{6}$	$\frac{2}{3}$
25	ν_R	$\stackrel{03}{(+i)}\stackrel{12}{(+)}\vert\stackrel{56}{(+)}\stackrel{78}{(+)}\Vert\stackrel{9\,10}{(+)}\stackrel{11\,12}{[+]}\stackrel{13\,14}{[+]}$	1	$\frac{1}{2}$	0	$\frac{1}{2}$	0	0	$-\frac{1}{2}$	0	0
26	ν_R	$\stackrel{03}{[-i]}\stackrel{12}{[-]}\vert\stackrel{56}{(+)}\stackrel{78}{(+)}\Vert\stackrel{9\,10}{(+)}\stackrel{11\,12}{[+]}\stackrel{13\,14}{[+]}$	1	$-\frac{1}{2}$	0	$\frac{1}{2}$	0	0	$-\frac{1}{2}$	0	0
27	e_R	$\stackrel{03}{(+i)}\stackrel{12}{(+)}\vert\stackrel{56}{[-]}\stackrel{78}{[-]}\Vert\stackrel{9\,10}{(+)}\stackrel{11\,12}{[+]}\stackrel{13\,14}{[+]}$	1	$\frac{1}{2}$	0	$-\frac{1}{2}$	0	0	$-\frac{1}{2}$	-1	-1
28	e_R	$\stackrel{03}{[-i]}\stackrel{12}{[-]}\vert\stackrel{56}{[-]}\stackrel{78}{[-]}\Vert\stackrel{9\,10}{(+)}\stackrel{11\,12}{[+]}\stackrel{13\,14}{[+]}$	1	$-\frac{1}{2}$	0	$-\frac{1}{2}$	0	0	$-\frac{1}{2}$	-1	-1
29	e_L	$\stackrel{03}{[-i]}\stackrel{12}{(+)}\vert\stackrel{56}{[-]}\stackrel{78}{(+)}\Vert\stackrel{9\,10}{(+)}\stackrel{11\,12}{[+]}\stackrel{13\,14}{[+]}$	-1	$\frac{1}{2}$	$-\frac{1}{2}$	0	0	0	$-\frac{1}{2}$	$-\frac{1}{2}$	-1
30	e_L	$\stackrel{03}{(+i)}\stackrel{12}{[-]}\vert\stackrel{56}{[-]}\stackrel{78}{(+)}\Vert\stackrel{9\,10}{(+)}\stackrel{11\,12}{[+]}\stackrel{13\,14}{[+]}$	-1	$-\frac{1}{2}$	$-\frac{1}{2}$	0	0	0	$-\frac{1}{2}$	$-\frac{1}{2}$	-1
31	ν_L	$\stackrel{03}{[-i]}\stackrel{12}{(+)}\vert\stackrel{56}{(+)}\stackrel{78}{[-]}\Vert\stackrel{9\,10}{(+)}\stackrel{11\,12}{[+]}\stackrel{13\,14}{[+]}$	-1	$\frac{1}{2}$	$\frac{1}{2}$	0	0	0	$-\frac{1}{2}$	$-\frac{1}{2}$	0
32	ν_L	$\stackrel{03}{(+i)}\stackrel{12}{[-]}\vert\stackrel{56}{(+)}\stackrel{78}{[-]}\Vert\stackrel{9\,10}{(+)}\stackrel{11\,12}{[+]}\stackrel{13\,14}{[+]}$	-1	$-\frac{1}{2}$	$\frac{1}{2}$	0	0	0	$-\frac{1}{2}$	$-\frac{1}{2}$	0

(Continued)

Table 1. (*Continued*)

i		$\vert^a\psi_i>,\ \Gamma^{(7,1)} = (-1)\,1\,,\ \Gamma^{(6)} = (1)\,-1$ (Anti)octet of (anti)quarks and (anti)leptons	$\Gamma^{(3,1)}$	S^{12}	τ^{13}	τ^{23}	τ^{33}	τ^{38}	τ^4	Y	Q
33	$\bar{d}^{\bar{c}1}_L$	$\stackrel{03}{[-i]}\stackrel{12}{(+)}\vert\stackrel{56}{(+)}\stackrel{78}{(+)}\Vert\stackrel{9\,10}{[-]}\stackrel{11\,12}{[+]}\stackrel{13\,14}{[+]}$	-1	$\frac{1}{2}$	0	$\frac{1}{2}$	$-\frac{1}{2}$	$-\frac{1}{2\sqrt{3}}$	$-\frac{1}{6}$	$\frac{1}{3}$	$\frac{1}{3}$
34	$\bar{d}^{\bar{c}1}_L$	$\stackrel{03}{(+i)}\stackrel{12}{[-]}\vert\stackrel{56}{(+)}\stackrel{78}{(+)}\Vert\stackrel{9\,10}{[-]}\stackrel{11\,12}{[+]}\stackrel{13\,14}{[+]}$	-1	$-\frac{1}{2}$	0	$\frac{1}{2}$	$-\frac{1}{2}$	$-\frac{1}{2\sqrt{3}}$	$-\frac{1}{6}$	$\frac{1}{3}$	$\frac{1}{3}$
35	$\bar{u}^{\bar{c}1}_L$	$\stackrel{03}{[-i]}\stackrel{12}{(+)}\vert\stackrel{56}{[-]}\stackrel{78}{[-]}\Vert\stackrel{9\,10}{[-]}\stackrel{11\,12}{[+]}\stackrel{13\,14}{[+]}$	-1	$\frac{1}{2}$	0	$-\frac{1}{2}$	$-\frac{1}{2}$	$-\frac{1}{2\sqrt{3}}$	$-\frac{1}{6}$	$-\frac{2}{3}$	$-\frac{2}{3}$
36	$\bar{u}^{\bar{c}1}_L$	$\stackrel{03}{(+i)}\stackrel{12}{[-]}\vert\stackrel{56}{[-]}\stackrel{78}{[-]}\Vert\stackrel{9\,10}{[-]}\stackrel{11\,12}{[+]}\stackrel{13\,14}{[+]}$	-1	$-\frac{1}{2}$	0	$-\frac{1}{2}$	$-\frac{1}{2}$	$-\frac{1}{2\sqrt{3}}$	$-\frac{1}{6}$	$-\frac{2}{3}$	$-\frac{2}{3}$
37	$\bar{d}^{\bar{c}1}_R$	$\stackrel{03}{(+i)}\stackrel{12}{(+)}\vert\stackrel{56}{(+)}\stackrel{78}{[-]}\Vert\stackrel{9\,10}{[-]}\stackrel{11\,12}{[+]}\stackrel{13\,14}{[+]}$	1	$\frac{1}{2}$	$\frac{1}{2}$	0	$-\frac{1}{2}$	$-\frac{1}{2\sqrt{3}}$	$-\frac{1}{6}$	$-\frac{1}{6}$	$\frac{1}{3}$
38	$\bar{d}^{\bar{c}1}_R$	$\stackrel{03}{[-i]}\stackrel{12}{[-]}\vert\stackrel{56}{(+)}\stackrel{78}{[-]}\Vert\stackrel{9\,10}{[-]}\stackrel{11\,12}{[+]}\stackrel{13\,14}{[+]}$	1	$-\frac{1}{2}$	$\frac{1}{2}$	0	$-\frac{1}{2}$	$-\frac{1}{2\sqrt{3}}$	$-\frac{1}{6}$	$-\frac{1}{6}$	$\frac{1}{3}$
39	$\bar{u}^{\bar{c}1}_R$	$\stackrel{03}{(+i)}\stackrel{12}{(+)}\vert\stackrel{56}{[-]}\stackrel{78}{(+)}\Vert\stackrel{9\,10}{[-]}\stackrel{11\,12}{[+]}\stackrel{13\,14}{[+]}$	1	$\frac{1}{2}$	$-\frac{1}{2}$	0	$-\frac{1}{2}$	$-\frac{1}{2\sqrt{3}}$	$-\frac{1}{6}$	$-\frac{1}{6}$	$-\frac{2}{3}$
40	$\bar{u}^{\bar{c}1}_R$	$\stackrel{03}{[-i]}\stackrel{12}{[-]}\vert\stackrel{56}{[-]}\stackrel{78}{(+)}\Vert\stackrel{9\,10}{[-]}\stackrel{11\,12}{[+]}\stackrel{13\,14}{[+]}$	1	$-\frac{1}{2}$	$-\frac{1}{2}$	0	$-\frac{1}{2}$	$-\frac{1}{2\sqrt{3}}$	$-\frac{1}{6}$	$-\frac{1}{6}$	$-\frac{2}{3}$
41	$\bar{d}^{\bar{c}2}_L$	$\stackrel{03}{[-i]}\stackrel{12}{(+)}\vert\stackrel{56}{(+)}\stackrel{78}{(+)}\Vert\stackrel{9\,10}{(+)}\stackrel{11\,12}{(-)}\stackrel{13\,14}{[+]}$	-1	$\frac{1}{2}$	0	$\frac{1}{2}$	$\frac{1}{2}$	$-\frac{1}{2\sqrt{3}}$	$-\frac{1}{6}$	$\frac{1}{3}$	$\frac{1}{3}$
42	$\bar{d}^{\bar{c}2}_L$	$\stackrel{03}{(+i)}\stackrel{12}{[-]}\vert\stackrel{56}{(+)}\stackrel{78}{(+)}\Vert\stackrel{9\,10}{(+)}\stackrel{11\,12}{(-)}\stackrel{13\,14}{[+]}$	-1	$-\frac{1}{2}$	0	$\frac{1}{2}$	$\frac{1}{2}$	$-\frac{1}{2\sqrt{3}}$	$-\frac{1}{6}$	$\frac{1}{3}$	$\frac{1}{3}$
43	$\bar{u}^{\bar{c}2}_L$	$\stackrel{03}{[-i]}\stackrel{12}{(+)}\vert\stackrel{56}{[-]}\stackrel{78}{[-]}\Vert\stackrel{9\,10}{(+)}\stackrel{11\,12}{(-)}\stackrel{13\,14}{[+]}$	-1	$\frac{1}{2}$	0	$-\frac{1}{2}$	$\frac{1}{2}$	$-\frac{1}{2\sqrt{3}}$	$-\frac{1}{6}$	$-\frac{2}{3}$	$-\frac{2}{3}$
44	$\bar{u}^{\bar{c}2}_L$	$\stackrel{03}{(+i)}\stackrel{12}{[-]}\vert\stackrel{56}{[-]}\stackrel{78}{[-]}\Vert\stackrel{9\,10}{(+)}\stackrel{11\,12}{(-)}\stackrel{13\,14}{[+]}$	-1	$-\frac{1}{2}$	0	$-\frac{1}{2}$	$\frac{1}{2}$	$-\frac{1}{2\sqrt{3}}$	$-\frac{1}{6}$	$-\frac{2}{3}$	$-\frac{2}{3}$
45	$\bar{d}^{\bar{c}2}_R$	$\stackrel{03}{(+i)}\stackrel{12}{(+)}\vert\stackrel{56}{(+)}\stackrel{78}{[-]}\Vert\stackrel{9\,10}{(+)}\stackrel{11\,12}{(-)}\stackrel{13\,14}{[+]}$	1	$\frac{1}{2}$	$\frac{1}{2}$	0	$\frac{1}{2}$	$-\frac{1}{2\sqrt{3}}$	$-\frac{1}{6}$	$-\frac{1}{6}$	$\frac{1}{3}$
46	$\bar{d}^{\bar{c}2}_R$	$\stackrel{03}{[-i]}\stackrel{12}{[-]}\vert\stackrel{56}{(+)}\stackrel{78}{[-]}\Vert\stackrel{9\,10}{(+)}\stackrel{11\,12}{(-)}\stackrel{13\,14}{[+]}$	1	$-\frac{1}{2}$	$\frac{1}{2}$	0	$\frac{1}{2}$	$-\frac{1}{2\sqrt{3}}$	$-\frac{1}{6}$	$-\frac{1}{6}$	$\frac{1}{3}$
47	$\bar{u}^{\bar{c}2}_R$	$\stackrel{03}{(+i)}\stackrel{12}{(+)}\vert\stackrel{56}{[-]}\stackrel{78}{(+)}\Vert\stackrel{9\,10}{(+)}\stackrel{11\,12}{(-)}\stackrel{13\,14}{[+]}$	1	$\frac{1}{2}$	$-\frac{1}{2}$	0	$\frac{1}{2}$	$-\frac{1}{2\sqrt{3}}$	$-\frac{1}{6}$	$-\frac{1}{6}$	$-\frac{2}{3}$
48	$\bar{u}^{\bar{c}2}_R$	$\stackrel{03}{[-i]}\stackrel{12}{[-]}\vert\stackrel{56}{[-]}\stackrel{78}{(+)}\Vert\stackrel{9\,10}{(+)}\stackrel{11\,12}{(-)}\stackrel{13\,14}{[+]}$	1	$-\frac{1}{2}$	$-\frac{1}{2}$	0	$\frac{1}{2}$	$-\frac{1}{2\sqrt{3}}$	$-\frac{1}{6}$	$-\frac{1}{6}$	$-\frac{2}{3}$

i		$\vert^a\psi_i>$, $\Gamma^{(7,1)} = (-1)\,1\,,\,\Gamma^{(6)} = (1)\,-1$ (Anti)octet of (anti)quarks and (anti)leptons	$\Gamma^{(3,1)}$	S^{12}	τ^{13}	τ^{23}	τ^{33}	τ^{38}	τ^{4}	Y	Q
49	$\bar{d}_{L}^{\bar{c}3}$	$\stackrel{03}{[-i]}\stackrel{12}{(+)}\vert\stackrel{56}{(+)}\stackrel{78}{(+)}\vert\vert\stackrel{9\,10}{(+)}\;\stackrel{11\,12}{[+]}\;\stackrel{13\,14}{(-)}$	-1	$\frac{1}{2}$	0	$\frac{1}{2}$	0	$\frac{1}{\sqrt{3}}$	$-\frac{1}{6}$	$\frac{1}{3}$	$\frac{1}{3}$
50	$\bar{d}_{L}^{\bar{c}3}$	$\stackrel{03}{(+i)}\stackrel{12}{[-]}\vert\stackrel{56}{(+)}\stackrel{78}{(+)}\vert\vert\stackrel{9\,10}{(+)}\;\stackrel{11\,12}{[+]}\;\stackrel{13\,14}{(-)}$	-1	$-\frac{1}{2}$	0	$\frac{1}{2}$	0	$\frac{1}{\sqrt{3}}$	$-\frac{1}{6}$	$\frac{1}{3}$	$\frac{1}{3}$
51	$\bar{u}_{L}^{\bar{c}3}$	$\stackrel{03}{[-i]}\stackrel{12}{(+)}\vert\stackrel{56}{[-]}\stackrel{78}{[-]}\vert\vert\stackrel{9\,10}{(+)}\;\stackrel{11\,12}{[+]}\;\stackrel{13\,14}{(-)}$	-1	$\frac{1}{2}$	0	$-\frac{1}{2}$	0	$\frac{1}{\sqrt{3}}$	$-\frac{1}{6}$	$-\frac{2}{3}$	$-\frac{2}{3}$
52	$\bar{u}_{L}^{\bar{c}3}$	$\stackrel{03}{(+i)}\stackrel{12}{[-]}\vert\stackrel{56}{[-]}\stackrel{78}{[-]}\vert\vert\stackrel{9\,10}{(+)}\;\stackrel{11\,12}{[+]}\;\stackrel{13\,14}{(-)}$	-1	$-\frac{1}{2}$	0	$-\frac{1}{2}$	0	$\frac{1}{\sqrt{3}}$	$-\frac{1}{6}$	$-\frac{2}{3}$	$-\frac{2}{3}$
53	$\bar{d}_{R}^{\bar{c}3}$	$\stackrel{03}{(+i)}\stackrel{12}{(+)}\vert\stackrel{56}{(+)}\stackrel{78}{[-]}\vert\vert\stackrel{9\,10}{(+)}\;\stackrel{11\,12}{[+]}\;\stackrel{13\,14}{(-)}$	1	$\frac{1}{2}$	$\frac{1}{2}$	0	0	$\frac{1}{\sqrt{3}}$	$-\frac{1}{6}$	$-\frac{1}{6}$	$\frac{1}{3}$
54	$\bar{d}_{R}^{\bar{c}3}$	$\stackrel{03}{[-i]}\stackrel{12}{[-]}\vert\stackrel{56}{(+)}\stackrel{78}{[-]}\vert\vert\stackrel{9\,10}{(+)}\;\stackrel{11\,12}{[+]}\;\stackrel{13\,14}{(-)}$	1	$-\frac{1}{2}$	$\frac{1}{2}$	0	0	$\frac{1}{\sqrt{3}}$	$-\frac{1}{6}$	$-\frac{1}{6}$	$\frac{1}{3}$
55	$\bar{u}_{R}^{\bar{c}3}$	$\stackrel{03}{(+i)}\stackrel{12}{(+)}\vert\stackrel{56}{[-]}\stackrel{78}{(+)}\vert\vert\stackrel{9\,10}{(+)}\;\stackrel{11\,12}{[+]}\;\stackrel{13\,14}{(-)}$	1	$\frac{1}{2}$	$-\frac{1}{2}$	0	0	$\frac{1}{\sqrt{3}}$	$-\frac{1}{6}$	$-\frac{1}{6}$	$-\frac{2}{3}$
56	$\bar{u}_{R}^{\bar{c}3}$	$\stackrel{03}{[-i]}\stackrel{12}{[-]}\vert\stackrel{56}{[-]}\stackrel{78}{(+)}\vert\vert\stackrel{9\,10}{(+)}\;\stackrel{11\,12}{[+]}\;\stackrel{13\,14}{(-)}$	1	$-\frac{1}{2}$	$-\frac{1}{2}$	0	0	$\frac{1}{\sqrt{3}}$	$-\frac{1}{6}$	$-\frac{1}{6}$	$-\frac{2}{3}$
57	$\bar{e}_{L}$	$\stackrel{03}{[-i]}\stackrel{12}{(+)}\vert\stackrel{56}{(+)}\stackrel{78}{(+)}\vert\vert\stackrel{9\,10}{[-]}\;\stackrel{11\,12}{(-)}\;\stackrel{13\,14}{(-)}$	-1	$\frac{1}{2}$	0	$\frac{1}{2}$	0	0	$\frac{1}{2}$	1	1
58	$\bar{e}_{L}$	$\stackrel{03}{(+i)}\stackrel{12}{[-]}\vert\stackrel{56}{(+)}\stackrel{78}{(+)}\vert\vert\stackrel{9\,10}{[-]}\;\stackrel{11\,12}{(-)}\;\stackrel{13\,14}{(-)}$	-1	$-\frac{1}{2}$	0	$\frac{1}{2}$	0	0	$\frac{1}{2}$	1	1
59	$\bar{\nu}_{L}$	$\stackrel{03}{[-i]}\stackrel{12}{(+)}\vert\stackrel{56}{[-]}\stackrel{78}{[-]}\vert\vert\stackrel{9\,10}{[-]}\;\stackrel{11\,12}{(-)}\;\stackrel{13\,14}{(-)}$	-1	$\frac{1}{2}$	0	$-\frac{1}{2}$	0	0	$\frac{1}{2}$	0	0
60	$\bar{\nu}_{L}$	$\stackrel{03}{(+i)}\stackrel{12}{[-]}\vert\stackrel{56}{[-]}\stackrel{78}{[-]}\vert\vert\stackrel{9\,10}{[-]}\;\stackrel{11\,12}{(-)}\;\stackrel{13\,14}{(-)}$	-1	$-\frac{1}{2}$	0	$-\frac{1}{2}$	0	0	$\frac{1}{2}$	0	0
61	$\bar{\nu}_{R}$	$\stackrel{03}{(+i)}\stackrel{12}{(+)}\vert\stackrel{56}{[-]}\stackrel{78}{(+)}\vert\vert\stackrel{9\,10}{[-]}\;\stackrel{11\,12}{(-)}\;\stackrel{13\,14}{(-)}$	1	$\frac{1}{2}$	$-\frac{1}{2}$	0	0	0	$\frac{1}{2}$	$\frac{1}{2}$	0
62	$\bar{\nu}_{R}$	$\stackrel{03}{[-i]}\stackrel{12}{[-]}\vert\stackrel{56}{[-]}\stackrel{78}{(+)}\vert\vert\stackrel{9\,10}{[-]}\;\stackrel{11\,12}{(-)}\;\stackrel{13\,14}{(-)}$	1	$-\frac{1}{2}$	$-\frac{1}{2}$	0	0	0	$\frac{1}{2}$	$\frac{1}{2}$	0
63	$\bar{e}_{R}$	$\stackrel{03}{(+i)}\stackrel{12}{(+)}\vert\stackrel{56}{(+)}\stackrel{78}{[-]}\vert\vert\stackrel{9\,10}{[-]}\;\stackrel{11\,12}{(-)}\;\stackrel{13\,14}{(-)}$	1	$\frac{1}{2}$	$\frac{1}{2}$	0	0	0	$\frac{1}{2}$	$\frac{1}{2}$	1
64	$\bar{e}_{R}$	$\stackrel{03}{[-i]}\stackrel{12}{[-]}\vert\stackrel{56}{(+)}\stackrel{78}{[-]}\vert\vert\stackrel{9\,10}{[-]}\;\stackrel{11\,12}{(-)}\;\stackrel{13\,14}{(-)}$	1	$-\frac{1}{2}$	$\frac{1}{2}$	0	0	0	$\frac{1}{2}$	$\frac{1}{2}$	1

the state $\overset{03}{[-i]}\overset{12}{(+)} \mid \overset{56}{(+)}\overset{78}{[-]} \mid\mid \overset{9\,10}{(+)}\;\;\overset{11\,12}{(-)}\;\;\overset{13\,14}{(-)}$, which is the state in the seventh line of Table 1. All the states of one representation can be obtained from the starting state by applying on the starting state the generators S^{ab}. From the first state, for example, we obtain the seventh one by the application of $S^{0\,7}$ (or of $S^{0\,8}$, $S^{3\,7}$, $S^{3\,8}$).

Let us make a few examples to get inside how can one read the quantum numbers of states from 7 products of nilpotents and projectors. All nilpotents and projectors are eigen states, Eq. (11), of Cartan sub-algebra, Eq. (8).

Example 1.: Let us calculate properties of the two states: The first state — $\overset{03}{(+i)}\overset{12}{(+)} \mid \overset{56}{(+)}\overset{78}{(+)} \mid\mid \overset{9\,10}{(+)}\overset{11\,12}{(-)}\overset{13\,14}{(-)}\,|\psi_0\rangle$ — and the seventh state — $\overset{03}{[-i]}\overset{12}{(+)} \mid \overset{56}{(+)}\overset{78}{[-]} \mid\mid \overset{9\,10}{(+)}\overset{11\,12}{(-)}\overset{13\,14}{(-)}\,|\psi_0\rangle$ — of Table 1.

The handedness of the whole one Weyl representation (64 states) follows from Eqs. (9,8): $\Gamma^{(14)} = i^{13}2^7 S^{03}S^{12}\cdots S^{13\,14}$. This operator gives, when applied on the first state of Table 1, the eigenvalue $= i^{13}2^7\frac{i}{2}(\frac{1}{2})^4(-\frac{1}{2})^2 = -1$ (since the operator S^{03} applied on the nilpotent $\overset{03}{(+i)}$ gives the eigenvalue $\frac{k}{2} = \frac{i}{2}$, the rest four operators have the eigenvalues $\frac{1}{2}$, and the last two $-\frac{1}{2}$, Eq. (11)).

In an equivalent way we calculate the handedness $\Gamma^{(3,1)}$ of these two states in $d = (3+1)$: The operator $\Gamma^{(3,1)} = i^3 2^2 S^{03}S^{12}$, applied on the first state, gives 1 — the right handedness, while $\Gamma^{(3,1)}$ is for the seventh state -1 — the left handedness.

The weak charge operator $\tau^{13}(= \frac{1}{2}(S^{56} - S^{78}))$, Eq. (5), applied on the first state, gives the eigenvalue 0: $\frac{1}{2}(\frac{1}{2} - \frac{1}{2})$, The eigenvalue of τ^{13} is for the seventh state $\frac{1}{2}$: $\frac{1}{2}(\frac{1}{2} - (-\frac{1}{2}))$, τ^{23} $(= \frac{1}{2}(S^{56} + S^{78}))$, applied on the first state, gives as its eigenvalue $\frac{1}{2}$, while when τ^{23} applies on the seventh state gives 0. The "fermion charge" operator τ^4 $(= -\frac{1}{3}(S^{9\,10} + S^{11\,12} + S^{13\,14})$, Eq. (6)) gives, when applied on any of these two states, the eigenvalues $-\frac{1}{3}(\frac{1}{2} - \frac{1}{2} - \frac{1}{2}) = \frac{1}{6}$. Correspondingly is the hyper charge Y $(= \tau^{23} + \tau^4)$ of these two states $Y = (\frac{2}{3}, \frac{1}{6})$, respectively, what the *standard model* assumes for u_R and u_L, respectively.

One finds for the colour charge of these two states, (τ^{33}, τ^{38}) $(= (\frac{1}{2}(S^{9\,10} - S^{11\,12}), \frac{1}{\sqrt{3}}(S^{9\,10} + S^{11\,12} - 2S^{13\,14}))$ the eigenvalues $(1/2, 1/(2\sqrt{3}))$.

The first and the seventh states differ in the handedness $\Gamma^{(3,1)} = (1, -1)$, in the weak charge τ^{13} $=(0, \frac{1}{2})$ and the hyper charge $Y = (\frac{2}{3}, \frac{1}{6})$, respectively. All the states of this octet — $SO(7,1)$ — have the same colour charge and the same "fermion charge" (the difference in the hyper charge Y is caused by the difference in $\tau^{23} = (\frac{1}{2}, 0)$).

The states for the d_R-quark and d_L-quark of the same octet follow from the state u_R and u_L, respectively, by the application of S^{57} (or S^{58}, S^{67}, S^{68}).

All the $SO(7,1)$ $(\Gamma^{(7,1)} = 1)$ part of the $SO(13,1)$ spinor representation are the same for either quarks of all the three colours (quarks states appear in Table 1 from the first to the 24th line) or for the colourless leptons (leptons appear in Table 1 from the 25th line to the 32nd line).

Leptons distinguish from quarks in the part represented by nilpotents and projectors, which is determined by the eigenstates of the Cartan subalgebra of $(S^{9\,10}, S^{11\,12}, S^{13\,14})$. Taking into account Eq. (11) one calculates that (τ^{33}, τ^{38}) is for the colourless part of the lepton states $(\nu_{R,L}, e_{R,L})$ — $(\cdots || \stackrel{9\,10}{(+)}\; \stackrel{11\,12}{[+]}\; \stackrel{13\,14}{[+]})$ — equal to $= (0, 0)$, while the "fermion charge" τ^4 is for these states equal to $-\frac{1}{2}$ (just as assumed by the *standard model*).

Let us point out that the octet $SO(7,1)$ manifests how the spin and the weak and hyper charges are related.

Example 2.: Let us look at the properties of the anti-quark and anti-lepton states of one fundamental representation of the $SO(13,1)$ group. These states are presented in Table 1 from the 33rd line to the 64th line, representing anti-quarks (the first three octets) and anti-leptons (the last octet).

Again, all the anti-octets, the $SO(7,1)$ $(\Gamma^{(7,1)} = -1)$ part of the $SO(13,1)$ representation, are the same either for anti-quarks or for anti-leptons. The last three products of nilpotents and projectors (the part appearing in Table 1 after "$||$") determine anti-colours for the anti-quarks states and the anti-colourless state for anti-leptons.

Let us add that all the anti-spinor states are reachable from the spinor states (and opposite) by the application of the operator[12] $\mathbb{C}_{\mathcal{N}} \mathcal{P}_{\mathcal{N}}$ [d]. The part of this operator, which operates on only the spinor part of the state (presented in Table 1), is $\mathbb{C}_{\mathcal{N}} \mathcal{P}_{\mathcal{N}}|_{spinor} = \gamma^0 \prod^{d}_{\Im \gamma^s, s=5} \gamma^s$. Taking into account Eq. (12) and this operator one finds that $\mathbb{C}_{\mathcal{N}} \mathcal{P}_{\mathcal{N}}|_{spinor}$ transforms u^{c1}_{R} from the first line of Table 1 into $\bar{u}^{\bar{c1}}_{L}$ from the 35th line of the same table. When the operator $\mathbb{C}_{\mathcal{N}} \mathcal{P}_{\mathcal{N}}|_{spinor}$ applies on ν_R (the 25th line of the same table, with the colour chargeless part equal to $\cdots || \stackrel{9\,10}{(+)}\; \stackrel{11\,12}{[+]}\; \stackrel{13\,14}{[+]})$, transforms ν_R into $\bar{\nu}_L$ (the 59th line of the table, with the colour anti-chargeless part equal to $(\cdots || \stackrel{9\,10}{[-]}\; \stackrel{11\,12}{(-)}\; \stackrel{13\,14}{(-)})$.

2.2. *A short inside into families in the spin-charge-family theory*

The operators $\tilde{S}^{ab}$, commuting with S^{ab} (Eq. (7)), transform any spinor state, presented in Table 1, to the same state of another family, orthogonal to the starting state and correspondingly to all the states of the starting family.

[d] Discrete symmetries in $d = (3+1)$ follow from the corresponding definition of these symmetries in d- dimensional space[12]. This operator is defined as: $\mathbb{C}_{\mathcal{N}} \mathcal{P}_{\mathcal{N}} = \gamma^0 \prod^{d}_{\Im \gamma^s, s=5} \gamma^s \, I_{\vec{x}_3} \, I_{x^6, x^8, \dots, x^d}$, where γ^0 and γ^1 are real, γ^2 imaginary, γ^3 real, γ^5 imaginary, γ^6 real, alternating imaginary and real up to γ^d, which is in even dimensional spaces real. γ^a's appear in the ascending order. Operators I operate as follows: $I_{x^0} x^0 = -x^0$; $I_x x^a = -x^a$; $I_{x^0} x^a = (-x^0, \vec{x})$; $I_{\vec{x}} \vec{x} = -\vec{x}$; $I_{\vec{x}_3} x^a = (x^0, -x^1, -x^2, -x^3, x^5, x^6, \dots, x^d)$; $I_{x^5, x^7, \dots, x^{d-1}} \, (x^0, x^1, x^2, x^3, x^5, x^6, x^7, x^8, \dots, x^{d-1}, x^d) = (x^0, x^1, x^2, x^3, -x^5, x^6, -x^7, \dots, -x^{d-1}, x^d)$; $I_{x^6, x^8, \dots, x^d} \, (x^0, x^1, x^2, x^3, x^5, x^6, x^7, x^8, \dots, x^{d-1}, x^d) = (x^0, x^1, x^2, x^3, x^5, -x^6, x^7, -x^8, \dots, x^{d-1}, -x^d)$, $d = 2n$.

Applying the operator $\tilde{S}^{01}$ $(= \frac{i}{2}\tilde{\gamma}^0\tilde{\gamma}^1)$, for example, on ν_R (the 25th line of Table 1 and the last line on Table 2), one obtains, taking into account Eq. (12), the ν_{R7} state belonging to another family, presented in the seventh line of Table 2.

Operators S^{ab} transform ν_R (the 25th line of Table 1, presented in Table 2 in the eighth line, carrying the name ν_{R8}) into all the rest of the 64 states of this eighth family, presented in Table 1. The operator $S^{11\,13}$, for example, transforms ν_{R8} into u_{R8} (presented in the first line of Table 1), while it transforms ν_{R7} into u_{R7}.

Table 2 represents eight families of neutrinos, which distinguish among themselves in the family quantum numbers: ($\tilde{\tau}^{13}$, $\tilde{N}_L$, $\tilde{\tau}^{23}$, $\tilde{N}_R$, $\tilde{\tau}^4$). These family quantum numbers can be expressed by $\tilde{S}^{ab}$ as presented in Eqs. (4, 5, 6), if S^{ab} are replaced by $\tilde{S}^{ab}$.

Eight families decouples into two groups of four families, one (I) is a doublet with respect to ($\vec{\tilde{N}}_L$ and $\vec{\tilde{\tau}}^1$) and a singlet with respect to ($\vec{\tilde{N}}_R$ and $\vec{\tilde{\tau}}^2$), the other (II) is a singlet with respect to ($\vec{\tilde{N}}_L$ and $\vec{\tilde{\tau}}^1$) and a doublet with with respect to ($\vec{\tilde{N}}_R$ and $\vec{\tilde{\tau}}^2$).

All the families follow from the starting one by the application of the operators ($\tilde{N}^{\pm}_{R,L}$, $\tilde{\tau}^{(2,1)\pm}$), Eq. (A.3). The generators ($N^{\pm}_{R,L}$, $\tau^{(2,1)\pm}$), Eq. (A.3), transform ν_{R1} to all the members belonging to the $SO(7,1)$ group of one family, S^{st}, $(s,t) = (9\cdots,14)$ transform quarks of one colour to quarks of other colours or to leptons.

Table 2. Eight families of the right handed neutrino ν_R (appearing in the 25th line of Table 1), with spin $\frac{1}{2}$. $\nu_{Ri}, i = (1,\cdots,8)$, carries the family quantum numbers $\tilde{\tau}^{13}$, $\tilde{N}^3_L$, $\tilde{\tau}^{23}$, $\tilde{N}^3_R$ and $\tilde{\tau}^4$. Eight families decouple into two groups of four families.

			$\tilde{\tau}^{13}$	$\tilde{\tau}^{23}$	$\tilde{N}^3_L$	$\tilde{N}^3_R$	$\tilde{\tau}^4$
I	ν_{R1}	$\stackrel{03}{(+i)}\stackrel{12}{[+]}\,\vert\,\stackrel{56}{[+]}\stackrel{78}{(+)}\,\Vert\,\stackrel{9\,10}{(+)}\;\stackrel{11\,12}{[+]}\;\stackrel{13\,14}{[+]}$	$-\frac{1}{2}$	0	$-\frac{1}{2}$	0	$-\frac{1}{2}$
I	ν_{R2}	$\stackrel{03}{[+i]}\stackrel{12}{(+)}\,\vert\,\stackrel{56}{[+]}\stackrel{78}{(+)}\,\Vert\,\stackrel{9\,10}{(+)}\;\stackrel{11\,12}{[+]}\;\stackrel{13\,14}{[+]}$	$-\frac{1}{2}$	0	$\frac{1}{2}$	0	$-\frac{1}{2}$
I	ν_{R3}	$\stackrel{03}{(+i)}\stackrel{12}{[+]}\,\vert\,\stackrel{56}{(+)}\stackrel{78}{[+]}\,\Vert\,\stackrel{9\,10}{(+)}\;\stackrel{11\,12}{[+]}\;\stackrel{13\,14}{[+]}$	$\frac{1}{2}$	0	$-\frac{1}{2}$	0	$-\frac{1}{2}$
I	ν_{R4}	$\stackrel{03}{[+i]}\stackrel{12}{(+)}\,\vert\,\stackrel{56}{(+)}\stackrel{78}{[+]}\,\Vert\,\stackrel{9\,10}{(+)}\;\stackrel{11\,12}{[+]}\;\stackrel{13\,14}{[+]}$	$\frac{1}{2}$	0	$\frac{1}{2}$	0	$-\frac{1}{2}$
II	ν_{R5}	$\stackrel{03}{[+i]}\stackrel{12}{[+]}\,\vert\,\stackrel{56}{[+]}\stackrel{78}{[+]}\,\Vert\,\stackrel{9\,10}{(+)}\;\stackrel{11\,12}{[+]}\;\stackrel{13\,14}{[+]}$	0	$-\frac{1}{2}$	0	$-\frac{1}{2}$	$-\frac{1}{2}$
II	ν_{R6}	$\stackrel{03}{(+i)}\stackrel{12}{(+)}\,\vert\,\stackrel{56}{[+]}\stackrel{78}{[+]}\,\Vert\,\stackrel{9\,10}{(+)}\;\stackrel{11\,12}{[+]}\;\stackrel{13\,14}{[+]}$	0	$-\frac{1}{2}$	0	$\frac{1}{2}$	$-\frac{1}{2}$
II	ν_{R7}	$\stackrel{03}{[+i]}\stackrel{12}{[+]}\,\vert\,\stackrel{56}{(+)}\stackrel{78}{(+)}\,\Vert\,\stackrel{9\,10}{(+)}\;\stackrel{11\,12}{[+]}\;\stackrel{13\,14}{[+]}$	0	$\frac{1}{2}$	0	$-\frac{1}{2}$	$-\frac{1}{2}$
II	ν_{R8}	$\stackrel{03}{(+i)}\stackrel{12}{(+)}\,\vert\,\stackrel{56}{(+)}\stackrel{78}{(+)}\,\Vert\,\stackrel{9\,10}{(+)}\;\stackrel{11\,12}{[+]}\;\stackrel{13\,14}{[+]}$	0	$\frac{1}{2}$	0	$\frac{1}{2}$	$-\frac{1}{3}$

All the families of Table 2 and the family members of the eighth family in Table 1 are in the massless basis.

The scalar fields, which are the gauge scalar fields of $\vec{\tilde{N}}_R$ and $\vec{\tilde{\tau}}^2$, couple only to the four families which are doublets with respect to these two groups. The scalar fields which are the gauge scalars of $\vec{\tilde{N}}_L$ and $\vec{\tilde{\tau}}^1$ couple only to the four families which are doublets with respect to these last two groups.

After the electroweak phase transition, caused by the scalar fields with the space index $(7,8)$[3–5,11], the two groups of four families become massive. The lowest of the two groups of four families contains the observed three, while the fourth family remains to be measured. The lowest of the upper four families is the candidate to form the dark matter[4,10].

2.3. *Vector gauge fields and scalar gauge fields in the spin-charge-family theory*

In the *spin-charge-family* theory[2–4], like in all the Kaluza-Klein like theories, either vielbeins or spin connections can be used to represent the vector gauge fields in $d=(3+1)$ space, when space with $d \geq 5$ has large enough symmetry and no strong spinor source is present. This is proven in Ref. 1 and the references therein. There are the superposition of $\omega_{stm}, m=(0,1,2,3), (s,t) \geq 5$, which are used in the *spin-charge-family* theory to represent vector gauge fields — $A^{Ai}_{m} (= \sum_{s,t} c^{Ai}{}_{st}\, \omega^{st}{}_{m})$ — in $d=(3+1)$ in the low energy regime. Here Ai represent the quantum numbers of the corresponding subgroups, expressed by the operators S^{st} in Eqs. (5, 6). Coefficients $c^{Ai}{}_{st}$ can be read from Eqs. (5,6). These vector gauge fields manifest the properties of all the directly and indirectly observed gauge fields[e].

In the *spin-charge-family* theory also the scalar fields[1–4,9,11] have the origin in the spin connection field, in $\omega_{sts'}$ and $\tilde{\omega}_{sts'}, (s,t,s') \geq 5$. These scalar fields offer the explanation for the Higgs's scalar and the Yukawa couplings of the *standard model*[4,9].

Both, scalar and vector gauge fields, follow from the simple starting action of the *spin-charge-family* presented in Eq. (1).

The Lagrange function for the vector gauge fields follows from the action for the curvature R in Eq. (1) and manifests in the case of the flat $d=(3+1)$ space as assumed by the *standard model*: $L_v = -\frac{1}{4} \sum_{A,i,m,n} F^{Ai}{}_{mn} F^{Ai\,mn}$, $F^{Ai}{}_{mn} = \partial_m A^{Ai}_n - \partial_n A^{Ai}_m - i f^{Aijk} A^{Aj}_m A^{Ak}_n$, with

$$
\begin{aligned}
A^{Ai}{}_{m} &= \sum_{s,t} c^{Ai}{}_{st}\, \omega^{st}{}_{m}\,, \\
\tau^{Ai} &= \sum_{s,t} c^{Aist}\, M_{st}\,, \qquad M_{st} = S_{st} + L_{st}\,.
\end{aligned} \tag{13}
$$

In the low energy regime only S_{st} manifest. These expressions can be found in Ref. 1, Eq. (25), for example, and the references therein.

[e]In the *spin-charge-family* theory there are, besides the vector gauge fields of $(\vec{\tau}^1, \vec{\tau}^3)$, Eqs. (5,6), also the vector gauge fields of $\vec{\tau}^2$, Eq. (5), and τ^4, Eq. (6). The vector gauge fields of τ^{21}, τ^{22} and $Y' = \tau^{23} - \tan\theta_2 \tau^4$ gain masses when interacting with the condensate[4] (and the references therein) at around 10^{16} GeV, while the vector gauge field of the hyper charge $Y = \tau^{23} + \tau^4$ remains massless, together with the gauge fields of $\vec{\tau}^1$ and $\vec{\tau}^3$, manifesting at low energies properties, postulated by the *standard model*.

From Eq. (1) we read the interaction between fermions, presented in Table 1, and the corresponding vector gauge fields in flat $d = (3 + 1)$ space.

$$\mathcal{L}_{fv} = \bar{\psi}\gamma^m (p_m - \sum_{A,i} \tau^{Ai} A^{Ai}_m)\psi \,. \tag{14}$$

Particular superposition of spin connection fields, either $\omega_{sts'}$ or $\tilde{\omega}_{abs'}$, $(s,t,s') \geq 5, (a,b) = (0,\cdots,8)$, with the scalar space index $s' = (7,8)$, manifest at low energies as the scalar fields, which contribute to the masses of the family members. The superposition of the scalar fields $\omega_{stt"}$ with the space index $t'' = (9,\cdots,14)$ contribute to the transformation of matter into antimatter and back, causing in the presence of the condensate[2,4] the matter-antimatter asymmetry of our universe. The interactions of all these scalar fields with fermions follow from Eq. (1)

$$\mathcal{L}_{fs} = \left\{ \sum_{s=7,8} \bar{\psi}\gamma^s p_{0s}\, \psi \right\} + \left\{ \sum_{t=5,6,9,\dots,14} \bar{\psi}\gamma^t p_{0t}\, \psi \right\}, \tag{15}$$

where $p_{0s} = p_s - \frac{1}{2}S^{s's"}\omega_{s's"s} - \frac{1}{2}\tilde{S}^{ab}\tilde{\omega}_{abs}$, $p_{0t} = p_t - \frac{1}{2}S^{t't"}\omega_{t't"t} - \frac{1}{2}\tilde{S}^{ab}\tilde{\omega}_{abt}$, with $m \in (0,1,2,3)$, $s \in (7,8)$, $(s', s") \in (5,6,7,8)$, (a,b) (appearing in $\tilde{S}^{ab}$) run within either $(0,1,2,3)$ or $(5,6,7,8)$, t runs $\in (5,\dots,14)$, $(t', t")$ run either $\in (5,6,7,8)$ or $\in (9,10,\dots,14)$. The spinor function ψ represents all family members of all the $2^{\frac{7+1}{2}-1} = 8$ families presented in Table 2.

There are the superposition of the scalar fields $\omega_{s's"s}$ — $(A^Q_{\pm}, A^{Q'}_{\pm}, A^{Y'}_{\pm})$ [f] — and the superposition of $\tilde{\omega}_{s's''s}$ — $(\vec{\tilde{A}}^{\tilde{N}_L}_{\pm}, \vec{\tilde{A}}^{\tilde{1}}_{\pm}, \vec{\tilde{A}}^{\tilde{N}_R}_{\pm}, \vec{\tilde{A}}^{\tilde{2}}_{\pm})$ [g] — which determine mass terms of family members of spinors after the electroweak break. I shall use $A^{Ai}_{\pm}$ to represent all the scalar fields, which determine masses of family members, the Yukawa couplings and masses of the weak boson vector fields, $A^{Ai}_{\pm} = (\sum_{A,i,a,b} c^{Aist}(\omega_{st7} \pm i\omega_{st8})$ as well as $= \sum_{A,i,a,b} c^{Aist}(\tilde{\omega}_{ab7} \pm i\omega_{ab8})$.

The part of the first term of Eq. (15), in which summation runs over the space index $s = (7,8)$ — $\sum_{s=7,8} \bar{\psi}\gamma^s p_{0s}\psi$ — determines after the electroweak break masses of the two groups of four families. The highest of the lower four families is predicted to be observed at the L(arge)H(adron)C(ollider)[11], the lowest of the higher four families is explaining the origin of the dark matter[10].

[f] $Q := \tau^{13} + Y$, $Q' := -Y \tan^2\vartheta_1 + \tau^{13}$, $Q' := -\tan^2\vartheta_1 Y + \tau^{13}$, $Y := \tau^4 + \tau^{23}$, $Y' := -\tan^2\vartheta_2 \tau^4 + \tau^{23}$, $Q := \tau^{13} + Y$, and correspondingly $A^Q_s = \sin\vartheta_1\, A^{13}_s + \cos\vartheta_1\, A^Y_s$, $A^{Q'}_s = \cos\vartheta_1\, A^{13}_s - \sin\vartheta_1\, A^Y_s$, $A^{Y'}_s = \cos\vartheta_2\, A^{23}_s - \sin\vartheta_2\, A^4_s$, $A^4_s = -(\omega_{9\,10\,s} + \omega_{11\,12\,s} + \omega_{13\,14\,s})$, $A^{13}_s = (\omega_{56s} - \omega_{78s})$, $A^{23}_s = (\omega_{56s} + \omega_{78s})$, with $(s \in (7,8))$ (Re.[3], Eq. (A9)).

[g] $\vec{\tilde{A}}^{\tilde{1}}_s = (\tilde{\omega}_{\tilde{5}\tilde{8}s} - \tilde{\omega}_{\tilde{6}\tilde{7}s}, \tilde{\omega}_{\tilde{5}\tilde{7}s} + \tilde{\omega}_{\tilde{6}\tilde{8}s}, \tilde{\omega}_{\tilde{5}\tilde{6}s} - \tilde{\omega}_{\tilde{7}\tilde{8}s})$, $\vec{\tilde{A}}^{\tilde{N}_L}_s = (\tilde{\omega}_{\tilde{2}\tilde{3}s} + i\,\tilde{\omega}_{\tilde{0}\tilde{1}s}, \tilde{\omega}_{\tilde{3}\tilde{1}s} + i\,\tilde{\omega}_{\tilde{0}\tilde{2}s}, \tilde{\omega}_{\tilde{1}\tilde{2}s} + i\,\tilde{\omega}_{\tilde{0}\tilde{3}s})$, $\vec{\tilde{A}}^{\tilde{2}}_s = (\tilde{\omega}_{\tilde{5}\tilde{8}s} + \tilde{\omega}_{\tilde{6}\tilde{7}s}, \tilde{\omega}_{\tilde{5}\tilde{7}s} - \tilde{\omega}_{\tilde{6}\tilde{8}s}, \tilde{\omega}_{\tilde{5}\tilde{6}s} + \tilde{\omega}_{\tilde{7}\tilde{8}s})$ and $\vec{\tilde{A}}^{\tilde{N}_R}_s = (\tilde{\omega}_{\tilde{2}\tilde{3}s} - i\,\tilde{\omega}_{\tilde{0}\tilde{1}s}, \tilde{\omega}_{\tilde{3}\tilde{1}s} - i\,\tilde{\omega}_{\tilde{0}\tilde{2}s}, \tilde{\omega}_{\tilde{1}\tilde{2}s} - i\,\tilde{\omega}_{\tilde{0}\tilde{3}s})$, where $(s \in (7,8))$ (Ref. 3, Eq. (A8)).

The scalar fields in the part of the second term of Eq. (15), in which summation runs over the space index $t = (9, \cdots, 14)$ — $\sum_{t=9,\cdots,14} \bar{\psi}\gamma^t p_{0t}\psi$ — cause transitions from anti-leptons into quarks and anti-quarks into quarks and back, transforming antimatter into matter and back. In the expanding universe the condensate of two right handed neutrinos breaks this matter-antimatter symmetry, explaining the matter-antimatter asymmetry of our universe[2].

Spin connection fields $\omega_{sts'}$ and $\tilde{\omega}_{sts'}$ interact also with vector gauge fields and among themselves[1]. These interactions can be red from Eq. (1).

3. Discussions and Open Problems

The *spin-charge-family* theory is offering the next step beyond both *standard models*, by explaining:
i. The origin of charges of the (massless) family members and the relation between their charges and spins. The theory, namely, starts in $d = (13 + 1)$ with the simple action for spinors, which interact with the gravity only (Eq.1) (through the vielbeins and the two kinds of the spin connection fields), while one fundamental representation of $SO(13,1)$ contains, if analyzed with respect to the subgroups $SO(3,1), SU(3), SU(2)_I, SU(2)_{II}$ and $U(1)_{II}$ of the group $SO(13,1)$, all the quarks and anti-quarks and all the leptons and anti-leptons with the properties assumed by the *standard model*, relating handedness and charges of spinors as well as of anti-spinors (Table 1).
ii. The origin of families of fermions, since spinors carry two kinds of spins (Eq. (2)) — the Dirac γ^a and $\tilde{\gamma}^a$. In $d = (3 + 1)$ γ^a take care of the observed spins and charges, $\tilde{\gamma}^a$ take care of families (Table 2).
iii. The origin of the massless vector gauge fields of the observed charges, represented by the superposition of the spin connection fields $\omega_{stm}, (s,t) \geq 5, m \leq 3$[1,3,4].
iv. The origin of masses of family members and of heavy bosons. The superposition of $\omega_{sts'}, (s,t) \geq 5, s' = (7,8)$ and the superposition of $\tilde{\omega}_{abs'}, (a,b) = (0, \cdots, 8), s' = (7,8)$ namely gain at the electroweak break constant values, determining correspondingly masses of the spinors (fermions) and of the heavy bosons, explaining[3,4,11] the origin of the Higgs's scalar and the Yukawa couplings of the *standard model.*
v. The origin of the matter-antimatter asymmetry[2], since the superposition of $\omega_{sts'}, s' \geq 9$, cause transitions from anti-leptons into quarks and anti-quarks into quarks and back, while the appearance of the scalar condensate in the expanding universe breaks the CP symmetry, enabling the existence of matter-antimatter asymmetry.
vi. The origin of the dark matter, since there are two groups of decoupled four families in the low energy regime. The neutron made of quarks of the stable of the upper four families explains the appearance of the dark matter[10] [h].

[h] We followed in Ref. 10 freezing out of the fifth family quarks and anti-quarks in the expanding universe to see whether baryons of the fifth family quarks are the candidates for the dark matter.

vii. The origin of the triangle anomaly cancellation in the standard model. All the quarks and anti-quarks and leptons and anti-leptons, left and right handed, appear within one fundamental representation of $SO(13,1)$[3,4].
viii. The origin of all the gauge fields. The *spin-charge-family* theory unifies the gravity with all the vector and scalar gauge fields, since in the starting action there is only gravity (Eq. (1)), represented by the vielbeins and the two kinds of the spin connection fields, which in the low energy regime manifests in $d = (3+1)$ as the ordinary gravity and all the directly and indirectly observed vector and scalar gauge fields[1]. If there is no spinor condensate present, only one of the three fields is the propagating field (both spin connections are expressible with the vielbeins). In the presence of the spinor fields the two spin connection fields differ among themselves (Ref. 1, Eq. (4), and the references therein).

The more work is done on the *spin-charge-family* theory, the more answers to the open questions of both *standard models* is the theory offering.

There are, of course, still open questions (mostly common to all the models) like:

a. How has our universe really started? The *spin-charge-family* theory assumes $d = (13+1)$, but how "has the universe decided" to start with $d = (13+1)$? If starting at $d = \infty$, how can it come to (13+1) with the massless Weyl representation of only one handedness? We have studied in a toy model the break of symmetry from $d = (5+1)$ into (3+1)[14], finding that there is the possibility that spinors of one handedness remain massless after this break. This study gives a hope that breaking the symmetry from $(d-1)+1$, where d is even and ∞, could go, if the jump of $(d-1)+1$ to $((d-4)-1)+1$ would be repeated as twice the break suggested in Ref. 14. These jumps should then be repeated all the way from $d = \infty$ to $d = (13+1)$.

b. What did "force" the expanding universe to break the symmetry of $SO(13,1)$ to $SO(7,1) \times SU(3) \times U(1)_{II}$ and then further to $SO(3,1) \times SU(2) \times SU(3) \times U(1)_I$ and finally to $SO(3,1) \times SU(3) \times U(1)$?

From phase transitions of ordinary matter we know that changes of temperature and pressure lead a particular matter into a phase transition, causing that constituents of the matter (nuclei and electrons) rearrange, changing the symmetry of space.
In expanding universe the temperature and pressure change, forcing spinors to make condensates (like it is the condensate of the two right handed neutrinos in the *spin-charge-family* theory[2–4], which gives masses to vector gauge fields of $SU(2)_{II}$, breaking $SU(2)_{II} \times U(1)_{II}$ into $U(1)_I$). There might be also vector gauge fields causing a change of the symmetry (like does the colour vector gauge fields, which "dress" quarks and anti-quarks and bind them to massive colourless baryons and mesons of the ordinary, mostly the first family, matter). Also scalar gauge fields might cause the break of the symmetry of the space (as this do the superposition of $\omega_{s't's}$ and the superposition of $\tilde{\omega}_{abs}$, $s = (7,8), (s',t") \geq 5, (a,b) = (0,\cdots,8)$ in the *spin-charge-family* theory[3,4] by gaining constant values in $d = (3+1)$ and breaking correspondingly also the symmetry of the coordinate space in $d \geq 5$).

All these remain to be studied.

c. What is the scale of the electroweak phase transition? How higher is this scale in comparison with the colour phase transition scale? If the colour phase transition scale is at around 1 GeV (since the first family quarks contribute to baryons masses around 1 GeV), is the electroweak scale at around 1 TeV (of the order of the mass of Higgs's scalar) or this scale is much higher, possibly at the unification scale (since the *spin-charge-family* theory predicts two decoupled groups of four families and several scalar fields — twice two triplets and three singlets[3,4,11])?

d. There are several more open questions. Among them are the origin of the dark energy, the appearance of fermions, the origin of inflation of the universe, quantization of gravity, and several others. Can the *spin-charge-family* theory be — while predicting the fourth family to the observed three, several scalar fields, the fifth family as the origin of the dark matter, the scalar fields transforming anti-leptons into quarks and anti-quarks into quarks and back and the condensate which break this symmetry — the first step, which can hopefully show the way to next steps?

Appendix A. Some Useful Formulas and Relations are Presented[4,5]

$$
\begin{aligned}
S^{ac}\stackrel{ab}{(k)}\stackrel{cd}{(k)} &= -\frac{i}{2}\eta^{aa}\eta^{cc}\stackrel{ab}{[-k]}\stackrel{cd}{[-k]}\,, & \tilde{S}^{ac}\stackrel{ab}{(k)}\stackrel{cd}{(k)} &= \frac{i}{2}\eta^{aa}\eta^{cc}\stackrel{ab}{[k]}\stackrel{cd}{[k]}\,,\\
S^{ac}\stackrel{ab}{[k]}\stackrel{cd}{[k]} &= \frac{i}{2}\stackrel{ab}{(-k)}\stackrel{cd}{(-k)}\,, & \tilde{S}^{ac}\stackrel{ab}{[k]}\stackrel{cd}{[k]} &= -\frac{i}{2}\stackrel{ab}{(k)}\stackrel{cd}{(k)}\,,\\
S^{ac}\stackrel{ab}{(k)}\stackrel{cd}{[k]} &= -\frac{i}{2}\eta^{aa}\stackrel{ab}{[-k]}\stackrel{cd}{(-k)}\,, & \tilde{S}^{ac}\stackrel{ab}{(k)}\stackrel{cd}{[k]} &= -\frac{i}{2}\eta^{aa}\stackrel{ab}{[k]}\stackrel{cd}{(k)}\,,\\
S^{ac}\stackrel{ab}{[k]}\stackrel{cd}{(k)} &= \frac{i}{2}\eta^{cc}\stackrel{ab}{(-k)}\stackrel{cd}{[-k]}\,, & \tilde{S}^{ac}\stackrel{ab}{[k]}\stackrel{cd}{(k)} &= \frac{i}{2}\eta^{cc}\stackrel{ab}{(k)}\stackrel{cd}{[k]}\,.
\end{aligned}
\tag{A.1}
$$

$$
\begin{aligned}
&\stackrel{ab}{(k)}\stackrel{ab}{(k)} = 0\,, && \stackrel{ab}{(k)}\stackrel{ab}{(-k)} = \eta^{aa}\stackrel{ab}{[k]}\,, && \stackrel{ab}{(-k)}\stackrel{ab}{(k)} = \eta^{aa}\stackrel{ab}{[-k]}\,, && \stackrel{ab}{(-k)}\stackrel{ab}{(-k)} = 0\,,\\
&\stackrel{ab}{[k]}\stackrel{ab}{[k]} = \stackrel{ab}{[k]}\,, && \stackrel{ab}{[k]}\stackrel{ab}{[-k]} = 0\,, && \stackrel{ab}{[-k]}\stackrel{ab}{[k]} = 0\,, && \stackrel{ab}{[-k]}\stackrel{ab}{[-k]} = \stackrel{ab}{[-k]}\,,\\
&\stackrel{ab}{(k)}\stackrel{ab}{[k]} = 0\,, && \stackrel{ab}{[k]}\stackrel{ab}{(k)} = \stackrel{ab}{(k)}\,, && \stackrel{ab}{(-k)}\stackrel{ab}{[k]} = \stackrel{ab}{(-k)}\,, && \stackrel{ab}{(-k)}\stackrel{ab}{[-k]} = 0\,,\\
&\stackrel{ab}{(k)}\stackrel{ab}{[-k]} = \stackrel{ab}{(k)}\,, && \stackrel{ab}{[k]}\stackrel{ab}{(-k)} = 0, && \stackrel{ab}{[-k]}\stackrel{ab}{(k)} = 0\,, && \stackrel{ab}{[-k]}\stackrel{ab}{(-k)} = \stackrel{ab}{(-k)}\,.
\end{aligned}
\tag{A.2}
$$

$$
\begin{aligned}
&N^{\pm}_{+} = N^{1}_{+} \pm i\,N^{2}_{+} = -\stackrel{03}{(\mp i)}\stackrel{12}{(\pm)}\,, \quad N^{\pm}_{-} = N^{1}_{-} \pm i\,N^{2}_{-} = \stackrel{03}{(\pm i)}\stackrel{12}{(\pm)}\,,\\
&\tilde{N}^{\pm}_{+} = -\stackrel{03}{\tilde{(\mp i)}}\stackrel{12}{\tilde{(\pm)}}\,, \quad \tilde{N}^{\pm}_{-} = \stackrel{03}{\tilde{(\pm i)}}\stackrel{12}{\tilde{(\pm)}}\,,\\
&\tau^{1\pm} = (\mp)\stackrel{56}{(\pm)}\stackrel{78}{(\mp)}\,, \quad \tau^{2\mp} = (\mp)\stackrel{56}{(\mp)}\stackrel{78}{(\mp)}\,,\\
&\tilde{\tau}^{1\pm} = (\mp)\stackrel{56}{\tilde{(\pm)}}\stackrel{78}{\tilde{(\mp)}}\,, \quad \tilde{\tau}^{2\mp} = (\mp)\stackrel{56}{\tilde{(\mp)}}\stackrel{78}{\tilde{(\mp)}}\,.
\end{aligned}
\tag{A.3}
$$

References

1. D. Lukman and N. S. Mankoč Borštnik, "Gauge fields with respect to $d = (3 + 1)$ in the Kaluza-Klein theories and in the *spin-charge-family* theory", *Eur. Phys. J. C* (2017) Doi:10.1140/epjc/s10052-017-4804-y.
2. N. S. Mankoč Borštnik, "Can spin-charge-family theory explain baryon number non conservation?", *Phys. Rev.* **D 91**, 6 (2015) 065004 ID: 0703013; doi:10.1103 [arXiv: 1409.7791, arXiv:1502.06786v1].
3. N. S. Mankoč Borštnik, "The explanation for the origin of the Higgs's scalar and for the Yukawa couplings by the *spin-charge-family* theory", *J. of Mod. Phys.* **6** (2015) 2244-2274, doi:10.4236/jmp.2015.615230 [arXiv:1409.4981].
4. N. S. Mankoč Borštnik, "Spin-charge-family theory is offering next step in understanding elementary particles and fields and correspondingly universe", in Proceedings to The 10th Biennial Conference on Classical and Quantum Relativistic Dynamics of Particles and Fields, IARD conference, Ljubljana 6-9 of June 2016 [arXiv:1607.01618v2], *J. Phys. Conf. Ser.* **845**, 012017 (2017).
5. A. Borštnik Bračič and N. S. Mankoč Borštnik, "Origin of families of fermions and their mass matrices", *Phys. Rev.* **D 74**, 073013 (2006) [hep-ph/0301029; hep-ph/9905357, p. 52-57; hep-ph/0512062, p.17-31; hep-ph/0401043, p. 31-57].
6. N. S. Mankoč Borštnik, "Spin connection as a superpartner of a vielbein", *Phys. Lett.* **B 292**, 25-29 (1992).
7. N. S. Mankoč Borštnik, "Spinor and vector representations in four dimensional Grassmann space", *J. Math. Phys.* **34**, 3731-3745 (1993).
8. N. S. Mankoč Borštnik, "Unification of spins and charges", *Int. J. Theor. Phys.* **40**, 315-338 (2001).
9. G. Bregar, M. Breskvar, D. Lukman and N. S. Mankoč Borštnik, "On the origin of families of quarks and leptons — predictions for four families", *New J. of Phys.* **10**, 093002 (2008), [arXiv:0606159, arXiv:07082846, arXiv:0612250, p.25-50].
10. G. Bregar and N. S. Mankoč Borštnik, "Does dark matter consist of baryons of new stable family quarks?", *Phys. Rev.* **D 80**, 1 (2009), 083534 (2009).
11. G. Bregar and N. S. Mankoč Borštnik, "The new experimental data for the quarks mixing matrix are in better agreement with the *spin-charge-family* theory predictions", Proceedings to the 17th Workshop "What comes beyond the standard models", Bled, 20-28 of July, 2014, Ed. N. S. Mankoč Borštnik, H. B. Nielsen and D. Lukman, DMFA Založništvo, Ljubljana December 2014, p.20-45 [arXiv:1502.06786v1, arXiv:1412.5866].
12. N. S. Mankoč Borštnik and H. B. Nielsen, "Discrete symmetries in the Kaluza-Klein-like theories", *Jour. of High Energy Phys.* **04** (2014) 165-174, doi:10.1007 [arXiv:1212.2362v3].
13. N. S. Mankoč Borštnik and H. B. Nielsen, "How to generate spinor representations in any dimension in terms of projection operators", *J. of Math. Phys.* **43** (2002) 5782 [hep-th/0111257].
14. D. Lukman, N. S. Mankoč Borštnik and H. B. Nielsen, "An effective two dimensionality" cases bring a new hope to the Kaluza-Klein-like theories", *New J. Phys.* **13** (2011) 103027 [hep-th/1001.4679v5].
15. N. S. Mankoč Borštnik and H. B. Nielsen, "How to generate families of spinors", *J. of Math. Phys.* **44** (2003) 4817, [hep-th/0303224].

Acceleration of Dust-Ball Expansion due to GR Gravitational Time Dilation

S. K. Kauffmann

Retired (APS Senior Life Member)
Email: skkauffmann@gmail.com

Though the expansion of a simple FLRW dust ball would always decelerate in Newtonian gravitational dynamics, in GR, when the dust ball's radius insufficiently exceeds the Schwarzschild value, its expansion instead accelerates because the dominant gravitational time-dilation braking of its expansion speed weakens as it expands. But in "comoving coordinates" the fixing of the 00 component of the metric tensor to unity completely eliminates gravitational time dilation, which is reflected by the purely Newtonian Friedmann equation of motion in those "coordinates" for the dust-ball. For a particular dust-ball initial condition Oppenheimer and Snyder remedied the GR-inconsistent Newtonian behavior in "comoving coordinates" by their famed tour-de-force analytic transformation to GR-compatible "standard" coordinates. Recent extension of their transformation to arbitrary dust-ball initial conditions enables the derivation of GR-consistent equations of motion in "standard" coordinates for all shell radii of any simple FLRW dust ball. These non-Newtonian equations of motion not only show that a dust ball's expansion always accelerates when its radius insufficiently exceeds the Schwarzschild value, but also that for a range of initial conditions the dust ball's expansion never ceases to accelerate (although that acceleration asymptotically decreases toward zero), apparently eliminating any need for a nonzero "dark energy" cosmological constant.

Keywords: Gravitational time dilation; "Comoving coordinates"; Friedmann equation; Extended Oppenheimer-Snyder transformation; Acceleration of FLRW dust-ball expansion; "Dark energy"; Cosmological constant.

Force and acceleration in the presence of relativistic time dilation

Assuming purely radial motion, conservation of *Newtonian kinetic energy* plus gravitational potential energy, namely,

$$\tfrac{1}{2}m\dot{r}^2 - GmM/r = T_0 + V_0,$$

implies Newtonian Second Law *proportionality of acceleration to the gravitational force* $-GmM/r^2$, specifically that $\ddot{r} = -GM/r^2$. But conservation of *special relativistic kinetic energy* plus gravitational potential energy, namely,

$$mc^2[(1 - \dot{r}^2/c^2)^{-\frac{1}{2}} - 1] - GmM/r = T_0 + V_0,$$

upends proportionality of acceleration to the gravitational force, instead implying,

$$\ddot{r} = \frac{-GM/r^2}{[1 + (T_0 + V_0)/(mc^2) + (GM)/(c^2 r)]^3}.$$

We note in particular that as $r \to 0$ *and the gravitational force becomes infinite*, the acceleration $\ddot{r} \to 0$! That occurs because *special relativistic time dilation* overpowers gravitational force *to stop* $|\dot{r}|$ *from reaching* c *by driving* $\ddot{r}$ *toward zero.*

The *gravitational time dilation* intrinsic *to GR* is capable of going *beyond* such mere *cancellation* of the acceleration which is due to the gravitational *force*: under appropriate circumstances gravitational time dilation delivers net acceleration *opposite in sign to the gravitational force.* For example, as the surface radius $\bar{r}_a$ of a simple FLRW dust ball *contracts toward its Schwarzschild value* r_S, the gravitational time-dilation factor[1] at $\bar{r}_a$ *increases toward infinity,* driving the contracting dust ball's *negative surface radial velocity* $d\bar{r}_a/d\bar{t}_a$ *toward zero.* Thus a *contracting* dust ball's surface radius $\bar{r}_a$ will *ultimately* undergo *positive* (i.e., *outward*) acceleration when that radius gets close enough to its Schwarzschild value r_S—*despite* the immense *negative* (i.e., *inward*) gravitational *force* on the dust at that radius.

Unfortunately this fascinating consequence of gravitational time dilation *cannot* be reflected *by the only known direct analytic solution of the Einstein equation for the simple FLRW dust ball*[2] since that solution's metric-tensor component g_{00} *is fixed to unity,* which *extinguishes* gravitational time dilation *because* $(g_{00})^{-\frac{1}{2}}$ *is the gravitational time-dilation factor*[1]. *Fixing* g_{00} *to unity,* however, is well-known *to require the clock readings of an infinite number of different observers*[3], a requirement which *neither* can be physically fulfilled *nor* is compatible with Einstein's observer-to-coordinate-system paradigm!

That spherically-symmetric $g_{00} = 1$ *GR-unphysical* Einstein-equation solution for the FLRW dust ball was, for a *particular* initial condition, *transcended* in 1939 by Oppenheimer and Snyder[4], who, in an analytical tour-de-force, *worked out the transformation*[4,5] of the (r, t) "comoving coordinates" of that solution's *GR-unphysical* metric,

$$ds^2 = (cdt)^2 - U(r,t)(dr)^2 - V(r,t)\left((d\theta)^2 + (\sin\theta d\phi)^2\right),$$

to the $(\bar{r}, \bar{t})$ "standard coordinates" of the *GR-physical* metric,

$$ds^2 = B(\bar{r},\bar{t})(cd\bar{t})^2 - A(\bar{r},\bar{t})(d\bar{r})^2 - \bar{r}^2\left((d\theta)^2 + (\sin\theta d\phi)^2\right),$$

which doesn't extinguish of gravitational time dilation because its g_{00} metric component $B(\bar{r}, \bar{t})$ *isn't constrained.*

Preparatory to presentation of the Oppenheimer-Snyder transformation $(\bar{r}(r,t), \bar{t}(r,t))$, as *extended* to *arbitrary* dust-ball initial conditions, we note that the $g_{00} = 1$ *GR-unphysical* Einstein-equation solution's metric functions $U(r,t)$ and $V(r,t)$ above come out to be[2,6],

$$U(r,t) = (R(t))^2/[1 + \gamma\omega^2(r/c)^2] \quad \text{and} \quad V(r,t) = r^2(R(t))^2,$$

where the dimensionless function $R(t)$ is given in terms of the dust's *always-uniform* time-dependent energy density $\rho(t)$ by,

$$R(t) = (\rho(t_0)/\rho(t))^{\frac{1}{3}},$$

defined so that $R(t_0) = 1$. In the $g_{00} = 1$ *GR-unphysical* "comoving coordinates" (r, t), the dust not only has *always-uniform* energy density $\rho(t)$, it *also* has everywhere *zero* dust-particle *velocity*[6,7], so the dust ball's "comoving radius" a doesn't change with "comoving time" t.

The Einstein equation *within the dust ball in "comoving coordinates"* (r, t), i.e., *within* the "comoving region" $0 \leq r \leq a$, yields that $R(t)$ satisfies *the Friedmann equation*, which is,

$$(\dot{R}(t))^2 = \omega^2((1/R(t)) + \gamma), \tag{1}$$

where, of course, $R(t_0) = 1$. The constants ω^2 and γ above, which *also* occur in the metric function $U(r, t)$ given further above, are related to $\rho(t_0)$ and $\dot{\rho}(t_0)$ as follows,

$\omega^2 \stackrel{\text{def}}{=} (8\pi/3)G\rho(t_0)/c^2$ and $\gamma = [(\dot{R}(t_0))^2/\omega^2] - 1 = [\dot{\rho}(t_0)/(3\omega\rho(t_0))]^2 - 1 \geq -1.$

The Friedmann Eq. (1) is readily shown to be *isomorphic* to the elementary equation of conservation of a test particle's strictly *Newtonian* kinetic energy plus gravitational potential energy for purely radial motion which was given at the very beginning of this article, namely,

$$(\dot{r}(t))^2 - 2GM/r(t) = 2(T_0 + V_0)/m, \tag{2}$$

where, of course, $(2T_0/m) = (\dot{r}(t_0))^2$ and $(2V_0/m) = -2GM/r(t_0)$. The isomorphism of the Friedmann Eq. (1) to the strictly *Newtonian* Eq. (2) can be established by, for example, associating M to the dust ball's nominal total mass at the initial "comoving time" t_0,

$$M = (4\pi/3)a^3\rho(t_0)/c^2, \tag{3}$$

and *simultaneously* associating $r(t)$ to $aR(t)$,

$$r(t) = aR(t). \tag{4}$$

Equation (4) of course implies that,

$$R(t) = r(t)/a, \tag{5}$$

and since $R(t_0) = 1$, Eq. (4) *also* implies that,

$$a = r(t_0). \tag{6}$$

Equation (3), together with the above-given definition of ω^2 as $(8\pi/3)G\rho(t_0)/c^2$, implies that,

$$\omega^2 = 2GM/a^3, \tag{7}$$

while Eq. (5), together with the above-given rendition of γ as $[(\dot{R}(t_0))^2/\omega^2] - 1$, implies that,

$$\gamma = [(\dot{r}(t_0))^2/(a^2\omega^2)] - 1. \tag{8}$$

The insertion into the Friedmann Eq. (1) of Eqs. (5), (6), (7) and (8) establishes the *isomorphism* of Friedmann Eq. (1) to the strictly *Newtonian* Eq. (2).

Oppenheimer and Snyder solved for the transformation $(\bar{r}(r, t), \bar{t}(r, t))$ from the $g_{00} = 1$ *GR-unphysical* metric's (r, t) "comoving coordinates" to the GR-physical metric's $(\bar{r}, \bar{t})$ "standard coordinates" *only when the particular initial condition* $\dot{\rho}(t_0) = 0$ *holds*, i.e., *only when* $\dot{R}(t_0) = 0$ *and* $\gamma = -1$.[4,5] But the arduously intricate techniques they developed to obtain their limited transformation *are adequate to extend it to all values of* $\gamma \geq -1$.[6]

The extended Oppenheimer-Snyder transformation

The presentation of the extended Oppenheimer-Snyder transformation is more compact when all occurrences of the two constants ω^2 and γ are systematically replaced by occurrences of the Schwarzschild radius r_S in "standard" coordinates and the dimensionless constant α, which are defined as follows (note that the relation of ω^2 to G, M and a is given by Eq. (7)),

$$r_S \stackrel{\text{def}}{=} \omega^2 a^3/c^2 = 2GM/c^2 \;\;\&\;\; \alpha \stackrel{\text{def}}{=} \gamma(r_S/a).$$

With r_S and α superseding ω^2 and γ, the extended Oppenheimer-Snyder transformation $(\bar{r}(r,t), \bar{t}(r,t))$ for $0 \leq r \leq a$ comes out to be[6],

$$\bar{r}(r,t) = rR(t) \quad \text{and}$$
$$\bar{t}(r,t) = \bar{t}(a,t_0) \pm (a/c)(1+\alpha)^{\frac{1}{2}} \int_1^{S(r,t)} \frac{ds}{((r_S/(as)) + \alpha)^{\frac{1}{2}}(1 - (r_S/(as)))}, \qquad (9)$$

where the $\pm$ is the *sign* of $\dot{R}(t)$ if $\dot{R}(t) \neq 0$, but $\pm = -1$ if $\dot{R}(t) = 0$, and $S(r,t)$ is the following expression,

$$S(r,t) = R(t)\left(\frac{1+(r/a)^2\alpha}{1+\alpha}\right)^{\frac{1}{2}} - \left(\frac{r_S}{a\alpha}\right)\left[1 - \left(\frac{1+(r/a)^2\alpha}{1+\alpha}\right)^{\frac{1}{2}}\right]. \qquad (10)$$

Note that the integral in the Eq. (9) expression for $\bar{t}(r,t)$ *diverges to infinity* unless *both* $a > r_S$ (and therefore $\alpha > -1$, since $\alpha = \gamma(r_S/a)$ and $\gamma \geq -1$) *and* $S(r,t) > (r_S/a)$. Thus a subset of "comoving space-time" is transformed to *infinite* "standard" time, *which makes that subset inaccessible in GR-physical "standard" space-time.* (This space-time subset inaccessibility arises from *the singular character of the transformation from GR-unphysical "comoving coordinates" to GR-physical "standard" coordinates*—it patently *couldn't occur for the nonsingular transformation that necessarily obtains between two GR-physical metrics.*)

We now utilize this extended Oppenheimer-Snyder transformation to obtain the motion of the dust ball's interior-shell radii in "standard" coordinates.

Behavior of dust-ball interior-shell radii in "standard" coordinates

In *GR-unphysical* "comoving coordinates" (r,t), the dust ball's *interior-shell radii* ϵa, where $0 < \epsilon \leq 1$, *have zero velocity*, so *their world lines are simply* $(\epsilon a, t)$. The extended Oppenheimer-Snyder transformation given above maps these *GR-unphysical* "comoving" world lines $(\epsilon a, t)$ to GR-physical "standard" world lines

$(\bar{r}_{\epsilon a}(t), \bar{t}_{\epsilon a}(t)) \stackrel{\text{def}}{=} (\bar{r}(\epsilon a, t), \bar{t}(\epsilon a, t))$, which are obtained from Eqs. (9) and (10),

$$\bar{r}_{\epsilon a}(t) = \epsilon a R(t) \quad \text{and}$$

$$\bar{t}_{\epsilon a}(t) = \bar{t}(a, t_0) \pm (a/c)(1+\alpha)^{\frac{1}{2}} \int_1^{S(\epsilon a, t)} \frac{ds}{((r_S/(as)) + \alpha)^{\frac{1}{2}}(1 - (r_S/(as)))} =$$

$$\bar{t}(a, t_0) \pm (c\epsilon)^{-1}(1+\alpha)^{\frac{1}{2}} \int_{\epsilon a}^{\rho_{\epsilon a}(\bar{r}_{\epsilon a}(t))} \frac{d\rho}{((\epsilon r_S/\rho) + \alpha)^{\frac{1}{2}}(1 - (\epsilon r_S/\rho))} \stackrel{\text{def}}{=}$$

$$\bar{t}_{\epsilon a}(\rho_{\epsilon a}(\bar{r}_{\epsilon a}(t))), \tag{11}$$

where the *initial* integration variable s in Eq. (11) *was subsequently changed to* $\rho = (\epsilon a)s$, and consequently,

$$\rho_{\epsilon a}(\bar{r}_{\epsilon a}(t)) = (\epsilon a)S(\epsilon a, t) = \bar{r}_{\epsilon a}(t)\left(\frac{1+\epsilon^2\alpha}{1+\alpha}\right)^{\frac{1}{2}} - \left(\frac{\epsilon r_S}{\alpha}\right)\left[1 - \left(\frac{1+\epsilon^2\alpha}{1+\alpha}\right)^{\frac{1}{2}}\right]. \tag{12}$$

The Eq. (11) integral expression for $\bar{t}_{\epsilon a}(\rho_{\epsilon a}(\bar{r}_{\epsilon a}))$ *diverges to infinity* unless $\rho_{\epsilon a}(\bar{r}_{\epsilon a}) > \epsilon r_S$. Taking care to respect that *caveat*, the $d\rho$ integration can be carried out in closed form to produce an intricate *analytic result for* $\bar{t}_{\epsilon a}(\rho_{\epsilon a}(\bar{r}_{\epsilon a}))$.[6] That analytic result is useful for creating *numerical plots of* $\bar{r}_{\epsilon a}(\bar{t}_{\epsilon a})$, but its intricacy *resists direct interpretation.*

However the Eq. (11) integral result for $\bar{t}_{\epsilon a}(\rho_{\epsilon a}(\bar{r}_{\epsilon a}))$ *also* readily yields *a first-order differential equation for* $\bar{r}_{\epsilon a}(\bar{t}_{\epsilon a})$, namely,

$$d\bar{r}_{\epsilon a}/d\bar{t}_{\epsilon a} = (d\bar{t}_{\epsilon a}(\rho_{\epsilon a}(\bar{r}_{\epsilon a}))/d\bar{r}_{\epsilon a})^{-1} =$$

$$\pm c\epsilon\left(\frac{(\epsilon r_S/\rho_{\epsilon a}(\bar{r}_{\epsilon a})) + \alpha}{1+\epsilon^2\alpha}\right)^{\frac{1}{2}}(1 - (\epsilon r_S/\rho_{\epsilon a}(\bar{r}_{\epsilon a}))), \tag{13}$$

which of course carries with it the Eq. (11) *caveat* that $\rho_{\epsilon a}(\bar{r}_{\epsilon a}) > \epsilon r_S$.

Equation (13) and its *caveat* $\rho_{\epsilon a}(\bar{r}_{\epsilon a}) > \epsilon r_S$ imply an *upper bound* for the shell-radius *speed* $|d\bar{r}_{\epsilon a}/d\bar{t}_{\epsilon a}|$,

$$|d\bar{r}_{\epsilon a}/d\bar{t}_{\epsilon a}| < c\left(\frac{\epsilon^2 + \epsilon^2\alpha}{1+\epsilon^2\alpha}\right)^{\frac{1}{2}}(1 - (\epsilon r_S/\rho_{\epsilon a}(\bar{r}_{\epsilon a}))). \tag{14}$$

Since the relationship of $\rho_{\epsilon a}(\bar{r}_{\epsilon a})$ to $\bar{r}_{\epsilon a}$ *is a linear one* (see Eq. (12)), this upper bound drives $|d\bar{r}_{\epsilon a}/d\bar{t}_{\epsilon a}|$ *linearly toward zero* as $\rho_{\epsilon a}(\bar{r}_{\epsilon a}) \to \epsilon r_S+$, so it isn't possible for $\rho_{\epsilon a}(\bar{r}_{\epsilon a})$ to become equal to or smaller than ϵr_S *in any finite interval of "standard" local shell time* $\Delta\bar{t}_{\epsilon a}$. This *linear zeroing of the approach speed to the GR-unphysical configuration forbidden by the caveat reinforces that caveat*, which illustrates *the crucial role of gravitational time dilation.* (For the dust ball's *surface shell*, i.e., for $\epsilon = 1$, the caveat $\rho_{\epsilon a}(\bar{r}_{\epsilon a}) > \epsilon r_S$ *simplifies* to $\bar{r}_a > r_S$ [see Eq. (12)], so a dust ball's radius *always exceeds its Schwarzschild radius*, which implies that a dust ball *can't produce an event horizon.*)

Since $r_S = 2GM/c^2$ and $\alpha = \gamma(r_S/a)$, in the $c \to \infty$ *nonrelativistic limit*, $\rho_{\epsilon a}(\bar{r}_{\epsilon a}) \to \bar{r}_{\epsilon a}$, and the Eq. (13) dust-ball shell-radii equations of motion reduce to,

$$d\bar{r}_{\epsilon a}/d\bar{t}_{\epsilon a} = \pm\epsilon^{\frac{3}{2}}(2GM)^{\frac{1}{2}}((1/\bar{r}_{\epsilon a}) + (\gamma/(\epsilon a)))^{\frac{1}{2}}. \tag{15}$$

These nonrelativistic-limit shell-radii equations of motion are *devoid* of both the *speed* and the *configuration* constraints which featured so prominently in Eq. (14). Squaring *both sides* of *each* Eq. (15) nonrelativistic-limit shell-radius equation of motion reveals that it *corresponds* to the Eq. (1) *Newtonian Friedmann equation* $(\dot{R}(\bar{t}_{\epsilon a}))^2 = \omega^2((1/R(\bar{t}_{\epsilon a}))+\gamma)$, where $\omega^2 = 2GM/a^3$, *via the simple scaling relationship* $\bar{r}_{\epsilon a}(\bar{t}_{\epsilon a}) = \epsilon a R(\bar{t}_{\epsilon a})$, $0 < \epsilon \leq 1$. The *second-order form* of the nonrelativistic-limit shell-radius Eq. (15) above is also readily worked out. The result is,

$$d^2\bar{r}_{\epsilon a}/d\bar{t}_{\epsilon a}^2 = -\epsilon^3 GM/\bar{r}_{\epsilon a}^2, \tag{16}$$

which reflects *the nonrelativistic-limit* fact that the shell radius' acceleration arises from *the spherical effective mass* $\epsilon^3 M$, $0 < \epsilon \leq 1$, *which is the source of net Newtonian gravitational force on that shell.*

The *full* Eq. (13) shell-radius *first-order* equation of motion in "standard" coordinates *also* has a *second-order form*, which illuminates *the modification of shell-radius acceleration caused by gravitational time dilation.* One differentiates both sides of the Eq. (13) with respect to $\bar{t}_{\epsilon a}$, and then replaces the factor of $d\bar{r}_{\epsilon a}/d\bar{t}_{\epsilon a}$ on that result's right-hand side by the right-hand side of the Eq. (13) to obtain,

$$d^2\bar{r}_{\epsilon a}/d\bar{t}_{\epsilon a}^2 = \tfrac{1}{2}(\epsilon c^2/r_S) \times$$
$$\left[(-1+2\alpha)(\epsilon r_S/\rho_{\epsilon a}(\bar{r}_{\epsilon a}))^2 + 3(\epsilon r_S/\rho_{\epsilon a}(\bar{r}_{\epsilon a}))^3\right]\left[\frac{1-(\epsilon r_S/\rho_{\epsilon a}(\bar{r}_{\epsilon a}))}{(1+\epsilon^2\alpha)^{\frac{1}{2}}(1+\alpha)^{\frac{1}{2}}}\right], \tag{17}$$

which has *the same caveat* $\rho_{\epsilon a}(\bar{r}_{\epsilon a}) > \epsilon r_S$ as Eq. (13). Since $3 > (1-2\alpha)$ (because $\alpha > -1$), there *always* exists a *range of* $\rho_{\epsilon a}(\bar{r}_{\epsilon a})$ *values* which *both* satisfy the caveat $\rho_{\epsilon a}(\bar{r}_{\epsilon a}) > \epsilon r_S$ *and* produce *positive* (i.e., *outward*) shell-radius acceleration $d^2\bar{r}_{\epsilon a}/d\bar{t}_{\epsilon a}^2$—that is so *despite* the fact that the $c \to \infty$ *nonrelativistic limit* of Eq. (17) is Eq. (16), which implies *always negative* (i.e., *always inward*) acceleration. Furthermore, for all initial conditions such that $\alpha \geq \frac{1}{2}$, every shell-radius acceleration $d^2\bar{r}_{\epsilon a}/d\bar{t}_{\epsilon a}^2$ is *positive* (i.e., *outward*) at *all finite "standard" local times* $\bar{t}_{\epsilon a}$ (at any *finite* "standard" local time $\bar{t}_{\epsilon a}(\rho_{\epsilon a}(\bar{r}_{\epsilon a}))$, Eq. (11) implies that the caveat $\rho_{\epsilon a}(\bar{r}_{\epsilon a}) > \epsilon r_S$ *is automatically satisfied*). That *every* shell-radius acceleration can be *positive* (i.e., *outward*) at *all* finite "standard" local times apparently *eliminates any need to postulate a nonzero "dark energy" cosmological constant.*

References

1. S. Weinberg, *Gravitation and Cosmology: Principles and Applications of the General Theory of Relativity* (John Wiley & Sons, New York, 1972), Eq. (3.5.2), p. 79.
2. S. Weinberg, op. cit., Eqs. (15.1.1), (15.1.2), (15.1.20) and (15.1.22), pp. 471–472.
3. S. Weinberg, op. cit., Section 11.8, second paragraph, p. 338.
4. J. R. Oppenheimer and H. Snyder, Phys. Rev. **56**, 455 (1939).
5. S. Weinberg, op. cit., Eqs. (11.9.28)–(11.9.30), p. 345.
6. S. K. Kauffmann, "Is 'Dark Energy' Just an Effect of Gravitational Time Dilation?", viXra:1605.0196 (2016), `http://vixra.org/abs/1605.0196`.
7. S. Weinberg, op. cit., Eq. (11.8.9), p. 340 and Eq. (11.9.3), p. 342.

Nonlinear-supersymmetric General Relativity Theory

K. Shima

Laboratory of Physics, Saitama Institute of Technology,
Fukaya, Saitama 369-0293, Japan
**shima@sit.ac.jp*

Physical and methematical implications of nonlinear-supersymmetric general relativity theory(NLSUSYGR) are presented. They offer a new paradigm for the supersymmetric unification of space-time(Einstein general relativity theory(EGR)) and matter(the SM of particle physics), which gives new insight into the unsolved problems in EGR and SM.

Keywords: Linear and nonlinear SUSY; Einstein-Hilbert type action; goldstone fermion; compositeness; unified theory.

1. Introduction

The symmetry and its spontaneous breaking are key notions for the description of the rationale of being. Supersymmetry (SUSY)[1–3] related naturally to space-time symmetry is promising for the unification of general relativity and the low enegy SM in *one* single irreducible representation of the symmetry group. Therefore, the evidences of SUSY and its spontaneous breakdown[4–6] should be studied not only in (low energy) particle physics but also in cosmology, i.e. in the framework necessarily accomodating graviton[7]. And we have found by group theoretical arguments that among all $SO(N)$ super-Poincaré (sP) groups the $SO(10)$ sP group decomposed as $N = \underline{10} = \underline{5} + \underline{5^*}$ under $SO(10) \supset SU(5)$ may be a unique and minimal group which accomodates all observed particles including graviton in *a single* irreducible representation of $N = 10$ *linear(L)* SUSY[8], In that case 10 supercharges $Q^I, (I = 1, 2, \cdots .10)$ are embedded as follows: $\underline{10}_{SO(10)} = \underline{5}_{SU(5)} + \underline{5}^*{}_{SU(5)}$, $\underline{5}_{SU(5)} = [\ \underline{3}^{*c}, \underline{1}^{ew}, (\frac{e}{3}, \frac{e}{3}, \frac{e}{3}) : Q_a(a = 1, 2, 3)\ ] + [\ \underline{1}^c, \underline{2}^{ew}, (-e, 0) : Q_m(m = 4, 5)\ ]$, i.e., $\underline{5}_{SU(5)GUT}$ are $[Q_a$:$\bar{d}$-type, Q_m:Lepton-type] supercharges, And the massless helicity state $|h>$ of gravity multiplet of SO(10) sP with CPT conjugation are specified by the helicity $h = (2 - \frac{n}{2})$ and the dimension $\underline{d}_{[n]} = \frac{10!}{n!(10-n)!}$: $|h> = Q^n Q^{n-1} \cdots Q^2 Q^1 |2>$, Q^n $(n = 0, 1, 2, \cdots, 10)$. Here we assume a *maximal*

$\lvert h \rvert$	3	$\frac{5}{2}$	2	$\frac{3}{2}$	1	$\frac{1}{2}$	0
			$\underline{1}_{[0]}$	$\underline{10}_{[1]}$	$\underline{45}_{[2]}$	$\underline{120}_{[3]}$	$\underline{210}_{[4]}$
$\underline{d}_{[n]}$	$\underline{1}_{[10]}$	$\underline{10}_{[9]}$	$\underline{45}_{[8]}$	$\underline{120}_{[7]}$	$\underline{210}_{[6]}$	$\underline{252}_{[5]}$	$\underline{210}_{[4]}$

$SU(3) \times SU(2) \times U(1)$ invariant superHiggs mechanism among helicity states, i.e., all redundant high spin states for SM become massive by absorbing lower helicity states in SM invariant way. The results are interesting:

In the fermionic sector, just three generations of quark and lepton states survive. In the bosonic sector, gauge fields of SM in vector states and one Higgs scalar field of SM survive. Besides those observed states, one color-singlet neutral massive vector state and double-charge color-singlet massive spin $\frac{1}{2}$ states are predicted, which can be tested. Spin $\frac{1}{2}$ state survivors after superHiggs mechanism are shown in the table (tentatively as Dirac particles).

$SU(3)$	Q_e	$SU(2) \otimes U(1)$
$\underline{1}$	0, -1	$\begin{pmatrix} \nu_e \\ e \end{pmatrix} \begin{pmatrix} \nu_\mu \\ \mu \end{pmatrix} \begin{pmatrix} \nu_\tau \\ \tau \end{pmatrix}$
	-2	$(E)\,(M)$
$\underline{3}$	$5/3$, $2/3$	$\begin{pmatrix} a \\ f \end{pmatrix} \begin{pmatrix} g \\ m \end{pmatrix}$
	$2/3$, $-1/3$	$\begin{pmatrix} u \\ d \end{pmatrix} \begin{pmatrix} c \\ s \end{pmatrix} \begin{pmatrix} t \\ b \end{pmatrix}$
	$-1/3$, $-4/3$	$\begin{pmatrix} h \\ o \end{pmatrix}$
	$2/3$, $-1/3$, $-4/3$	$\begin{pmatrix} r \\ i \\ n \end{pmatrix}$
$\underline{6}$	$4/3$, $1/3$, $-2/3$	$\begin{pmatrix} P \\ Q \\ R \end{pmatrix} \begin{pmatrix} X \\ Y \\ Z \end{pmatrix}$
$\underline{8}$	0, -1	$\begin{pmatrix} N_1 \\ E_1 \end{pmatrix} \begin{pmatrix} N_2 \\ E_2 \end{pmatrix}$

We will show in the next section that no-go theorem for constructing non-trivial $SO(N>8)$SUGRA can be circumvented by adopting the *nonliner (NL)* representation of SUSY[12], i.e. by introducing *the degeneracy of space-time* through NLSUSY degrees of freedom. Also the NL representation of SUSY gives the (unique) action describing the *robust* spontaneous breakdown of SUSY. Therefore the NLSUSY invariant generalization of the general relativity theory may give the fundamental theory in our scenario.

2. Nonlinear-Supersymmetric General Relativity (NLSUSYGR)

For simplicity we discuss $N=1$ without the loss of the generality. The extension to $N>1$ is straightforward.

The fundamental action *nonlinear supersymmetric general relativity theory (NLSUSYGR)* has been constructed by extending the geometric arguments of Einstein general relativity (EGR) on Riemann space-time to new space-time inspired by NLSUSY, where tangent space-time is specified not only by the Minkowski coordinate

x_a for $SO(1,3)$ but also by the Grassmann coordinate ψ_α for $SL(2,C)$ related to NLSUSY[9,13]. They are coordinates of the coset space $\frac{superGL(4R)}{GL(4R)}$ and can be interpreted as NG fermions associated with the spontaneous breaking of *super*-$GL(4R)$ down to $GL(4R)$. (The noncompact isomorphic groups $SO(1,3)$ and $SL(2,C)$ for tangent space-time symmetry on curved space-time can be regarded as the generalization of the compact isomorphic groups $SU(2)$ and $SO(3)$ for the gauge symmetry of 't Hooft-Polyakov monopole on flat space-time.)

The NLSUSYGR action[9,13], is given by

$$L_{NLSUSYGR}(w) = -\frac{c^4}{16\pi G}|w|\{\Omega(w) + \Lambda\}, \tag{1}$$

$$|w| = \det w^a{}_\mu = \det\{e^a{}_\mu + t^a{}_\mu(\psi)\},$$

$$t^a{}_\mu(\psi) = \frac{\kappa^2}{2i}(\bar{\psi}\gamma^a\partial_\mu\psi - \partial_\mu\bar{\psi}\gamma^a\psi), \tag{2}$$

where G is the Newton gravitational constant, Λ is a (*small*) cosmological term and κ is an arbitrary constant of NLSUSY with the dimemsion $(\text{mass})^{-2}$. $w^a{}_\mu(x) = e^a{}_\mu + t^a{}_\mu(\psi)$ and $w^\mu{}_a = e^\mu{}_a - t^\mu{}_a + t^\mu{}_\rho t^\rho{}_a - t^\mu{}_\sigma t^\sigma{}_\rho t^\rho{}_a + t^\mu{}_\kappa t^\kappa{}_\sigma t^\sigma{}_\rho t^\rho{}_a$ which terminate at $O(t^4)$ for $N=1$ are the invertible *unified vierbeins* of new space-time. $e^a{}_\mu$ is the ordinary vierbein of EGR for the local $SO(1,3)$ and $t^a{}_\mu(\psi)$ is the mimic vierbein analogue (actually the stress-energy-momentum tensor) of NG fermion $\psi(x)$ for the local $SL(2,C)$. (We call $\psi(x)$ *superon* as the hypothetical fundamental spin $\frac{1}{2}$ particle carrying the supercharge of the supercurrent[15] of the global NLSUSY.) $\Omega(w)$ is the the unified scalar curvature of new space-time computed in terms of the *unified vierbein* $w^a{}_\mu(x)$. Interestingly Grassmann degrees of freedom induce the imaginary part of the unified vierbein $w^a{}_\mu(x)$, which represents straightforwardly the fermionic matter contribution. Note that $e^a{}_\mu$ and $t^a{}_\mu(\psi)$ contribute equally to the curvature of spac-time, which may be regarded as the Mach's principle in ultimate space-time. (The second index of mimic vierbein t, e.g. μ of $t^a{}_\mu$, means the derivative ∂_μ.) $s_{\mu\nu} \equiv w^a{}_\mu \eta_{ab} w^b{}_\nu$ and $s^{\mu\nu}(x) \equiv w^\mu{}_a(x) w^{\nu a}(x)$ are *unified metric tensors* of new spacetime.

NLSUSY GR action (1) possesses promissing large symmetries isomorphic to $SO(N)$ ($SO(10)$) SP group[14,16]; namely, $L_{NLSYSYGR}(w)$ is invariant under

$$[\text{new NLSUSY}] \otimes [\text{local GL(4, R)}] \otimes [\text{local Lorentz}] \tag{3}$$

for spacetime symmetries and

$$[\text{global}SO(N)] \otimes [\text{local}U(1)^N] \tag{4}$$

for internal symmetries in case of N superons $\psi^i, i = 1, 2, \cdots, N$.

For example, $L_{NLSUSYGR}(w)$ (2) is invariant under the following NLSUSY transformations:

$$\delta^{NL}\psi = \frac{1}{\kappa^2}\zeta + i\kappa^2(\bar{\zeta}\gamma^\rho\psi)\partial_\rho\psi, \quad \delta^{NL}e^a{}_\mu = i\kappa^2(\bar{\zeta}\gamma^\rho\psi)\partial_{[\rho}e^a{}_{\mu]}, \tag{5}$$

where ζ is a constant spinor parameter and $\partial_{[\rho}e^a{}_{\mu]} = \partial_\rho e^a{}_\mu - \partial_\mu e^a{}_\rho$, which induce the following GL(4R) transformations on the unified vierbein $w^a{}_\mu$

$$\delta_\zeta w^a{}_\mu = \xi^\nu \partial_\nu w^a{}_\mu + \partial_\mu \xi^\nu w^a{}_\nu, \quad \delta_\zeta s_{\mu\nu} = \xi^\kappa \partial_\kappa s_{\mu\nu} + \partial_\mu \xi^\kappa s_{\kappa\nu} + \partial_\nu \xi^\kappa s_{\mu\kappa}, \tag{6}$$

where $\xi^\rho = i\kappa^2(\bar{\zeta}\gamma^\rho\psi)$,

The commutators of two new NLSUSY transformations (5) on ψ and $e^a{}_\mu$ are $GL(4R)$, i.e. new NLSUSY (5) is the square-root of $GL(4R)$;

$$[\delta_{\zeta_1}, \delta_{\zeta_2}]\psi = \Xi^\mu \partial_\mu \psi, \quad [\delta_{\zeta_1}, \delta_{\zeta_2}]e^a{}_\mu = \Xi^\rho \partial_\rho e^a{}_\mu + e^a{}_\rho \partial_\mu \Xi^\rho, \tag{7}$$

where $\Xi^\mu = 2i\kappa(\bar{\zeta}_2\gamma^\mu\zeta_1) - \xi_1^\rho \xi_2^\sigma e_a{}^\mu(\partial_{[\rho}e^a{}_{\sigma]})$. The algebra closes. The ordinary local GL(4R) invariance is trivial by the construction.

Also NLSUSYGR is invariant under the following local Lorentz transformation: on $w^a{}_\mu$

$$\delta_L w^a{}_\mu = \epsilon^a{}_b w^b{}_\mu \tag{8}$$

or equivalently on ψ and $e^a{}_\mu$

$$\delta_L \psi = -\frac{i}{2}\epsilon_{ab}\sigma^{ab}\psi, \quad \delta_L e^a{}_\mu = \epsilon^a{}_b e^b{}_\mu + \frac{\kappa^4}{4}\varepsilon^{abcd}\bar{\psi}\gamma_5\gamma_d\psi(\partial_\mu \epsilon_{bc}), \tag{9}$$

with the local parameter $\epsilon_{ab} = (1/2)\epsilon_{[ab]}(x)$. The local Lorentz transformation forms a closed algebra, for example, on $e^a{}_\mu$

$$[\delta_{L_1}, \delta_{L_2}]e^a{}_\mu = \beta^a{}_b e^b{}_\mu + \frac{\kappa^4}{4}\varepsilon^{abcd}\bar{\psi}\gamma_5\gamma_d\psi(\partial_\mu \beta_{bc}), \tag{10}$$

where $\beta_{ab} = -\beta_{ba}$ is defined by $\beta_{ab} = \epsilon_{2ac}\epsilon_1{}^c{}_b - \epsilon_{2bc}\epsilon_1{}^c{}_a$.

Note that the no-go theorem is overcome (circumvented) in a sense that the nontivial N-extended SUSY gravitational theory with $N > 8$ has been constructed in the NLSUSY invariant way.

New *empty* space-time described by NLSUSYGR action $L_{NLSUSYGR}(w)$ of the *vacuum* EH type is unstable due to NLSUSY structure of tangent space-time and collapses (called *Big Collapse*[16]) spontaneously to ordinary Riemann space-time with the cosmological term and fermionic matter *superon* (called *superon-graviton model (SGM)*). Note that Big Collapse induces inststantaneously the rapid expansion of three dimensional space-time due to the Pauli principle of NG fermion *superon*. SGM action, whose gravitational evolution ignites Big Bang of the present observed universe, is the following;

$$\begin{aligned}
L_{SGM}(e,\psi) = & -\frac{c^4\Lambda}{16\pi G}e|w| - \frac{c^4}{16\pi G}e|w|R(e) + \frac{c^4}{16\pi G}e|w|[\, 2t^{(\mu\nu)}R_{\mu\nu}(e) \\
& +\frac{1}{2}\{g^{\mu\nu}\partial^\rho\partial_\rho t_{(\mu\nu)} - t_{(\mu\nu)}\partial^\rho\partial_\rho g^{\mu\nu} + g^{\mu\nu}\partial^\rho t_{(\mu\sigma)}\partial^\sigma g_{\rho\nu} - 2g^{\mu\nu}\partial^\rho t_{(\mu\nu)}\partial^\sigma g_{\rho\sigma} + \ldots\} \\
& +(t^\mu{}_\rho t^{\rho\nu} + t^\nu{}_\rho t^{\rho\mu} + t^{\mu\rho}t^\nu{}_\rho)R_{\beta\mu}(e) - \{2t^{(\mu\rho)}t^{(\nu}{}_{\rho)}R_{\mu\nu} + t^{(\mu\rho)}t^{(\nu\sigma)}R_{\mu\nu\rho\sigma}(e) \\
& +\frac{1}{2}t^{(\mu\nu)}(g^{\rho\sigma}\partial_\mu\partial_\nu t_{(\rho\sigma)} - g^\rho{}_\sigma\partial^\rho\partial_\mu t_{(\sigma\nu)} + \ldots)\} + \{O(t^3)\} + \cdots],
\end{aligned} \tag{11}$$

where $(\psi)^5 \equiv 0$(for $N = 1$), $e = dete^a{}_\mu$, $t^{(\mu\nu)} = t^{\mu\nu} + t^{\nu\mu}$, $t_{(\mu\nu)} = t_{\mu\nu} + t_{\nu\mu}$, and $|w| = detw^a{}_b = det(\delta^a{}_b + t^a{}_b)$ is the flat space NLSUSY action of VA[2] containing up to $O(t^4)$ and $R(e)$, $R_{\mu\nu}(e)$, and $R_{\mu\nu\rho\sigma}(e)$ are the ordinary Riemann curvature tensors of GR. Remarkably the first term should reduces to NLSUSY action[2] in Riemann-flat $e_a{}^\mu(x) \to \delta_a{}^\mu$ space-time, i.e. the arbitrary constant κ of NLSUSY is fixed to

$$\kappa^{-2} = \frac{c^4}{8\pi G}\Lambda. \tag{12}$$

$L_{SGM}(e,\psi)$ (11) can be recasted formally as the following famlliar form

$$L_{SGM}(e,\psi) = -\frac{c^4}{16\pi G}|e|\{R(e) + \Lambda + \tilde{T}(e,\psi)\}, \tag{13}$$

where $R(e)$ is the scalar curvature of ordinary EH action and $\tilde{T}(e,\psi)$ represents the kinetic term and the gravitational interaction of superons.

We have shown qualitatively that NLSUSY GR may potentially describe a new paradigm (SGM) for the SUSY unification of space-time and matter, where particular SUSY compositeness composed of superons for all (observed) particles emerges as an ultimate feature of nature behind the familiar LSUSY models (MSSM, SUSY GUTs)[9,17] and SM as well. That is, all (observed) low energy particles may be gravitational eigenstates of $SO(N)$ sP expressed uniquely as the SUSY composites of N superons. We examine explicitly these possibilities in the next section.

3. Linearizing NLSUSY and Vacuum of SGM

3.1. *Linearization of NLSUSY*

We anticipate that $SO(10)$ sP LSUSY algebra of space-time determines the particle configuration of superon-graviton system $L_{\rm SGM}(e,\psi)$, (like $O(4)$ for the relativistic. H-atom). extracting the (low energy) physical meaning of SGM directly on curved Riemann space-time is yet to be done.

Considering that the SGM action in the low energy reduces essentially to the N-extended NLSUSY action with $\kappa^2 = (\frac{c^4\Lambda}{8\pi G})^{-1}$ in Riemann-flat ($e^a{}_\mu \to \delta^a_\mu$) space-time, it is interesting from the viewpoint of the low energy physics on the tangent flat space-time to linearize the N-extended NLSUSY model and find the equivalent(related) N-extended LSUSY theory. We have shown explicitly in two dimensional space-time ($d = 2$)[22,23] for simplicity that $N = 2$ LSUSYQED is equivalent(related) to $N = 2$ NLSUSY model. We put three conditions: (i) *SUSY compositenes*, i.e. each field of LSUSY supermultiplet is expressed uniquely as the composite of the NLSUSY NG fermion which is the low energy leading order term of the superchage, (ii) *LSUSY algebra base*, i.e. LSUSY transformation for the supermultiplet and the invariance of the LSUSY action are reproduced for the composite supermultiplet under the NLSUSY transformations of the constituent NG fermion, (iii) *NL/L SUSY relation(equivalence)*, i.e. LSUSY action L_{LSUSY} reduces to NLSUSY action L_{NLSUSY} when *SUSY compositeness* is substituted into LSUSY field of LSUSY supermultiplet. (Note that the minimal realistic SUSY QED in SGM composite scenario is given by $N = 2$ SUSY[21].) Considering SUSY compositeness (ii) seriously we are tempted to imagine some composite structure (far) behind the familiar LSUSY unified models, e.g. MSSM and SUSY GUT may be in a sense a low energy effective theory of SGM/NLSUSYGR. The condition (iii) guarantees the equivalence irrespective of the renormalizability of basic theories.

In this section we study explicitly the vacuum structure of $N = 2$ LSUSY QED in the SGM scenario in $d = 2$[23]. $N = 2$ NLSUSY action for two superons (NG fermions) ψ^i $(i = 1, 2)$ in $d = 2$ is written as follows,

$$
\begin{aligned}
& L_{N=2\text{NLSUSY}} \\
& = -\frac{1}{2\kappa^2}|w| \\
& = -\frac{1}{2\kappa^2}\left\{1 + t^a{}_a + \frac{1}{2!}(t^a{}_a t^b{}_b - t^a{}_b t^b{}_a)\right\} \\
& = -\frac{1}{2\kappa^2}\left\{1 - i\kappa^2\bar{\psi}^i \partial\!\!\!/ \psi^i \frac{1}{2}\right. \\
& \left. -\frac{1}{2}\kappa^4(\bar{\psi}^i \partial\!\!\!/ \psi^i \bar{\psi}^j \partial\!\!\!/ \psi^j - \bar{\psi}^i \gamma^a \partial_b \psi^i \bar{\psi}^j \gamma^b \partial_a \psi^j)\right\}
\end{aligned} \tag{14}
$$

where κ is a constant whose dimension is $(\text{mass})^{-1}$ and $|w| = \det(w^a{}_b) = \det(\delta^a_b + t^a{}_b)$, $t^a{}_b = -i\kappa^2\bar{\psi}^i\gamma^a\partial_b\psi^i$. While, the helicity states contained formally in $(d = 2)$ $N = 2$ LSUSY QED are the vector supermultiplet containing $U(1)$ gauge field

$$
\begin{pmatrix} +1 \\ +\frac{1}{2},\ +\frac{1}{2} \\ 0 \end{pmatrix} + [\text{CPT conjugate}],
$$

and the scalar supermultiplet for matter fields

$$
\begin{pmatrix} +\frac{1}{2} \\ 0,\ 0 \\ -\frac{1}{2} \end{pmatrix} + [\text{CPT conjugate}].
$$

The most general $N = 2$ LSUSY QED action for the massless case in $d = 2$, is written as follows[23],

$$
\begin{aligned}
& L_{N=2\mathrm{SUSYQED}} \\
& = -\frac{1}{4}(F_{ab})^2 + \frac{i}{2}\bar{\lambda}^i \partial\!\!\!/ \lambda^i + \frac{1}{2}(\partial_a A)^2 + \frac{1}{2}(\partial_a \phi)^2 + \frac{1}{2}D^2 \\
& -\frac{1}{\kappa}\xi D + \frac{i}{2}\bar{\chi}\partial\!\!\!/ \chi + \frac{1}{2}(\partial_a B^i)^2 + \frac{i}{2}\bar{\nu}\partial\!\!\!/ \nu + \frac{1}{2}(F^i)^2 \\
& +f(A\bar{\lambda}^i\lambda^i + \epsilon^{ij}\phi\bar{\lambda}^i\gamma_5\lambda^j - A^2 D + \phi^2 D + \epsilon^{ab}A\phi F_{ab}) \\
& +e\left\{ iv_a\bar{\chi}\gamma^a\nu - \epsilon^{ij}v^a B^i\partial_a B^j + \bar{\lambda}^i\chi B^i + \epsilon^{ij}\bar{\lambda}^i\nu B^j\frac{1}{2} \right. \\
& \left. -\frac{1}{2}D(B^i)^2 + \frac{1}{2}(\bar{\chi}\chi + \bar{\nu}\nu)A - \bar{\chi}\gamma_5\nu\phi \right\} \\
& +\frac{1}{2}e^2({v_a}^2 - A^2 - \phi^2)(B^i)^2,
\end{aligned} \tag{15}
$$

where $(v^a, \lambda^i, A, \phi, D)$ $(F_{ab} = \partial_a v_b - \partial_b v_a)$ is the *minimal* off-shel vector supermultiplet containing v^a for a $U(1)$ vector field, λ^i for doublet (Majorana) fermions, A for a scalar field in addition to ϕ for another scalar field and D for an auxiliary scalar field, while $(\chi,\ B^i,\ \nu,\ F^i)$ is the *minimal* off-shell scalar supermultiplet containing (χ, ν) for two (Majorana) fermions, B^i for doublet scalar fields and F^i for auxiliary scalar fields. The linear term of F is forbidden by the gauge invariance[23]. Also ξ is an arbitrary demensionless parameter giving a magnitude of SUSY breaking mass, and f and e are Yukawa and gauge coupling constants with the dimension $(\mathrm{mass})^1$, respectively.

$N = 2$ LSUSY QED action (15) is invariant under the following LSUSY transformations parametrized by ζ^i,

$$
\delta_\zeta v^a = -i\epsilon^{ij}\bar{\zeta}^i\gamma^a\lambda^j,\ \delta_\zeta\lambda^i = (D - i\partial\!\!\!/ A)\zeta^i + \frac{1}{2}\epsilon^{ab}\epsilon^{ij}F_{ab}\gamma_5\zeta^j - i\epsilon^{ij}\gamma_5\partial\!\!\!/\phi\zeta^j, \tag{16}
$$

$$
\delta_\zeta A = \bar{\zeta}^i\lambda^i,\ \delta_\zeta\phi = -\epsilon^{ij}\bar{\zeta}^i\gamma_5\lambda^j,\ \delta_\zeta D = -i\bar{\zeta}^i\partial\!\!\!/\lambda^i.
$$

for the vector multiplet and

$$
\delta_\zeta\chi = (F^i - i\partial\!\!\!/ B^i)\zeta^i - e\epsilon^{ij}V^i B^j,\ \delta_\zeta\nu = \epsilon^{ij}(F^i + i\partial\!\!\!/ B^i)\zeta^j + eV^i B^i, \tag{17}
$$

$$
\delta_\zeta B^i = \bar{\zeta}^i\chi - \epsilon^{ij}\bar{\zeta}^j\nu,
$$

$$
\delta_\zeta F^i = -i\bar{\zeta}^i\partial\!\!\!/\chi - i\epsilon^{ij}\bar{\zeta}^j\partial\!\!\!/\nu - e\{\epsilon^{ij}\bar{V}^j\chi - \bar{V}^i\nu + (\bar{\zeta}^i\lambda^j + \bar{\zeta}^j\lambda^i)B^j - \bar{\zeta}^j\lambda^j B^i\} \tag{18}
$$

with $V^i = iv_a\gamma^a\zeta^i - \epsilon^{ij}A\zeta^j - \phi\gamma_5\zeta^i$ for the scalar multiplet.

For extracting the low energy physical contents of $N = 2$ SGM (NLSUSY GR) we consider SGM in asymptotic Riemann-flat space-time, where $N = 2$ SGM reduces to essentially $N = 2$ NLSUSY action. We will show the equivalence of $N = 2$ NLSUSY action to $N = 2$ SUSYQED action(called *NL/L SUSY relation.*), i.e.

$$
\begin{aligned}
& L_{N=2\mathrm{SGM}} \overset{{e^a}_\mu \to \delta^a_\mu}{\longrightarrow} \\
& L_{N=2\mathrm{NLSUSY}} + [\text{suface terms}] = f_\xi L_{N=2\mathrm{SUSYQED}},
\end{aligned} \tag{19}
$$

where f_ξ is the function of vacuum values of auxiliary fields. NL/L SUSY relation guarantees the equivalence(relation) of two theories irrespective of the renormalizability. NL/L SUSY relation shown explicitly by substituting the following SUSY compositeness[23] into the LSUSY QED theory.

The SUSY compositeness for the minimal vector supermultiplet $(v^a, \lambda^i, A, \phi, D)$ are

$$
\begin{aligned}
v^a &= -\frac{i}{2}\xi\kappa\epsilon^{ij}\bar{\psi}^i\gamma^a\psi^j|w|, \\
\lambda^i &= \xi\left[\psi^i|w| - \frac{i}{2}\kappa^2\partial_a\{\gamma^a\psi^i\bar{\psi}^j\psi^j(1-i\kappa^2\bar{\psi}^k\partial\!\!\!/\psi^k)\}\right], \\
A &= \frac{1}{2}\xi\kappa\bar{\psi}^i\psi^i|w|, \\
\phi &= -\frac{1}{2}\xi\kappa\epsilon^{ij}\bar{\psi}^i\gamma_5\psi^j|w|, \\
D &= \frac{\xi}{\kappa}|w| - \frac{1}{8}\xi\kappa^3\partial_a\partial^a(\bar{\psi}^i\psi^i\bar{\psi}^j\psi^j).
\end{aligned} \tag{20}
$$

While for the minimal scalar supermultiplet (χ, B^i, ν, F^i) the SUSY compositeness is

$$
\begin{aligned}
\chi &= \xi^i\left[\psi^i|w| + \frac{i}{2}\kappa^2\partial_a\{\gamma^a\psi^i\bar{\psi}^j\psi^j(1-i\kappa^2\bar{\psi}^k\partial\!\!\!/\psi^k)\}\right], \\
B^i &= -\kappa\left(\frac{1}{2}\xi^i\bar{\psi}^j\psi^j - \xi^j\bar{\psi}^i\psi^j\right)|w|, \\
\nu &= \xi^i\epsilon^{ij}\left[\psi^j|w| + \frac{i}{2}\kappa^2\partial_a\{\gamma^a\psi^j\bar{\psi}^k\psi^k(1-i\kappa^2\bar{\psi}^l\partial\!\!\!/\psi^l)\}\right], \\
F^i &= \frac{1}{\kappa}\xi^i\left\{|w| + \frac{1}{8}\kappa^3\partial_a\partial^a(\bar{\psi}^j\psi^j\bar{\psi}^k\psi^k)\right\} \\
&\quad -i\kappa\xi^j\partial_a(\bar{\psi}^i\gamma^a\psi^j|w|) - \frac{1}{4}e\kappa^2\xi\xi^i\bar{\psi}^j\psi^j\bar{\psi}^k\psi^k,
\end{aligned} \tag{21}
$$

where ξ and ξ^i are factors of the vacuum expectation value of auxiliary fields. SUSY compositeness (20) and (21) satisfy (i), (ii) and (iii). Furthermore substituting these relations into the $N = 2$ SUSYQED action (15) we can show directly that $N = 2$ SUSYQED action (15) is equivalent(related) to $N = 2$ NLSUSY action provided $f_\xi(\xi, \xi^i) = \xi^2 - (\xi^i)^2 = 1$.

Note that for the SUSY compositeness (21) of the scalar supermultiplet it is interesting that the four-fermion self-interaction term appearing only in the auxiliary fields F^i is the origin of the familiar local $U(1)$ gauge symmetry of LSUSY theory. The introduction of new auxiliary fields in the supermultiplet improves(clarifies) these situations. Especially the total derivative term in SUSY compositeness (20) and (21) disappears by introducing a new auxiliary fields composed of fermion self-interactions. (Note that in the straightforward linearization the commutator algebra does not contain U(1) gauge transformation even for the vector U(1) gauge field[21].)

NL/L SUSY relation of SGM scenario for the larger supermultiplet predicts the magnitude of the bare gauge coupling constant These situations are easily seen by using the superfield formulation of the linearizion of NLSUSY. We adopt for the $N = 2$ general vector supermultiplet of the general superfield

$$\mathcal{V}(x,\theta^i) = C(x) + \bar\theta^i\Lambda^i(x) + \frac{1}{2}\bar\theta^i\theta^j M^{ij}(x) - \frac{1}{2}\bar\theta^i\theta^i M^{jj}(x) + \frac{1}{4}\epsilon^{ij}\bar\theta^i\gamma_5\theta^j\phi(x)$$
$$-\frac{i}{4}\epsilon^{ij}\bar\theta^i\gamma_a\theta^j v^a(x) - \frac{1}{2}\bar\theta^i\theta^i\bar\theta^j\lambda^j(x) - \frac{1}{8}\bar\theta^i\theta^i\bar\theta^j\theta^j D(x), \tag{22}$$

and for the $N = 2$ scalar supermultiplet

$$\Phi^i(x,\theta^i) = B^i(x) + \bar\theta^i\chi(x) - \epsilon^{ij}\bar\theta^j\nu(x) - \frac{1}{2}\bar\theta^j\theta^j F^i(x) + \bar\theta^i\theta^j F^j(x) - i\bar\theta^i \partial\!\!\!/ B^j(x)\theta^j$$
$$+\frac{i}{2}\bar\theta^j\theta^j(\bar\theta^i\partial\!\!\!/\chi(x) - \epsilon^{ik}\bar\theta^k\partial\!\!\!/\nu(x)) + \frac{1}{8}\bar\theta^j\theta^j\bar\theta^k\theta^k\partial_a\partial^a B^i(x). \tag{23}$$

We take the following ψ^i-dependent supertranslations with $-\kappa\psi(x)$,

$$x'^a = x^a + i\kappa\bar\theta^i\gamma^a\psi^i, \quad \theta'^i = \theta^i - \kappa\psi^i, \tag{24}$$

and denote the resulting superfields on (x'^a, θ'^i) and their θ-expansions as

$$\mathcal{V}(x'^a,\theta'^i) = \tilde{\mathcal{V}}(x^a,\theta^i;\psi^i(x)), \quad \Phi(x'^a,\theta'^i) = \tilde\Phi(x^a,\theta^i;\psi^i(x)). \tag{25}$$

More generalized (Hybrid) global SUSY transformations $\delta^h{}_\varepsilon = \delta^L(x.\theta) + \delta^{NL}(\psi)$ of $\mathcal{V}(x'^a,\theta'^i)$ and $\Phi(x'^a,\theta'^i)$ on (x'^a,θ'^i) give:

$$\delta^h\tilde{\mathcal{V}}(x^a,\theta^i;\psi^i(x)) = \xi_\mu\partial^\mu\tilde{\mathcal{V}}(x^a,\theta^i;\psi^i(x)), \delta^h\tilde\Phi(x^a,\theta^i;\psi^i(x)) = \xi_\mu\partial^\mu\tilde\Phi(x^a,\theta^i;\psi^i(x)), \tag{26}$$

where $\xi_\mu = -i\bar\varepsilon^i\gamma_\mu\psi^i$. Therefore, we obtain the following generalized global SUSY invariant constraints for (the component fields in) $\tilde{\mathcal{V}}$ and $\tilde\Phi$,

$$\tilde\varphi^I_{\mathcal{V}}(x) = \xi^I_{\mathcal{V}}(\text{constant}) \quad \tilde\varphi^I_\Phi(x) = \xi^I_\Phi(\text{constant}), \tag{27}$$

which provide SUSY compositeness conditions.

From the physical viewpoint, i.e. the Lorentz invariance of the vacuum expectation value of auxiliary field, the following simple SUSY invariant constraint is the minimal one:

$$\tilde C = \tilde\Lambda^i = \tilde M^{ij} = \tilde\phi = \tilde v^a = \tilde\lambda^i = 0, \tilde D = \frac{\xi}{\kappa}, \tilde B^i = \tilde\chi = \tilde\nu = 0, \quad \tilde F^i = \frac{\xi^i}{\kappa}, \tag{28}$$

which produces the previous simple SUSY compositeness (20) and (21). More general and physical SUSY invariant constraint corresponding to vev of general 0^+ auxiliary fields:

$$\tilde C = \xi_c, \; \tilde\Lambda^i = \tilde M^{ij} = \tilde\phi = \tilde v^a = \tilde\lambda^i = 0, \; \tilde D = \frac{\xi}{\kappa}, \; \tilde B^i = \tilde\chi = \tilde\nu = 0, \; \tilde F^i = \frac{\xi^i}{\kappa}. \tag{29}$$

gives SUSY compositeness of largear supermultiplet and produces for NL/L SUSY relation

$$f_\xi(\xi, \xi^i, \xi_c) = \xi^2 - (\xi^i)^2 e^{-4e\xi_c} = 1, \quad i.e., e = \frac{\ln(\frac{\xi^{i2}}{\xi^2 - 1})}{4\xi_c}, \tag{30}$$

where e is the bare gauge coupling constant. This mechanism is natural and favorable for SGM scenario from the viewpoint of theory of everything.

By changing integration variables $(x^a, \theta^i) \to (x'^a, \theta'^i)$in superspace, we can confirm systematically NL/L SUSY relation by using the superfield formulation in superspace. Broken LSUSY(QED) gauge theory is encoded in the vacuum of $N = 2$ NLSUSY theory as composites of NG fermion, i.e. in the local flat space of NLSUSYGR with the constant energy density (cosmological term). Broken LSUSY(QED) gauge theory may be a low energy effective theory of NLSUSY.

3.2. *Vacuum structure of NLSUSY*

Now we study the vacuum structure of $N = 2$ SUSY QED action (15)[25] as the the physical configuration of the vacuum of NLSUSY theory. The vacuum is determined by the minimum of the potential $V(A, \phi, B^i, D)$,

$$\begin{aligned} &V(A, \phi, B^i, D) \\ &= -\frac{1}{2}D^2 + \left\{ \frac{\xi}{\kappa} + f(A^2 - \phi^2) + \frac{1}{2}e(B^i)^2 \right\} D. \end{aligned} \tag{31}$$

Substituting the solution of the equation of motion for the auxiliary field D we obtain

$$V(A, \phi, B^i) = \frac{1}{2}f^2 \left\{ A^2 - \phi^2 + \frac{e}{2f}(B^i)^2 + \frac{\xi}{f\kappa} \right\}^2 \geq 0. \tag{32}$$

The configurations of the fields corresponding to the vacua in (A, ϕ, B^i)-space, which are $SO(1,3)$ or $SO(3,1)$ symmetric, are classified according to the signatures of the parameters e, f, ξ, κ. We find two interesting vacua:
(I) For $ef > 0$, $\frac{\xi}{f\kappa} < 0$ case,

$$A^2 - \phi^2 + (\tilde{B}^i)^2 = k^2. \quad \left(\tilde{B}^i = \sqrt{\frac{e}{2f}} B^i, \ k^2 = \frac{-\xi}{f\kappa} \right) \tag{33}$$

(II) For $ef < 0$, $\frac{\xi}{f\kappa} > 0$ case,

$$-A^2 + \phi^2 + (\tilde{B}^i)^2 = k^2. \quad \left(\tilde{B}^i = \sqrt{\frac{-e}{2f}} B^i, \ k^2 = \frac{\xi}{f\kappa} \right) \tag{34}$$

Other cases of signatures with $SO(1,3)$ isometry in (A, ϕ, B^i)-space are unphysical, for they produce the pathological wrong sign kinetic terms for the fields induced around the vacuum. As for the above cases we find two different physical vacua. The physical particle spectrum is obtained by expanding the field (A, ϕ, B^i) around the vacuum with $SO(3,1)$ isometry. Furthermore (i) and (II) are essentially the same configurations due to the symmetry, we discuss case (I).

For case (I), the following expressions (a) and (b) are considered.:

$$\text{Case (Ia)}\ \begin{array}{l} A = -(k+\rho)\cos\theta\cos\varphi\cosh\omega,\ \ \phi = (k+\rho)\sinh\omega, \\ \tilde{B}^1 = (k+\rho)\sin\theta\cosh\omega,\ \ \tilde{B}^2 = (k+\rho)\cos\theta\sin\varphi\cosh\omega. \end{array} \quad \text{and}$$

$$\text{Case (Ib)}\ \begin{array}{l} A = (k+\rho)\sin\theta\cosh\omega,\ \ \phi = (k+\rho)\sinh\omega \\ \tilde{B}^1 = (k+\rho)\cos\theta\cos\varphi\cosh\omega,\ \ \tilde{B}^2 = (k+\rho)\cos\theta\sin\varphi\cosh\omega. \end{array}$$

Substituting these expressions into $L_{N=2\text{SUSYQED}}\ (A,\phi,B^i)$ and expanding the action around the vacuum configuration we obtain the physical particle contents. For the cases (a) we obtain

$$\begin{aligned} L_{N=2\text{SUSYQED}} = & \frac{1}{2}\{(\partial_a\rho)^2 - 4f^2k^2\rho^2\} + \frac{1}{2}\{(\partial_a\theta)^2 + (\partial_a\varphi)^2 - e^2k^2(\theta^2+\varphi^2)\} \\ & +\frac{1}{2}(\partial_a\omega)^2 - \frac{1}{4}(F_{ab})^2 + \frac{1}{2}(i\bar{\lambda}^i\partial\!\!\!/\lambda^i - 2fk\bar{\lambda}^i\lambda^i) \\ & +\frac{1}{2}\{i(\bar{\chi}\partial\!\!\!/\chi + \bar{\nu}\partial\!\!\!/\nu) - ek(\bar{\chi}\chi + \bar{\nu}\nu)\} + \cdots, \end{aligned}$$

and the following mass spectrum which indicates the spontaneous breakdown of $N=2$ SUSY;

$$\begin{aligned} & m_\rho^2 = m_{\lambda^i}^2 = 4f^2k^2 = \frac{-4\xi f}{\kappa},\ m_\theta^2 = m_\varphi^2 = m_\chi^2 = m_\nu^2 = e^2k^2 = \frac{-\xi e^2}{\kappa f}, \\ & m_{v_a} = m_\omega = 0, \end{aligned} \tag{35}$$

which can produce mass hierarchy by the factor $\frac{e}{f}$ independent of κ. ($\kappa^{-2} = \frac{c^4\Lambda}{16\pi G}$). Interestingly all fermions acquire masses through the spontaneous SUSY breaking. The local $U(1)$ gauge symmetry is not broken. The massless scalar ω is a NG boson for the degeneracy of the vacuum in $(A,\tilde{B}_2)$-space, which is gauged away provided the local gauge symmetry between the vector and the scalar multiplet is introduced. We have shown explicitly that N=2 LSUSY QED, i.e. the cosmological constant Λ term of $N=2$ SGM (in flat-space), possesses a true vacuum. The resulting model describes: one massive charged Dirac fermion ($\psi_D{}^c \sim \chi + i\nu$), one massive neutral Dirac fermion ($\lambda_D{}^0 \sim \lambda^1 - i\lambda^2$), one massless vector (a photon) (v_a), one charged scalar ($\hat{B}^1 + i\hat{B}^2$), one neutral complex scalar ($\hat{A} + i\hat{\phi}$), which are composites of superons.

Remakably, the lepton-Higgs sector of SM analogue $SU(2)_{gl}\times U(1)$ appears (without manifest superpartners). From these arguments we conclude that $N=2$ SUSY QED is equivalent(related) to $N=2$ NLSUSY action, i.e., to the matter sector of $N=2$ SGM produced by Big Collapse (phase transition) of $N=2$ NLSUSYGR (new space-time). And the true vacuum of $N=2$ NLSUSY model is achieved by the LSUSY model where all particles of th LSUSY supermultiplet are SUSY composites of NG fermions.

By the similar computations for (b) we obtain the following mass genaration

$$m_\rho^2 = m_\theta^2 = m_\omega^2 = m_{v_a}^2 = 2(ef)k^2 = -\frac{2\xi e}{\kappa},$$
$$m_{\lambda^i} = m_\chi = m_\nu = m_\varphi = 0.$$

The vacuum breaks both SUSY and the local $U(1)$ spontaneously.

3.3. *SU(5) Superon-quintet model (SQM) of particles and forces*

We have found that (i) the SM with just three generations of quarks and leptons are contained in SO(10) sP group, (ii) all physical states of SO(10) sP are uniquly constructed by multiplications of all possible combinations of the supercharge, (iii) the canonical quantization of superon(NG fermion) is carried out in compatible with the superalgebra[15], i.e. the commutation relation of the supercarge. and the leading order term of the supercharge in the low energy is superon(soft NG fermion), (iv) NL/L SUSY relation(the equivalence of NLSUSY with broken global LSUSY) for N-extended SUSY is possible. (iii) and (iv) are demonstraited by the field theory in the flat space. From these considerations we speculate that all (observed) particles are such composite objects of superons as constructed group theoretically by the multiplicatoion of the supercharge. And SM, GUT may be regarded as the low energy effective theory of $N = 10$ NLSUSYGR/SGM, where $N = \underline{10} = \underline{5} + \underline{5^*}$ under $SO(10) \supset SU(5)$ We call *superon-quintet model(SQM)* of particles. We have adoptted tentatively the follwing naive L-R symmetric assignment for the SM particles: for (ν_e, e), (ν_μ, μ), (ν_τ, τ) we choose $\epsilon^{lm} Q_l Q_m Q^*{}_n$, $Q_a Q^*{}_a Q_m$, $Q_a Q^*{}_a \epsilon^{lm} Q_l Q_m Q^*{}_n$, for (u, d), (c, s), (t, b) we choose $\epsilon^{abc} Q_b Q_c Q_m$, $\epsilon^{abc} Q_a Q_b Q_c Q^*_d Q_m$, $\epsilon^{lm} Q_l Q_m \epsilon^{abc} Q_b Q_c Q^*_n$, for neutral Higgs particle we have $\delta^{ab} Q_a Q^*{}_b \delta^{lm} Q_l Q^*{}_m$, for gauge bosons we have $Q_a Q^*{}_a$, $Q_l Q_m$, $Q_l Q^*{}_m$, $Q_a Q^*{}_l$, $Q_a Q_m$, $\cdots$, $(a, b, \cdots = 1, 2, 3,\ l, m, \cdots = 4, 5)$ By specifying the superon content of particles explicitly we interpret the Feynmann diagram of the SM/GUT in terms of the composite superon pictures, i.e. the single line of the propagator of the SM particle in the Feynmann diagram is replaced by the multiple lines of the constituent.superons. We find that most Feynmann diagrams of the observed physical process of SM/GUT are reproduced in terms of the composite picture of particles. However the Feynmann diagram of the leading order of the dangerous (phenomenologically forbidden) process in SM/GUT, e.g. FCNC and proton decay, etc. are not reproduced due to the selection rurle for the superon number conservation at the vertex. The dangerous process diagram is forbidden automatically in the superon-quintet composite model. Finally we just mention that there is no low energy (non-gravitational) excited states in SM/GUT particles in the SGM scenario.

4. Cosmology of NLSUSYGR

4.1. *Big collapse*

Now we discuss the cosmological implications of NLSUSYGR (or SGM via Big Collapse). NLSUSYGR space-time $L_{NLSUSYGR}$ is unstable and spontaneously collapses *Big Collapse* to L_{SGM} of familiar Riemann space-time *(graviton)* and Nambu-Goldstone fermion matter *(superon)*. The Big Collapse may induce the rapid expansion of three dimensional space-time by the quantum effect *Pauli exclusion principle*. NLSUSYGR scenario predicts the *four* dimension of space-time. Because we assume *space-time supersymmetry* as the origin of *Hilbert space(particle) supersymmetry*, i.e. we ask the isomorophism of $SO(1, D-1)$ and $SL(d, C)$:

$$\frac{D(D-1)}{2} = 2(d^2 - 1), \tag{36}$$

which holds for *only*

$$D = 4, \quad d = 2. \tag{37}$$

The variation of SGM action L_{SGM} with respect to $e^a{}_\mu$ gives the equation of motion for $e^a{}_\mu$ recasted as follows

$$R_{\mu\nu}(e) - \frac{1}{2} g_{\mu\nu} R(e) = \frac{8\pi G}{c^4} \left\{ \tilde{T}_{\mu\nu}(e, \psi^j) - g_{\mu\nu} \frac{c^4 \Lambda}{8\pi G} \right\}, \tag{38}$$

where $\tilde{T}_{\mu\nu}(e, \psi^i)$ $(i.j = 1, 2$ for $N = 2)$ abbreviates the stress-energy-momentum of superon(NG fermion) matter including the gravitational interactions. Note that $-\frac{c^4\Lambda}{8\pi G}$ can be interpreted as *the negative energy density of Riemann space-time*, i.e. *the dark energy density* ρ_D. (The negative sign is provided uniquely and produces simultaneously the correct kinetic term of superons in NLSUSY.) We anticipate that the graviton is the universal atrative force which dictates the evolution of the soperon-graviton world (SGM action) L_{SGM} and constitutes gravitatonal (massless) conposites eigenststes of space-time symmetry $SO(N)$ sP, which ignites Big Bang model of the universe and continues to the SMs. If superon-graviton phase(SGM action(11)) with the cosmological constant of space-time survives(swictched off) in the evolution after Big Collapsec and oexists after the ignition of Big Bang, such SGM supace-tme behaves as the dark side of the universe, i.e. the dark energy inducing the acceleration of the present universe, which is recognized only by the gravitational interaction.

4.2. *Cosmology and particle physics*

On tangent flat space-time, the low energy theorem of the particle physics gives the following superon(massless NG fermion)-vacuum coupling

$$< \psi^j_\alpha | S^{k\mu}{}_\beta | 0 > = i \sqrt{\frac{c^4 \Lambda}{8\pi G}} (\gamma^a)_{\alpha\beta} \delta^{jk}, \tag{39}$$

where $S^{k\mu} = i\sqrt{\frac{c^4\Lambda}{8\pi G}} e_a{}^\mu \gamma^a \psi^k + \cdots, j,k = 1,2$ is the conserved supercurrent obtained by applying the Noether theorem[15] and $\sqrt{\frac{c^4\Lambda}{8\pi G}}$ is *the coupling constant* g_{sv} *of superon with the vacuum via the supercurrent.* Further we have seen in the preceding section that the right hand side of (38) for N=2 is essentially N=2 NLSUSY VA action. And it is equivalent to the broken $N = 2$ LSUSYQED action (15) with the non-zero vacuum expectation value of the auxiliary field(Fayet-Iliopoolos term). For the vacuum of the case (a) it gives the SUSY breaking masse,

$$M_{SUSY}{}^2 \sim < D > \sim \sqrt{\frac{c^4\Lambda}{8\pi G}}, \tag{40}$$

to the component fields of the (massless) LSUSY supermultiplet, provided $-f\xi \sim O(1)$. We find NLSUSYGR(SGM) scenario gives interesting relations among the important quantities of the cosmology and the low energy particle physics, i.e.,

$$\rho_D \sim \frac{c^4\Lambda}{8\pi G} \sim < D >^2 \sim g_{sv}{}^2. \tag{41}$$

Suppose that in the (low energy) LSUSY supermultiplet the stable and the lightest massive particle retains the mass of the order of the spontaneous SUSY breaking. And if we identify the neutrino with such a particle and with $\lambda^i(x)$, i.e.

$$m_\nu{}^2 \sim \sqrt{\frac{c^4\Lambda}{8\pi G}}, \tag{42}$$

then SGM predicts the observed value of the (dark) energy density of the universe and naturally explains the mysterious numerical relations between m_ν and ρ_D^{obs}:

$$\rho_D^{obs} \sim (10^{-12} GeV)^4 \sim m_\nu{}^4 (\sim g_{sv}{}^2). \tag{43}$$

The tiny neutrino mass is the direct evidence of SUSY (breaking) in the NLSUSYGR scenario, i.e. Big collapse of space-time and the subsequent creation of graitational (massless) composite of the superon-graviton SGM space-time in advance of Big Bang. The large mass scales and the non-abelian gauge symmetry necessary for building the realistic broken LSUSY model will appear by the extension to the large N NLSUSY and by the linearization of $\tilde{T}_{\mu\nu}(e,\psi)$ which contains the mass scale in the higher order with ψ^{24}.

In Riemann flat space-time of SGM, ordinary LSUSY gauge theory with the spontaneous SUSY breaking emerges as (massless) composites of NG fermion from the NLSUSY cosmological constant of SGM SM and GUT may be regarded as the low energy effective composite theory of NLSUSYGR in flat space-time.

5. Summary

We have proposed NLSUSYGR(SGM) scenario for unity of nature. : The ultimate shape of nature is unstable $d = 4$ space-time specified by $[x^a, \psi_\alpha{}^N; x^\mu]$ and described by NLSUSYGR $L_{\rm NLSUSYGR}(w^a{}_\mu)$ with $\Lambda > 0$. Mach principle is encoded geometrically. Big Collapse of space-time occurs (due to false vacuum $V_{\rm P.E.} = \Lambda > 0$) to $L_{\rm SGM}(e.\psi)$ and creates Riemann space-time $[x^a; x^\mu]$ and massless fermionic matter *superon* ${\psi_\alpha}^N$, which is described by SGM action $[L_{\rm SGM} = L_{\rm EH}(e) - \Lambda + T(\psi.e)]$ containing Einstein GR with $V_{\rm P.E.} = \Lambda > 0$ and N superons. The phase transition towards the true vacuum $V_{\rm P.E} = 0$, is achieved by forming gravitational composite massless LSUSY supermultiplet, which ignites Big Bang Universe and the subsequent oscilations around the true vacuum. The graviton dictates the evolution of SGM world. We have shown in flat space-time that broken N-LSUSY theory emerges from the N-NLSUSY cosmological term of $L_{\rm SGM}(e,\psi)$ (NL/L SUSY relation). Interestingy SGM may be regarded as the superfluidity of space-time and matter. The vacua of composite SGM scenario created by the Big Collapse of new space-time possesses rich structures promissing for the unified description of nature. SGM gives new insights into the unsolved problems of cosmology and particle physics, for example, e.g. the origin of three-generations strcture for quarks andleptons, the tiny nutrino mass, proton decay, dark matter, dark energy, the space-time dimennsion *four*, etc. We have shown explicitly in 2 dimensional space-time that $N = 2$ LSUSY QED theory for the realistic $U(1)$ gauge theory appears as the physical field configurations on the vacuum of $N = 2$ NLSUSY theory on Minkowski space-time, which relate the particle physics with cosmology.

Establishing NL/L SUSY relation on curved space-time, i.e. linearizing SGM action $L_{SGM}(e,\psi)$ and the extension to large N, especially to $N = \underline{10} = \underline{5} + \underline{5^*}$ for equipping the SU(5) GUT structure[9], is essential for *superon quintet hypothesis* of SGM scenario.as a model of nature

Finally we just mention that NLSUSY GR and the subsequent SGM scenario for the spin $\frac{3}{2}$ NG fermion[14,27] is in the same scope.

References

1. J. Wess and B. Zumino, *Phys. Lett. B*49, 52 (1974).
2. D.V. Volkov and V.P. Akulov, *Phys. Lett. B*46, 109 (1973).
3. Yu. A. Golfand and E.S. Likhtman, *JETP Lett.*13, 323 (1971).
4. A. Salam and J. Strathdee, *Phys. Lett. B*49, 465 (1974).
5. P. Fayet and J. Iliopoulos, *Phys. Lett. B*51, 461 (1974).
6. L. O'Raifeartaigh, *Nucl. Phys. B*96, 331 (1975).
7. M. Gell-Mann, *Proceedings of supergravity workshop at Stony Brook*, eds. P. van Nieuwenhuisen and D. Z. Freedman(North Holland, Amsterdam, 1977).
8. K. Shima, *Z. Phys.* **C18**, 25(1983).
9. K. Shima, *European Phys. J.* **C7**, 341 (1999).
10. S. Coleman and J. Mandula, *Phys. Rev.*159, 1251 (1967).
11. R. Haag, J. Lopuszanski and M. Sohnius, *Nucl. Phys. B*88, 257 (1975).
12. J. Wess and J. Bagger, *Supersymmetry and Supergravity* (Princeton University Press, Princeton, 1992).
13. K. Shima, *Phys. Lett. B*501, 237 (2001).
14. K. Shima and M. Tsuda, *Phys. Lett. B*507, 260 (2001).

15. K. Shima, *Phys. Rev. D***15**, 2165 (1977).
16. K. Shima and M. Tsuda, *PoS HEP2005*, 011 (2006); K. Shima and M. Tsuda, *Phys. Lett. B***645**, 455 (2007). K. Shima, M. Tsuda and M. Sawaguchi, *Int. J. Mod. Phys.* **E13**, 539 (2004).
17. E.A. Ivanov and A.A. Kapustnikov, *J. Phys.A***11**, 2375 (1978).
18. M. Roček, *Phys. Rev. Lett.***41**, 451 (1978).
19. T. Uematsu and C.K. Zachos, *Nucl. Phys.B***201**, 250 (1982).
20. K. Shima, Y. Tanii and M. Tsuda, *Phys. Lett. B***525**, 183 (2002).
21. K. Shima, Y. Tanii and M. Tsuda, *Phys. Lett. B***546**, 162 (2002).
22. K. Shima and M. Tsuda, *Mod. Phys. Lett. A***22**, 1085 (2007).
23. K. Shima and M. Tsuda, *Mod. Phys. Lett. A***22**, 3027(2007).
24. K. Shima and M. Tsuda, *Class. Quant. Grav.* **19**, 5101 (2002).
25. K. Shima, M. Tsuda and W. Lang, *Phys. Lett. B***659**, 741(2007).
26. K. Shima and M. Kasuya, *Phys. Rev. D***22**, 290(1980).
27. N. S. Baaklini, *Phys. Lett. B***67**, 335 (1977).

Printed in the United States
By Bookmasters

Printed in the United States
By Bookmasters